Enemy Encounters in Modern Warfare

Holly Furneaux • Matilda Greig
Editors

Enemy Encounters in Modern Warfare

palgrave
macmillan

Editors
Holly Furneaux
Cardiff University
Cardiff, UK

Matilda Greig
National Army Museum
London, UK

ISBN 978-3-031-56747-6 ISBN 978-3-031-56748-3 (eBook)
https://doi.org/10.1007/978-3-031-56748-3

Cover illustration: Courtesy Everett Collection

This Palgrave Macmillan imprint is published by the registered company Springer Nature Switzerland AG.
The registered company address is: Gewerbestrasse 11, 6330 Cham, Switzerland

Foreword

In Wilfred Owen's 'Strange Meeting', two young soldiers-poets—fighting on opposite sides but both dead now—meet in the darkness of the underworld:

> I am the enemy you killed, my friend.
> I knew you in this dark: for so you frowned
> Yesterday through me as you jabbed and killed.
> I parried; but my hands were loath and cold.
> Let us sleep now...[1]

The fact that this iconic war poem would be about an enemy encounter shows how central the concept has been to the psychology of warfare. Enemy encounters, real and fictive, have a long and illustrious history, but what makes Owen's account so powerful is its combination of the intimacy of the address and the violence of the act: the pity of 'I knew you' and 'Let us sleep now' is all the more poignant when juxtaposed with the visceral shock of the monosyllabic 'jabbed and killed'. Pity, tenderness, brutality, and eros are fused and confused in this encounter as the enemy is reconfigured as friend, the friend as another young man, the young man as another poet. Where does the 'I' end and 'you' begin? But what happens if, instead of this romance between the self and the other, the

[1] 'Strange Meeting', in *Wilfred Owen, Complete Poems and Fragments*, ed. by Jon Stallworthy (Oxford: Oxford University Press, 1983), pp. 148–149.

'enemies' are separated by the exigencies of race, gender, and age, or the weight of colonial and postcolonial histories?

The First World War provides a particularly rich archive for 'enemy encounters' because of having, for the first time in history, a largely literate civilian-army (at least from Europe) who left records and traces of such encounters as in no previous war; second, the sustained nature of warfare in close proximity, particularly on the Western Front, fostered ideal conditions, mythologised in the famous Christmas truce and football match of 1914. Owen's poetic fantasy finds its real-life echo in the memoir of his friend and mentor Siegfried Sassoon who recalls meeting a dead German soldier, and cleaning the face of the young blond boy who had just been killed: 'Propped against the bank, his blond face was undisfigured, except by mud which I wiped away from his eyes and mouth with my coat sleeve'.[2] Henri Barbusse's *Le Feu* (1916), one of the earliest novels about the war, concludes with the image of a French and German soldier locked in a death-embrace while, in Erich Maria Remarque's *All Quiet On the Western Front* (1929), Paul realises that the Frenchman he had just stabbed and killed is remarkably like him. The drama of touch and intimacy that each of these scenes involves is gathered up into a lyric reverie in the Canadian poet Robert Service's poem 'Only a Boche' as the Allied soldier observes that 'the dying Boche on the stretcher there has a queer resemblance to me'.[3] Such moments were, however, by no means the preserve of European soldier-poets. In a hospital in Baghdad, an Indian POW-turned-nurse recalls a conversation with a wounded Turkish soldier who refers to him as 'Kardes', meaning 'brother':

> We spoke of our lands, our joys and sorrows. ... One thing that they always used to say was, 'This war that we are fighting—what is our stake in this? Why are we slashing each others' throats? You stay in Hindustan, we in Turkey, we do not know each other, share no enmity, and yet we became enemies overnight because one or two people deemed it so.' Is this on the mind of every soldier of every nation?[4]

[2] Siegfried Sassoon, *Memoirs of An Infantry Officer* (1930; London: Faber, 1997), p. 57.

[3] Robert Service, 'Only a Boche', in *Rhymes of a Red Cross Man* (New York: Barse & Hopkins, 1916), p. 111.

[4] Sisir Prasad Sarbadhikari, *Abhi Le Baghdad* [*On the Road to Baghdad*] (Kolkata: privately printed, 1957), p. 158.

Is there a Global south–south identification born out of common disenchantment with European imperialism, or is it a cosmopolitan brotherhood of the dispossessed, based on a shared sense of exploitation and vulnerability?

What the chapters of this remarkable collection do is to spectacularly expand the lines of enquiry—conceptually, historically and politically; in the process, they redefine the very scope of the term 'enemy encounters'. Rigorously researched and delicately textured, the chapters range from representations of guerrilla warfare in the Napoleonic era through colonial warfare, the two world wars and Vietnam to memorialisation and museum culture of today. Cumulatively, they alert us to the extraordinary variety as well as specificity of these encounters on one hand and the deep, often competing, histories—geopolitical, ideological, and emotional—that inform the actual moments and processes. Equally importantly, they investigate the role of narrative: the way narratives of enemy encounters have been produced to repair the image of a coercive colonial state and how they continue to be manufactured for propaganda purposes (to strike terror at or appease particular imagined audiences); to justify and legitimise the use of violence against a 'savage enemy'; and how they often end up shaping the actual experiences.

With contexts, meanings shift; boundaries blur. 'Enemy', the volume shows, is not so much a stable concept but is elastic, evolving and shapeshifting. An 'enemy' in the battlefield, as in the above example from Sarbadhikari, can become a friend and brother in a POW camp or a hospital; one set of enemies can be created to forget or mask another. Gender, age, and, more importantly for this volume, race and racism can determine definitions of enemy and the limits of empathy: indeed, in the First World War, some European combatants shared a deeper imaginative bond across enemy lines than with their African or Indian brothers-in-arms. Moreover, if relationships with the enemy have ranged from self-identification to violent othering, in today's world the word 'enemy' has expanded its scope to encompass a wide range of non-human agents, from pathogens and natural disasters to prohibited drugs. Indeed, in some of the essays, 'enemy' emerges less as a concept and more as a pressure-point, providing insights into our intimate selves and raising questions about competing affiliations (national, ethnic, and religious), concepts of just and unjust wars, and what lives we consider grievable.

Encounters prove no less capacious or complex a concept: mediated, in person, deferred, forced, and unforced, occurring through poems scrawled on the door of a dugout or through faked photographs. The essays alert us to the vast, complex, and sophisticated media technologies through which 'enemy encounters' increasingly occur, from self-othering by Islamic State terrorists to encountering the 'enemy' in the galleries of the Imperial War Museum, without ever letting go of the possibility of face-to-face attacks. And they ask fundamental questions: why are face-to-face encounters so uncanny? Can encounters happen long after the enemy has left or died? How does context (battlefield, hospital, POW camp, alive, injured, or dead) change the very conceptualisation of the enemy? At the same time, these carefully historized essays never let us forget that, in spite of all the fantasy and poetry we may invest in such moments, they are often remarkably violent, destructive, and short-lived.

As the book goes into print, the Russia-Ukraine war is into its third year; here, we have seen some of the most strenuous and far-fetched construction of the 'enemy' by the Russian state and some of the most brutish—and, occasionally, some heart-rendering—'enemy encounters'. As I write, some enemy encounter is happening somewhere in Ukraine. While academic research, like Auden's famous definition of poetry, cannot make anything 'happen' in the short term, these essays are instructive for our times, making us think afresh about the varied and complex lives of 'enemy encounters', with their roots in geo-political histories and propaganda machineries on one hand and the innermost recesses of our psyche on the other.

University of Oxford Santanu Das

Acknowledgements

This edited collection forms part of an AHRC-funded project at Cardiff University, 'Strange Meetings: Enemy Encounters 1800–2020' and builds on an international conference organised in partnership with Imperial War Museums in July 2021. As the dates suggest, work on this book spanned the global COVID-19 pandemic and would not have been possible without the help and support of many different people. During several years of working through Zoom windows, email chains, Dropbox folders, and an intimidating Excel file that we fondly nicknamed 'the spreadsheet of dreams', this book has been a current of connection, collegiality, and warmth for which we are immensely grateful.

First, we would like to thank the team at IWM who worked with us on the *Enemy Encounters* conference, to which they contributed time and creativity over the course of a year in discussing plans, finalising programmes, advising on speakers and topics, and both chairing and giving talks on panels. Thank you, Suzanne Bardgett, Chris Cooper, Vikki Hawkins, Becky Newell, Iris Veysey, James Taylor, and Hannah Wills. An extra thank you to Hannah: the smooth running of the online conference across all three days was entirely due to your incredible technical work behind the scenes. Thanks also to the team who worked with us on producing a webpage and a follow-up edition of the IWM Institute newsletter. We are extremely grateful, too, to the external speakers, chairs, and attendees who participated in the conference and created such thought-provoking discussions.

Thank you to the Cardiff University research team, and especial thanks to Martin Willis for his great support of this project. From Matilda, a thank you to Glyn Prysor and Justin Maciejewski at the National Army Museum for supporting my continued work on this project when I started there.

Our thanks are also due to our editor, Emily Russell, for supporting this book and guiding us through the publishing process, as well as to the production team at Palgrave Macmillan/Springer, and to the anonymous peer reviewer for their insightful comments. We are grateful to Emily also for her encouragement in pursuing the inclusion of a chapter on the war in Ukraine and to Catherine Baker and Lisa Gaufman for identifying potential contributors and generously giving their time and expertise to consult on the review process.

The generosity of our colleagues is continued in the kind and beautifully written preface provided by Santanu Das: thank you again for that piece, which lifted our hearts in a grey late-winter month anticipating the last of the editorial work! It is also reflected in the Afterword, where we hope we have done some justice to the sincerity, openness, and creativity of our friends at Re-Live: a huge thank you to Karin Diamond, Rach, Graham, Dave, and Robbie.

Finally, and most importantly of all, thank you to our contributors for all your hard work over the last two years. These chapters talk to each other, they complement each other, and they shine, and we are proud to be able to present them to the world.

Contents

Notes on Contributors

Stefan Aguirre Quiroga is a PhD student in history in the Department of Historical Studies, University of Göteborg. His research centers on the life and agency of the Kit Carson Scouts during the Vietnam War, exploring the reasons why South and North Vietnamese defectors volunteered to be employed by the United States to fight against their former comrades and how American soldiers perceived them during and after the war.

Jared Ahmad is Lecturer in Journalism, Politics and Communication at the University of Sheffield, UK. His research focuses on media discourses and (self)representations of terrorism and the way those portrayals are shaped by the dynamics of the contemporary 'hybrid' media environment. His work has been published in *International Journal of Communication*, *Critical Studies on Terrorism* and *Media, War and Conflict*, and he is the author of *The BBC, the War on Terror and the Discursive Construction of Terrorism: Representing 'Al-Qaeda'* (Palgrave Macmillan 2018). His next book is titled *Imagining the Caliphate: Analysing Public, Political and Popular Representations of the Islamic State.*

Jordyn Bailey is a PhD Candidate at the University of New Brunswick (Canada). She received her BA and MA from Western University in London, Ontario. Her MA thesis examined platonic and intimate encounters between Canadian women and German soldiers, sailors, and aviators during the Second World War. Bailey's doctoral dissertation examines connections between gender, sexuality, and espionage in the *Staatssicherheit*, or STASI, during the German Democratic Republic.

Catherine Bateson is Associate Lecturer in American History at the University of Kent. She researches and teaches the American Civil War, 1700s to early 1900s US social and cultural topics, slavery, civil rights and indigenous history, and American Studies. She has authored *Irish American Civil War Songs: Identity, Loyalty, and Nationhood* (LSU, 2022) and has also written articles about the Irish American wartime experience and the importance of music in military history. She is Associate Editor of the 'Irish in the American Civil War' project and co-founder of the Stuff of War Society.

Allison L. Bennett is a PhD candidate at Memorial University of Newfoundland in St. John's, Newfoundland and Labrador, Canada. She holds a Bachelor of Arts (Hons.), Master of Arts (History), and Bachelor of Education (Intermediate/Secondary) also from Memorial. Bennett was recently awarded the 2022–2023 Linda Souter Humanities Award from the Confederation of University Women for her doctoral research on soldier-civilian sexual encounters in First World War Greek Macedonia and the Middle East.

Rio Creech-Nowagiel is in the third year of an AHRC-funded collaborative doctoral PhD studentship with Imperial War Museums and the department of Journalism, Media & Culture at Cardiff University. Their PhD project explores the role of annotated photography in shaping public knowledge around Britain's decolonisation wars, focusing on framings and critical re-framings of the so-called Malayan Emergency (1948–1960). In 2020, Rio set up NDPR (@PhotoRschNet), an interdisciplinary research network for postgraduate and early career researchers working in areas related to photography. Alongside their PhD, they also work as a cultural producer and facilitator specialising in vernacular photography and community co-curation.

Holly Furneaux is Professor of Victorian Literature and Culture at Cardiff University. She currently leads the AHRC project 'Strange Meetings: Enemy Encounters 1800–2020' and her next project will explore Victorian international adoption. Her books include *Military Men of Feeling: Emotion, Touch and Masculinity in the Crimean War* (2016) and *Queer Dickens: Erotics, Families, Masculinities* (2009).

Tim Grady is Professor of Modern European History at the University of Chester. He is the author of the *German-Jewish Soldiers of the First*

World War in History and Memory (2012), *A Deadly Legacy: German Jews and the Great War* (2017), as well as numerous edited volumes on war, memory, and the contested legacies of conflict.

Matilda Greig is a Historian at the National Army Museum in London, specialising in the history of conflict in the Napoleonic period and the long nineteenth century. She received her PhD in 2018 from the European University Institute in Florence and has since held research and teaching fellowships at Sciences Po Reims, University College Dublin, Cardiff University, and Universitat Pompeu Fabra in Barcelona. Her first book, *Dead Men Telling Tales* (Oxford University Press, 2021) explored the stories of the several hundred military memoirs written by French, British and Spanish veterans of the Peninsular War. She is a managing editor for the *British Journal for Military History* and is co-editing collections on war and gender in modern Europe and on Napoleonic art and material culture.

Vikki Hawkins was a curator at the Imperial War Museum between 2016 and 2021, working on the £30.7 million transformation of the permanent Second World War Galleries at IWM London, which were awarded 'Permanent Exhibition of the Year 2022' at the Museum and Heritage Awards. She is the co-author of the book *Total War: A People's History of the Second World War* (Thames & Hudson, 2021) and is a Managing Editor of the *British Journal for Military History*. Vikki has published and contributed to various conferences on the topic of displaying marginalised histories of conflict in museums, focusing on narratives of the British Empire and war in Asia and the Pacific. Vikki is now a curator at the British Museum working on a project to develop new curatorial approaches to interpreting the collection and developing the narratives that will underpin a comprehensive redisplay of the galleries.

Mykola Homanyuk is an associate professor at the Kherson State University. He graduated from Kherson State Pedagogical Institute and defended his PhD thesis in Sociology at Karazin Kharkiv National University. In 2022–2023, he has been a member of the *Prisma Ukraine: War, Migration, Memory* research group (Forum Transregionale Studien in Berlin). Since 2022 Mykola is a member of *Contested Ukraine: Military Patriotism, Russian Influence and Implications for European Security* research group. In 2018, he won the ADAMI Media Prize for Cultural Diversity in Eastern Europe. At present, his main research interests are in the sphere of war commemoration, toponymy, and mental geography.

Valeria Khotsina is a translator and political refugee from Belarus, now based in Warsaw. She specialises in translating books and texts on radical politics, gender issues, and social change. Valeria is a member of the publishing cooperative Radical Theory and Practice and the Translation for Change project.

Tom Menger studied European Studies and History at the University of Amsterdam. He received his doctorate from the University of Cologne in 2021, with a thesis titled 'The Colonial Way of War: Extreme Violence in Knowledge and Practice of Colonial Warfare in the British, German and Dutch Colonial Empires, c. 1890–1914'. He is currently a postdoctoral researcher at the Käte Hamburger Research Centre *global dis:connect* at the Ludwig Maximilian University Munich. His main research interests lie in transimperial history, the history of colonial violence, and the history of imperial infrastructures and oil.

Christine Sylvester is Professor of Political Science at University of Connecticut and is affiliated with the School of Global Studies at the University of Gothenburg, Sweden. A native of Connecticut, she also holds Australian citizenship and has worked extensively outside the United States, including at the Australian National University (Canberra), The Institute of Social Studies (The Hague, Netherlands), and Lancaster University (UK). She was awarded the Swedish Research Council's Kerstin Hesselgren Professorship for Sweden for 2010–2011. Other awards include a Leverhulme fellowship at the School of Oriental and African Studies at The University of London; the Susan Northcutt Award of the International Studies Association (ISA); Eminent Scholar of the Feminist Theory and Gender Studies Section of the ISA; and the inaugural Ann Tickner Award of the ISA. She was also named one of *Fifty Key Thinkers in International Relations*, Martin Griffiths, Steven Roach, M. Scott Solomon, eds. (Routledge, 2008). Her most recent research and writings are on war as experience and art/museums and international relations. She is the editor of the Routledge book series: *War, Politics, Experience.* In 2014, Lund University awarded her an honorary doctorate in the social sciences, and in 2016, she was selected as a faculty fellow of the University of Connecticut Humanities Institute. Her latest book is *Curating and Re-Curating the American Wars in Vietnam and Iraq* (2019).

Lily Tekseng is a Rowan Williams Cambridge and Wolfson College PhD candidate at the Faculty of English, University of Cambridge. She works on the cultural productions on the Second World War in the borderland Indian Northeast. Her areas of interest include literatures from South Asia and Southeast Asia, indigeneity and race, and comparative borderlands. She has previously taught at the University of Delhi, India.

Julia Ribeiro S. C. Thomaz is a PhD student in French Literature at the Université Paris Nanterre and in History at the École des Hautes Études en Sciences Sociales, working on French poetry of the First World War. She is a collaborator of the Digital Humanities project *Poésie Grande Guerre* (https://pgg.parisnanterre.fr) and a founding member of the ECR network *Une Plus Grande Guerre/A Greater War*. She is a recipient of the Gerda Henkel Scholarship (Historial de la Grande Guerre) and of the *Prix d'Études des Mondes Contemporains* (*Association des Amis de* La Contemporaine).

Lisa M. Todd is Professor of History and Department Chair at the University of New Brunswick (Canada). She holds a PhD from the University of Toronto, an MA from Royal Holloway College, University of London, and a BA from the University of New Brunswick. She has authored *Sexual Treason in Germany During the First World War* with Palgrave Macmillan in 2017, and her current project examines German fears of 'miscegenation' from 1880 to 1945. Todd is also co-editing (with Cindy Brown), 'A Cultural History of War in the Modern Age, 1920–2000' and 'European Racism: A History in Documents' (with Gary Waite).

List of Figures

Dying with the Enemy: Prisoners of War Deaths in First World War Britain

Enemy Encounters in the Imperial War Museum's Second World War Galleries

'Children Made to Carry Bombs': Staging the Malayan 'Emergency' in Postwar Britain

'Islam is the religion of the sword not pacifism': Strategic Nostalgia and Self-Othering in Islamic State Propaganda

Memorializing the Enemy Three Ways: Australia, Japan, the USA

Afterword: Coming Home—Finding Form for Veterans' Life Stories of Enemy Encounter

Introduction: Fellow Feeling and Estrangement—Approaching Enemy Encounters in Modern Warfare

Holly Furneaux and Matilda Greig

Moments before a battle in 1811 in the Peninsular War, a French cavalry officer named Denis-Charles Parquin remembered galloping his horse across to the enemy British line to offer them a bottle of brandy. As he approached, he waved a white handkerchief to signal his peaceful intentions. When a British officer rode out to meet him, Parquin explained that he came bearing gifts, on his commander's orders: 'I've come to offer you a bottle of brandy, to drink with your comrades and mine, before we become acquainted in another manner'. The offer was readily accepted, and the two men signalled to their waiting regiments. Soon, about twenty enemy horsemen were mingling around on open ground between the lines, passing around the brandy, which did not last long. Parquin could

H. Furneaux (✉)
Cardiff University, Cardiff, UK
e-mail: furneauxh@cardiff.ac.uk

M. Greig
National Army Museum, London, UK
e-mail: mgreig@nam.ac.uk

H. Furneaux, M. Greig (eds.), *Enemy Encounters in Modern Warfare*, https://doi.org/10.1007/978-3-031-56748-3_1

not resist commenting on how impressed his opponents were with the quality of French distilling, as well as their behaviour: 'We found it excellent, especially the English officers, who thanked us for our conduct, which seemed to have made a strong impression on them'. Conversation sprang up about the course of the war and the reputations of various regiments. A British officer asked whether anyone on the French side might deliver a letter for him, to a friend of his who was a prisoner of war in Moulins in France, and a staff sergeant whose father serendipitously happened to be the prefect of that town 'took on the responsibility with pleasure'. The amiable group was only broken up by the battle itself starting, with the British artillery beginning to fire shells that landed not far away—an interruption that 'forced us to separate, not however without having agreed to drink the rum which the English officers had offered us in return'.[1]

The scene has the air of gentlemen being introduced to each other in a private club. It explicitly emphasises their similarities, rather than their differences; representatives of long-time enemy countries they may be, but what is more important to Parquin in his account of that moment is that they share similar customs of polite behaviour, educated conversation, even a cosmopolitan outlook—at least some, likely most, of the men present are able to converse easily in a foreign language. In a late nineteenth-century reprint, Parquin's story was accompanied by an artist's impression of the meeting on the field in Spain that adds an element of physical proximity and the visual blurring of enemy sides (see Fig. 1). Without a good knowledge of military uniform, the reader would not be able to distinguish the figures on horseback who intermingle on a green field, ignoring the discarded helmet and weaponry on the ground that might remind them of battles past and future. Class, rank and a shared sense of 'civility' are depicted as superseding national allegiance in this moment of peaceful enemy interaction shortly before conflict began. Such non-violent encounters, as this volume shows, have taken place in different forms in almost every instance of modern warfare. In fact, even within one memoir, this was far from an isolated incident: Parquin writes that it soon became a habit among his brigade to share rum and brandy with the English officers 'when we found ourselves in firing positions opposite them'.[2]

[1] Denis-Charles Parquin, *Récits de guerre: souvenirs du capitaine Parquin: 1803–1814*, ed. Frédéric Masson (Paris: Boussod, Valadon & Cie, 1892), pp. 94–95. Translations our own.

[2] Parquin, p. 109.

Fig. 1 L. Sergent, '... L'officier anglais accepta et fit signe à ses camarades; de mon côté, j'en fis autant au commandant de Vérigny qui me rejoignit avec une dizaine d'officiers. La bouteille passa à la ronde...', in Denis-Charles Parquin, *Récits de guerre: souvenirs du capitaine Parquin, 1803–1814*, ed. Frédéric Masson (Paris: Boussod, Valadon & Cie, 1892), opp. p. 94. Courtesy of the Bibliothèque nationale de France (BnF)

While Parquin's account suggests a gentlemanly understanding between officers, soldiers of the ranks also testified to enjoying hospitality across sides. John Morgan, another veteran of the Peninsular War, recalled that 'there was often a good and friendly feeling between English and French soldiers when they were in the field'. Separated from the enemy sentries by only fifty yards 'he would call out, "Bon soir". The Frenchmen would sing out in return "will you boire [drink]?" Then they would lay down their arms, meet in the middle space and drink together. Morgan liked drinking with the French sentries because they mostly had something hot' and risked it though he believed if discovered they would be punished by their officers and might have even been 'shot or hung'.[3] Moving forward to the First World War, Corporal Louis Barthas, a French infantryman, recalled similarly sustaining and pleasurable 'exchanges of gifts, like packets of tobacco from the Régie Française [French government tobacco monopoly] which went to fill the big German pipes, or delicious German cigarettes which came over to the French side. We also exchanged lighters, buttons, newspapers, bread'.[4] These were accompanied by friendly conversation, following which a German exclaimed: '"Frenchmen, Germans, soldiers, all comrades. Officers ..." —and here he raised a clenched fist— "No!"' Barthas's response makes the familiar point that men of the ranks, regardless of side, might feel more in common with those sharing their danger and adversity than with commanders positioned at a safe distance: 'Oh how right he was, this German. It's true that you shouldn't generalise. But how many of our officers were more distantly separated, more morally estranged from us soldiers than were the poor German devils, who were being led to the same slaughterhouse despite themselves'.[5] This agreement also points to the differences of politics and power between the soldiers suffering wars and those who stand to gain from them.

As these opening examples suggest, experiences of affinity and enmity were shifting, with feelings of commonality with the enemy often emerging through a sense of physical separation or 'moral estrange[ment]', as Barthas puts it, from another group, and facilitated by the exchange of information, food and drink, or gifts. A sense of brotherhood with the enemy might be forged through a sense of difference from other military ranks and classes, from civilians at home, especially women, and—repeatedly—from those deemed insufficiently civilised to participate in 'humane'

[3] Reverend Francis Kilvert, *Kilvert's Diary: Selections from the Diary of Rev. Francis Kilvert*, ed. William Plomer (Hay-on-Wye: O Donoghue, 2006), Vol. 1 (first published 1938), p. 212.

[4] Louis Barthas, *Poilu: The World War 1 Notebooks of Corporal Louis Barthas, Barrelmaker 1914–1918*, translated by Edward Strauss (New Haven and London: Yale University Press, 2014), 13 July–29 August 1916, p. 244.

[5] Barthas, 13 July–29 August 1916, p. 247.

warfare. Whether or not the opposing force was considered to have the same level of 'civilisation' as oneself became a determining factor of the character of enemy encounters during war in the nineteenth century, both with European and colonial adversaries, and forms a recurring thread throughout the collection. As the following chapters interrogate, civilisation was understood in a multifaceted sense applying both to race and nation and to class and behaviour. Parquin's account above also has to be read in the context of a developing Western imperial ideology and discourse around the laws of war that posed the northern European white man as the natural superior to all other peoples in the world.[6] The amiable, honourable and restrained manner in which ten northern European white officers met another ten of the same on a battlefield was conditioned by and could help to confirm racist assumptions of a hierarchy of expected behaviour.

The book's cover image illustrates the complexities—historical, political and personal—of enemy encounter. An image of apparent reconciliation in a handshake and partial embrace between a Confederate and Union veteran at the semi-centennial of the Battle of Gettysburg in 1913, the men's expressions, with compressed mouths, are difficult to read. Are they equally invested in the celebration of American unity which post Civil War reunions aimed to promote?[7] The whiteness of the image speaks to the exclusions of enemy encounter with which this book is concerned. Historians are divided as to whether Black veterans participated in the Gettysburg reunion, but certainly the public discourse of the event celebrated 'a triumph of the American spirit' through 'white reconciliation', reflecting the 'erasure of the legacy of emancipation and the unfulfilled

[6] Elizabeth Kolsky, *Colonial Justice in British India* (New York: Cambridge University Press, 2010), pp. 4–5. For an exploration of the body of scholarship on racial hierarchies of civilisation see Holly Furneaux, '"I was a stranger and ye took me in": Adopting the Enemy', Tom Menger, 'Racial Difference and the Limits of Fellow Feeling: British, German, and Dutch Soldiers Encounter the Enemy in Colonial Wars, c. 1900', and Jared Ahmad, '"Islam is the religion of the sword not pacifism": Strategic Nostalgia and Self-Othering in Islamic State Propaganda', all this volume.

[7] Caroline E. Janney considers the mixed feelings at Gettysburg in 1913: "Almost everyone revelled in the reconciliationist spirit of the day [...] A few, however, resolutely rejected the Blue-Gray lovefest [...] When the son of Confederate major Robert Randolph Henry began disparaging Lincoln at a local hotel, a Union veteran jumped to his feet to defend the martyred president, precipitating a raucous fight that left eight men suffering stab wounds." *Remembering the Civil War: Reunion and the Limits of Reconciliation* (Chapel Hill: The University of North Carolina Press, 2013), pp. 266–267.

promise of equality for African-Americans'.[8] While national and racial politics demand careful contextual reading of this photograph, the apparently gentle touch between these former adversaries also invites us to consider the shifts of feeling—tactile and emotional—as the category of enemy is superseded, whether by a shift in position as considered in Tim Grady's chapter here on the treatment of German prisoners of war who died in Britain, or by the passage of time. We return to the knotty question of reconciliation to consider its role in veterans' post-traumatic growth in the Afterword to this volume, written in partnership with former British soldiers and a therapeutic arts charity, Re-Live.

Proximate Enemies: Definitions, Distance, Shifts

Combatants' experiences of enmity could shift and change during war (as well as afterwards), compounding the longstanding difficulty of defining the enemy. The modern period is characterised by a continuing wrangle over who might be legitimately subjected to war violence, and which parts of a warring nation deserve protection. War has expanded from being predominantly the business of professional armies to forms of total war, in which civilians might be directly involved and are purposefully targeted, and diversified through a range of forms of irregular warfare in which individuals might move between the positions of soldier and civilian.

How far the entire population of an opposed state may be held an enemy and the duration of enmity of its forces—particularly when disarmed, wounded or captured—have long been debated and are central concerns within international law.[9] As Samuel Johnson, better known for his dictionary, put it in a 1758 essay in aid of the London Committee to Relieve the Wants of the French Prisoners of War, let 'no man be longer deemed an enemy than while his sword is drawn against us'.[10] Modes of

[8] Evan Preston, '"All May Visit the Big Camp": Race and the Lessons of the Civil War at the 1913 Gettysburg Reunion,' *The Gettysburg College Journal of the Civil War Era*, 2.5 (2011), pp. 66–85, p. 66. For different views on Black participation at this reunion see Carol Reardon, *Pickett's Charge in History and Memory* (Chapel Hill: University of North Carolina, 1997) and David Blight, *Race and Reunion: The Civil War in American Memory* (Cambridge: Belknap, 2001).

[9] See Geoffrey Best, *Humanity in Warfare: The Modern History of the International Law of Armed Conflicts* (London: Methuen, 1983), pp. 55–58.

[10] Samuel Johnson, 'French Prisoners of War', in *Johnson: Prose and Poetry* (Cambridge: Harvard University Press, 1963), p. 486.

possible encounter with the enemy have also shifted, as technologies have enabled killing at greater distance while not displacing direct contact between ground troops, as the continuing Russia/Ukraine war makes clear.

The extent to which drones and other modes of distanced killing enable desensitisation to enemy suffering is central to theoretical and creative explorations of the technology.[11] This equipment at once derealises the other side and makes possible grossly asymmetrical applications of force, which perhaps, as Jean Baudrillard suggests in the case of the Gulf War (1990–1991), disqualify such conflicts from the category of war and place them in the category of atrocity: 'Except in the New World order, war is born of an antagonistic, destructive but dual relation between two adversaries. This war is an asexual, surgical war, a matter of war processing in which the enemy only appears as a computerised target'.[12] While mourning the absence of 'any dual or personal relation' Baudrillard cautions against the romanticisation of earlier wars complete with passionate feeling between sides: 'the point is not to rehabilitate other wars'.[13] As Yvonne Chui puts it, 'generally the greater the distance and more impersonal (or massive) the method of killing, the more likely it is to be seen as unfair'.[14] This sense of the unfairness of killing at distance invokes a misleading nostalgia for closer range contact as fair. To demonstrate the problems of this idea we might recall the British designation of the extreme violence of imperial expansion in the nineteenth century as governed by ideas of 'fair-play'.[15] To offset a sense of the unfairness of killing by remotely piloted aircraft, contemporary Western culture has sought to repersonalise it through what Roger Stahl describes as a fetishization of the 'drone

[11] See, for example, Grégoire Chamayou, *A Theory of the Drone* (New York: The New Press, 2015), Nasser Hussain, 'The Sound of Terror' (2013), <https://www.bostonreview.net/articles/hussain-drone-phenomenology/> [accessed on 2 May 2023], and Omer Fast's film *Five Thousand Feet is the Best* (2011).

[12] Jean Baudrillard, *The Gulf War Did Not Take Place* (Sydney: Power, 2006, first published 1991), p. 62. See also Derek Gregory on the rhetoric of 'optic precision weapons and surgical strikes' by which 'death virtually disappears from the battle space', and war for nations operating those weapons may be re-enchanted, 'War and Peace', *Transactions of the Institute of British Geographers* 35 (2010), pp. 154–186, pp. 164–165.

[13] Baudrillard, pp. 44, 76.

[14] Yvonne Chui, *Conspiring with the Enemy: The Ethic of Cooperation in Warfare* (New York: Columbia University Press, 2019), p. 82.

[15] Lara Kriegel, 'The Strange Career of Fair Play, or Warfare and Gamesmanship in the Time of Victoria', in *The Oxford Handbook of Victorian Literary Culture*, ed. by Juliet John (Oxford: Oxford University Press, 2016), pp. 268–283.

operator as martyr', subject to 'emotional and mental trauma', 'compounded, not lessened, by the absence of bodily risk and the loss of the "warrior ethos"'.[16] This investment in the drone operator's suffering due to their feeling for the enemy has continuity with earlier endorsements of killing with 'right feeling' that sought to offset contradictions within the idea of the 'Christian soldier' as loving while killing his fellow man.[17] Announcements of feeling for the enemy at both close range and distance, then, might variously morally legitimate and enable war.

The range of conflicts scrutinised in the following chapters suggests that new technologies have not straightforwardly created barriers to enemy intimacy; in some ways, the enemy is closer than ever before, through social media on our screens in our bedrooms, as Jared Ahmad's chapter argues. Contributors scrutinise the ways in which the enemy has, often controversially, been defined in conflicts ranging across a broad geography from the early nineteenth century to the present. They consider which groups are excluded from being counted as enemies, particularly women and children, and where the lines become blurred in violence towards them. While civilian status clearly does not inevitably protect women from war violence, the historic symbolic association of combatancy with masculinity could also result in greater enmity towards women soldiers. Chapters also consider shifts between categories where individuals move from an enemy to allied force, become Prisoners of War, where civilian populations are purposely redefined as enemies, or post-war contexts of (sometimes uneasy) reconciliation.

Co-operation and Grievability in Contact Zones: The Critical Context of Enemy Encounter

The small field of existing work on interactions between enemies suggests that war depends on co-operation and reciprocity, so that formal and informal agreements on behavioural codes between opposing sides are commonplace. As Yvonne Chui argues, reciprocal restraint is a recurrent feature of warfare, elided by the popular emphasis on the First World War as exceptional: 'The live and let live systems in WW1 trenches could be the

[16] Roger Stahl, *Through the Crosshairs: War, Visual Culture, and the Weaponised Gaze* (New Brunswick: Rutgers University Press, 2018), pp. 78–79.

[17] Holly Furneaux, *Military Men of Feeling: Emotion, Touch and Masculinity in the Crimean War* (Oxford: Oxford University Press, 2016), pp. 68–74.

exception that proves the rule, but in fact, the very structure of modern war itself is a co-operative enterprise'.[18] As Chui summarises, 'during the modern period, from the sixteenth-century onward, the ethic of co-operation has become systematised in international laws and institutions', incorporating 'weapons bans, Prisoner of War regulations, requirements to wear uniforms into combat, and protections for clergy and medics'.[19] Planned, institutionalized co-operation is supplemented by individual instances in which hostility evaporates, when the enemy is spotted, for example, with trousers down, or enjoying a smoke, or when a treasured photograph of a loved one invites a different feeling.[20]

Chui's work extends the scope of Tony Ashworth's ground-breaking attention to the normalcy of 'live and let live' between British and German fronts in the First World War, by which the enemy's mealtimes, toilet facilities and even leisure pursuits might be protected through a 'process of reciprocal exchange among antagonists, where each diminished the other's risk of death, discomfort and injury by a deliberate restriction of aggressive activity, but only on the condition that the other requited the restraint'.[21] Though First World War fraternization has, historically, been more difficult to recognise in France as an invaded country and in Germany where it was denied as 'anti-German' under National Socialism, Rémy Cazals and Olaf Muller point to similarly extensive practices of live and let live in both armies, concluding that 'understandings and fraternization as part of trench warfare cannot be seen as exceptional or marginal'.[22]

Rather than a unique, unrepeatable event, as it is often presented in British popular culture, the Christmas Truce of 1914 is, instead, the best-known example of the ceasefires and often friendly interactions between sides that structure war violence. While care for enemies in truces might expose the illogic of war and be retold in hopes of peace, a growing body of scholarship demonstrates a history of intimacy and co-operation between sides that was no barrier to continuing violence, and might even enable the continuation of war. This complicates a widely held view that

[18] Chui, p. ix. Chui cites forms of live and let live in the American Civil War and the Korean War (1950–1953), pp. 15–16.

[19] Chui, pp. ix-x.

[20] See Chui, pp. 37, 39, 9.

[21] Tony Ashworth, *Trench Warfare 1914–1918: The Live and Let Live System* (Basingstoke: Macmillan, 1980), p. 19.

[22] Rémy Cazals, 'Good Neighbours', in *Meetings in No Man's Land*, pp. 78–166, p. 161.

dehumanization of the enemy enables war violence.[23] As Olaf Mueller suggests in the context of German experiences of truces in the First World War 'perhaps the chance to escape, however briefly, from the horror of static warfare enabled so many men to endure more than four years of conflict and prevented any genuinely effective resistance. Similarly, we cannot be certain that the agreements, often tacitly tolerated by the officers, did not in the end "help" to extend the war'.[24] In the different context of the American Civil War, Lauren Thompson suggests that frequent incidents of fraternisation were a means by which 'soldiers redefined themselves', 'constructing outlets to help them withstand mental and physical hardships'.[25] Friendliness between enemies might not only help to prolong war, it could also confirm rather than challenge inequalities, as shown by Thompson's findings that 'fraternization was a ritual denied to black men, and would be indicative of race relations in postwar America'.[26]

The modern period of warfare this book covers, from the nineteenth century to the present, is characterised by the codification of laws designed to reduce the evils of war accompanied by national claims to peculiarly humane ways of waging war and empire, which were and continue to be undercut by starkly uneven applications of international law and humanitarianism. While war relies on the co-operation documented by Chui, the forms and extent of reciprocity are clearly different depending on the nations involved, their relative military and economic power, their global position and their pre-existing attitudes about themselves and their opponents. Noting what she gently terms the 'irony of the contrast between attempts to codify laws of war during the American Civil War and the simultaneous horrific treatment of American Indians', Chui recognises that laws of war are also 'used to exclude certain people from such treatment and to both justify and perhaps unintentionally worsen their situation—although even then it is not clear that it would be better to have no

[23] See, for example, Margaret Macmillan's 2018 Reith Lecture on 'War and Humanity', in which she made the familiar argument that 'dehumanisation is absolutely critical in how people behave in war and will often lead them to war' citing the Nazi propagandistic languages of germs, viruses and animals to transform neighbours into aliens. 26 June 2018, BBC Radio 4.

[24] Mueller, 'Brother Boche', p. 169.

[25] Lauren Thompson, *Friendly Enemies: Soldier Fraternization throughout the American Civil War* (Lincoln: University of Nebraska Press, 2020), p. 3.

[26] Thompson, p. 7.

rules at all'.[27] While international laws and relief organisations seek to reduce the evils of war, they simultaneously support the continuation of war-legitimating claims that conflicts can be prosecuted virtuously and humanely, part of a wider insistence, always highly subjective as the following chapters show, that clear distinctions can be made between civilised and barbarous warfare and legitimate and illegitimate violence. Chapters in this collection reckon with the extreme violence enabled by such exclusionary logics of civilisation and indeed of humanity, and consider the asymmetries of 'co-operation'.

The book as a whole is concerned with ways in which nationhood and race structure differential valuations of enemy lives. In the context of the Black Lives Matter campaign, Judith Butler has described the 'racial structures that make such grotesque distinctions between which lives are valuable (and potentially grievable, if lost) and those which are not'.[28] She builds on her earlier work on uneven distributions of precarity through which some lives and deaths remain globally invisible while others are invested with great significance: 'only under conditions in which the loss would matter does the value of life appear'.[29] This collection reckons, variously, with the global inequalities of grievability and precarity that structure encounters between enemies.

Nations typically present their own actions as humane by establishing a barbarous enemy as their moral opposite. As Jared Ahmad argues in his essay in this collection, 'throughout history, nations have sought to define themselves by demonising, dehumanising and exaggerating the threat posed by "enemy" populations, something that, as a consequence, has served to bolster recruitment efforts and mobilise citizens in support for wars and unpopular foreign policies'. This collection scrutinises the instability of divisions between ally and enemy, considering ways in which this othering of the enemy is strained or breaks down through encounters that might reveal alternative lines of allegiance or raise questions about the national narrative advanced by one's own country.

Postcolonial studies have complicated straightforward ideas of power dynamics between oppressors and oppressed, arguing instead as Homi

[27] Chui, p. 89.

[28] Judith Butler, *The Force of Non-Violence: An Ethico-Political Bind* (New York and London: Verso, 2020), p. 11.

[29] Butler, *Frames of War: When is Life Grievable?* (London and New York: Verso, 2009), p. 14. See also Butler, *Precarious Life: The Powers of Mourning and Violence* (London and New York, Verso, 2004).

Bhabha does for a more hybrid understanding of dialogues between these groups that shape and form them both, albeit unevenly.[30] In her widely applied work on 'contact zones' as 'social spaces where disparate cultures meet, clash and grapple with each other, often in highly asymmetrical relations of domination and subordination', Mary Louise Pratt brings these ideas together with the linguistic concept of 'contact languages', forms of communication extemporised in encounters across ideologies and cultures.[31] She proposes a method that foregrounds 'the interactive, improvisational dimensions of colonial encounters so easily ignored or suppressed by diffusionist accounts of conquest and domination' to consider 'copresence, interaction, interlocking understandings and practices often within radically asymmetrical relations of power'.[32] In her use of this terminology to consider global encounters in the First World War, Anna Maguire posits that 'what contact zones did ... was to provide an extraordinary opportunity for individuals to elaborate on and confront the givens of colonialism'.[33]

The chapters of this book unfold in a range of contact zones. All are concerned with asymmetries of power in enemy encounters, including by gender, sexuality, economic security or precarity. Within these unequal power dynamics though, there are also opportunities for both sides to be shaped by the encounter and to reassess their own cultures and norms. Even in the most asymmetric conflicts, there are moments at which the value of enemy life can 'flicker' into view, to use Butler's term, and potentially provoke a reassessment of national narratives.[34] We follow Pratt in taking 'a "contact" perspective' that emphasises how enemies 'are

[30] Homi Bhabha argues that 'if the effect of colonial power is seen to be the *production* of hybridization rather than the noisy command of colonialist authority or the silent repression of native traditions, then an important change of perspective occurs'. 'Signs Taken for Wonders: Questions of Ambivalence and Authority Under a Tree Outside Delhi, May 1817' *Critical Inquiry* 12, 1985, pp. 144–165, p. 154.

[31] Mary Louise Pratt, *Imperial Eyes: Travel Writing and Transculturation* (London and New York: Routledge, 1992), p.4.

[32] Pratt, p. 7.

[33] Anna Maguire, *Contact Zones of the First World War: Cultural Encounters Across the British Empire* (Cambridge: Cambridge University Press, 2021), p. 4.

[34] Butler, *The Force of Non-Violence*, p. 73. In *Militarized Cultural Encounters in the Long Nineteenth Century: Making War, Mapping Europe*, Joseph Clarke and John Horne (eds), (Basingstoke: Palgrave, 2018), Clarke and Horne use the term 'counter-encounters' to recognise that even within the most asymmetric conflicts, 'encounters could never be unilateral', 'Introduction', pp. 1–22, p. 4.

constituted in and by their relations to each other'.[35] Here we extend the nuanced account given by Santanu Das, Anna Maguire and Daniel Steinbach of colonial encounters in the First World War, which they characterise as governed by a 'combination of different impulses—racism, suspicion, warmth, affection—coexisting or evolving from one to the other'.[36]

What's Enmity Got to Do with It? Contributions and Provocations

This collection aspires to a fully intersectional recognition of the ways in which constellations of class, gender and race informed relations between enemies. It features an extended, multi-authored chapter by Lisa Todd, Jordyn Bailey and Allison L. Bennett on global sexual encounters, considering the weaponization of rape as well as the power dynamics of more consensual cross-side intimacies. Vikki Hawkins continues the attention to military sexual violence as a key node of enemy encounter in her discussion of the Imperial War Museum's reworking of its presentation of the Second World War. Chapters by Holly Furneaux and Rio Creech-Nowagiel demonstrate that age is another significant category in interactions with the enemy, with children occupying a particularly emotive space in both state propaganda and individual experiences.

Language is a powerful force in forming these expectations and allowing for extreme violence, particularly against colonial or irregular opponents. Spanning two centuries, the chapters in this volume show the longevity of specific terms and tropes used to marginalise the enemy 'other'. In just one example, the word 'bandit' is used for insurgents in Matilda Greig's chapter on the Napoleonic era and recurs in Creech-Nowagiel's study of the Malayan 'Emergency' in the interwar period, a hundred and fifty years later. Contributors also highlight the hypocrisy this strategic language can hide: in both Greig and Creech-Nowagiel's chapters, a stress on the criminal or 'barbaric' conduct of insurgents masks similarly condemnable conduct by establishment forces and allows for an insistence on the honourable image of the regular army. By contrast, Jared Ahmad shows how Islamic State propaganda has exploited and repurposed

[35] Pratt, p. 7.

[36] Santanu Das, Anna Maguire and Daniel Steinbach (eds), *Colonial Encounters in a Time of Global Conflict*, 'Introduction' (London and New York: Routledge, 2022).

a long history of Western stereotypes around the 'East', Islam and terrorism to increase the emotive power of its messaging.

Narrative tropes and constructions of the enemy are often long-lasting, spanning centuries, though appropriated and modelled to fit each new conflict. Ahmad's chapter shows how these can be weaponised by both sides, and Hawkins's chapter suggests the stickiness of attitudes to the enemy. She recalls how members of the public in a recent people's forum for the IWM Second World War galleries found it hard to imagine seeing a German fighter pilot the same way as a British one—as a real person, with a name, bravery and their own personal and patriotic motivations. Christine Sylvester's reflective essay, meanwhile, shows how a range of war museums around the world continue to obscure the enemy entirely, whether through simple absence or by redirecting the visitor to be dazzled instead by military equipment, dioramas and flashing lights.

This collection also underlines the potential heterogeneity of experiences of enemy encounter within one army or 'side'. Catherine Bateson's chapter assesses the perception of fellow feeling between Irish-American soldiers fighting on opposing sides in the American Civil War, asking how far identification with a larger Irish nationalist fight or a sense of shared ethnic heritage might blur the category of enemy and ally for these troops. Meanwhile, Lily Tekseng's chapter unpicks shifting concepts of the 'enemy' in the Indo-Myanmar borderlands in the Second World War and post-independence periods, showing how ideologies of race, familiarity and cosmopolitanism were mobilised to cross enemy lines and generate both new alliances and new suspicions between the multiple groups involved. Perceptions of racial identity underscore all these encounters, with racist thought sometimes creating different reactions to separate subgroups within an opposing force: in Julia Ribeiro Thomaz's study of First World War poetry, German soldiers describe their hatred for the British as race traitors, for their use of colonial forces in Europe, and are therefore able to feel 'sympathy' for the enemy Russian and French troops perceived as under the British heel.

How far, then, is thinking of the enemy as an 'equal' in some way a precondition for friendly fraternisation in warfare? This is a question which the chapters in this volume all, variously, pose. Fellow feeling with the enemy is, in one interpretation, easy to build: it can be based on a presumption of shared ethnicity, gender, language, customs and adherence to the laws of war, rank or class. It can also be born in the moment of encounter itself, like the taking of a prisoner of war or the funeral of an enemy who died on domestic soil, as Grady's chapter explores, contradicting even previously strongly held preconceptions. The Afterword suggests that these preconceptions of the enemy might be revisited and reworked long after the conflict in affectively powerful meetings, by choice or chance, between former adversaries. If fellow feeling is quite easily found, though, what consequence does that have for soldier psyche and the continuation of war? Is work needed to keep shoring up and solidifying ideas of enmity and otherness, perhaps in the efforts of the state and military hierarchy or among soldiers themselves because otherwise it would decay?

Taken together, the chapters raise the question: Is a clear feeling of enmity necessary for war? Sources analysed in this book claim that groups of friendly, fraternal opponents fought each other without hesitation in the Napoleonic Wars and the American Civil War and the First World War. Friendships, even putting-life-in-their-hands trust—as Stefan Aguirre Quiroga's chapter shows—could be formed with former mortal enemies in the Vietnam War, while Tekseng's chapter traces how quickly and fluidly alliances and enmities could shift in a complex borderland region.

Not all attitudes to the enemy, however, are equally flexible—some, especially those underpinned by ideas about racial or colonial hierarchies, or the criminality of irregular warfare, persist and continue to authorise extreme violence. Tom Menger's study of the limits of fellow feeling in *fin-de-siècle* colonial warfare demonstrates that for representatives of the British, German and Dutch Empires, interactions with the colonial enemy were coloured by assumptions of their opponent's inferiority that had been inculcated from an early age. By contrast, Furneaux's chapter could be used to argue that there were other ways of forming sympathetic relationships across differences of race and allegiance, with feeling for the enemy both shoring up and destabilising imperial ideologies of superiority.

This collection shows the particular significance of the concept of civilisation in shaping enemy encounters through the nineteenth century to the present. Designations of the civilised or uncivilised enemy could both allow friendly fraternisation between apparent enemies and enable extreme violence. By spanning European and global conflicts, and by stretching through the modern period, the collection creates a new field in encounter studies, showing continuities across time, geographies and different forms of conflict. Far from being restricted to the Christmas Truces of 1914, intimate enemy encounters are prevalent across various modes of modern conflict, from pitched battlefields to punitive colonial expeditions to counterinsurgency campaigns.

Using a wide range of disciplinary approaches, the volume shows the significance of a wealth of different sources to understanding the nuances of enemy encounters. It incorporates literature, life writing, photography, poetry, song, theory and handbooks, diaries, letters, propaganda, material objects, museum displays, graves and funerals, monuments, the Internet and social media, even a wooden door. Many of the chapters reckon with the violence embedded in these historical materials and reflect on how these sources might be approached today without perpetuating that violence.

Chapter Summaries

The chapters in this collection have been arranged chronologically, starting with the early nineteenth century and continuing up to the present day with the ongoing war in Ukraine. What follows is a brief reader's guide, but each chapter has a full abstract at the start for further reference.

In her chapter 'Counterinsurgency for Readers at Home: Chilling Encounters with Guerrillas in Post-Napoleonic Novels' Matilda Greig examines how military and civilian encounters with Spanish guerrillas during the Peninsular War (1808–1814) were imagined as violent, dangerous and horrifying in novels and memoirs of the post-Napoleonic era. Unpicking the common tropes and narratives around the figure of the *guerrillero*, the chapter shows how irregular warfare itself was cast as a barbaric, criminal activity at the start of the nineteenth century, decades before the international laws of war were agreed.

The adoption of 'enemy' children by British troops in the eighteenth and nineteenth centuries forms the focus of '"I was a stranger and ye took me in": Adopting the Enemy', with Holly Furneaux exploring the emotive stories of the sons of Tipu Sultan, two Russian boys from the Crimean War and Ethiopian Prince Alemayehu, all brought under British guardianship and portrayed as 'good news stories' in the Georgian and Victorian press, despite the dubious benefit for the children themselves.

'"He said he was an Irishman also": Irish-American Union and Confederate Encounter Stories in the American Civil War' moves to the context of the US Civil War, where Catherine Bateson analyses real and fictional stories of encounters between Irish-American Confederate and Union soldiers. Using letters, memoirs, songs and newspaper articles, Bateson shows that references to these encounters reveal more about broader international desires for a common purpose, often connected to a larger Irish nationalist fight, than about shifting allegiances on the battlefield itself.

In 'Racial Difference and the Limits of Fellow Feeling: British, German, and Dutch Soldiers Encounter the Enemy in Colonial Wars, c. 1900', Tom Menger investigates how conceptions of racial difference informed moments of enemy encounter in *fin-de-siècle* colonial warfare, focusing on case studies from the British, German and Dutch Empires in Africa and Asia. Colonial soldiers understood their role as performing their supposed racial, civilisational and masculine superiority, which imposed limits on the potential for fraternisation and created a wide, elastic and sometimes paradoxical category of 'enemy', which could include civilians, including women and children.

'Global Sexual Encounters with Racialized 'Others' in War, Genocide, and Captivity, 1905–1945' is an extended, multi-authored chapter by Lisa M. Todd, Jordyn Bailey and Allison L. Bennett, focusing on the theme of soldier-civilian sexual encounters in global conflicts in the first half of the twentieth century. Running through each case study is the theme of how 'whiteness' worked to facilitate violence, coercion and consent, from the sexual violence perpetrated by occupying German soldiers against Indigenous Namibian women in 1904–1908 in Todd's analysis, to Bailey's examination of the cautions given to Australian servicemen in First World War Egypt to avoid sexual encounters with women in brothel areas for fear of venereal disease and the effects it might have on the 'health' of the white race. In Bennett's study of 1940s Canada, by contrast, wartime ideals of 'whiteness' and sexualised masculinity led many civilians to treat German prisoners of war with hospitality and affection, despite the men's status as 'enemies' and 'Nazis'.

Julia Ribeiro S C Thomaz's '"Yoors truly German": A Micro-History of Poetic Enemy Encounters of the 1917 Retreat' is a micro-history of enemy encounters during the First World War through the lens of poetry, anchored in the close reading of a 13-line poem carved onto a cupboard door in English by German soldiers, and left behind for British troops to find. The poem forms an example of deferred encounter, which happens through text and object rather than physical proximity, allowing the German authors to shape the narrative of the meeting and question their status as 'other'.

In 'Dying with the Enemy: Prisoners of War Deaths in First World War Britain', Tim Grady focuses on how civilians interacted with prisoners of war on home soil, examining the ways in which the enemy died in First World War Britain and how both the press and the local population responded. He argues that death unavoidably altered British perceptions of the enemy, blunting fears of Germans as a 'faceless mass' as German fighter pilots or prisoners gained an identity, received individual funerals and were buried in the countryside with grave markers bearing their names.

'Enemies, Subjects and Allies in Wartime Indian Northeast' by Lily Tekseng considers the borderland Indian Northeast in the Second World War, which transformed the region into a 'potpourri of murky categories: empires both old and new, colonial subjects, witnesses, friends, allies and enemies'. Using short stories by writers Kenzaburō Ōe and Temsüla Ao, Tekseng highlights the potential for changed expectations and surprises in the encounters between purported enemies and allies and the importance of ideas about race in shaping these meetings.

'Enemy Encounters in the Imperial War Museum's Second World War Galleries' takes the volume into dialogue with the heritage sector, with Vikki Hawkins analysing her critical and interpretative approach as one of the curators of the Imperial War Museum's new Second World War Galleries, opened in October 2021. The chapter forms a case study of how to bring together objects, themes and stories from around the world to challenge visitors to a national war museum to think critically about the boundary between enemy and ally.

In '"Children Made to Carry Bombs": Staging the Malayan "Emergency" in Postwar Britain', Rio Creech-Nowagiel examines the use of staged press imagery to convey a misleading impression of postwar Britain's counterinsurgency campaign in Malaya, during the so-called 'Emergency'. Continuing themes from earlier chapters about the way racial stereotypes informed colonial encounters with civilians and irregular combatants, as well as the blurry status of children as 'enemies', Creech-Nowagiel also challenges historians of colonial archives to think afresh about best practice when dealing with and reproducing difficult source material.

Confronting the question of what happens when a former enemy officially crosses over into being an ally instead, Stefan Aguirre Quiroga's 'Trusting Your Enemy: American Encounters with the Kit Carson Scouts During the Vietnam War, 1966–1973' tells the story of the Kit Carson Scout Program in the Vietnam War, in which South and North Vietnamese defectors were employed as auxiliaries by the United States army. For American soldiers, living and working closely with the Scouts, and often relying on them in life and death situations, caused a range of reactions, from entrenched distrust to camaraderie, greater understanding and personal friendship.

In the final full chapter in this collection, '"Islam is the religion of the sword not pacifism": Strategic Nostalgia and Self-Othering in Islamic State Propaganda', Jared Ahmad offers a provocative take on the theme of enemy encounters by showing how contemporary terrorist organisations, particularly Islamic State, deliberately invoke narratives and imagery in their propaganda that are designed to exploit existing Western stereotypes about the 'East' and Islam. Identifying two key messaging strategies, 'strategic nostalgia' and 'self-othering', Ahmad shows how debates around contemporary encounters with the enemy are shifting, and how new communication technologies are being weaponised.

The following two chapters take the form of reflections on two different aspects of the present-day commemoration and representation of the enemy. In 'Memorializing the Enemy Three Ways: Australia, Japan, the USA', inspired by discussions at the international conference we organised in conjunction with Imperial War Museums in 2021, Christine Sylvester reconsiders her visits to three separate sites of war memorialisation through the lens of enemy encounters: the Australian War Memorial, the American Vietnam Veterans Memorial and the Hiroshima Park and Museum. Her piece culminates by posing timely questions about the many-sided notions of the enemy that are emerging at a time of international conflict and rapidly advancing climate change. In 'USSR 2.0 vs. The Fourth Reich: Depicting the Enemy in the Official Occupation Media of Kherson Oblast', considerations of how 'the enemy' is formed are brought into sharp relief in Mykola Homanyuk's study of Russian and Ukrainian social media and press propaganda in occupied Kherson. Drawing on a dataset gathered over the course of seven months spent there under the occupation, Homanyuk's essay reveals the key Russian and Ukrainian narratives and tactics used to present the enemy.

Acting as a conclusion to this collection is an 'Afterword: Coming Home—Finding Form for Veterans' Life Stories of Enemy Encounter' co-written by the editors, inspired by the partnership between the Enemy Encounters AHRC-funded project and Re-Live, an arts in health charity acclaimed for its work with veterans. As part of this collaboration, we had the privilege of holding a series of conversations with Karin Diamond, Artistic Director at Re-Live, and a focus group composed of former British Army soldiers, in which the participants discussed the history of interactions across military sides, the transformative power of life-writing (including in the abridged form of a comic strip) and what 'the enemy' might mean today.

Key Bibliography on Enemy Encounters in Modern Warfare

Ashworth, Tony. 1980. *Trench Warfare 1914–1918: The Live and Let Live System.* Basingstoke: Macmillan.

Chui, Yvonne. 2019. *Conspiring with the Enemy: The Ethic of Cooperation in Warfare.* New York: Columbia University Press.

Clarke, Joseph, and John Horne, eds. 2018. *Militarized Cultural Encounters in the Long Nineteenth Century: Making War, Mapping Europe.* Palgrave Macmillan.

Das, Santanu, Anna Maguire, and Daniel Steinbach, eds. 2022. *Colonial Encounters in a Time of Global Conflict.* London and New York: Routledge.

Ferro, Marc, Malcolm Brown, Rémy Cazals, and Olaf Muller. 2007. *Meetings in No Man's Land: Christmas 1914 and Fraternisation in the Great War.* London: Constable and Robinson.

McCoppin, Rachel (ed.). 2019. 'War and Literature: Commiserating with the Enemy'. Special issue of *Humanities* 8.

Reiss, Matthias, and Brian K. Feltman, eds. 2022. *Prisoners of War and Local Women in Europe and the United States, 1914–1956: Consorting with the Enemy.* Cham: Springer International Publishing.

Thompson, Lauren. 2020. *Friendly Enemies: Soldier Fraternization Throughout the American Civil War.* Lincoln: University of Nebraska Press.

Counterinsurgency for Readers at Home: Chilling Encounters with Guerrillas in Post-Napoleonic Novels

Matilda Greig

In the early nineteenth century, readers of adventure stories, romances, and military memoirs from Britain and France were introduced to a new trope: the guerrilla ambush scene. While travelling in the wilds of a romanticised southern European country, your favourite character would wander off the path in some way. Perhaps they have been misled by a guide; perhaps they have become lost returning to their regiment; perhaps they are simply too occupied with their own thoughts to notice that the woods around them have become darker, thicker, more mysterious or that the route now leads through a narrow mountain pass. And in an instant, everything changes. Sinister figures materialise as if from nowhere, surrounding the hero or heroine and their party and taking them prisoner. Their faces bear cruel expressions, their clothes are rough and mismatched, their belts positively bristle with firearms, and, most important of all, they each carry a long, wicked knife. To their victims, their appearance marks

M. Greig (✉)
National Army Museum, London, UK
e-mail: mgreig@nam.ac.uk

H. Furneaux, M. Greig (eds.), *Enemy Encounters in Modern Warfare*, https://doi.org/10.1007/978-3-031-56748-3_2

them out immediately for what they are: guerrillas, scourge of the local countryside, brigands motivated only by greed. If the protagonist is really unlucky, there will even be an evil-minded monk leading the band, urging them on to ever greater atrocities. The prisoners will then always be taken off to some type of isolated evil hideout which, depending on the creativity of the author, might be in a ruined monastery or a moonlit river gorge, or might simply try to match its inhabitants in sinisterness by being called 'Cadaver Hamlet'.[1] There they are typically threatened with immediate execution, which is postponed long enough to build a little tension for the reader or to give the hero time to pledge his love or service to the heroine, before they either manage to escape or are saved by the timely arrival of honourable regular soldiers.

This scene, played out in similar forms in numerous French and British novels and memoirs from the first half of the nineteenth century, would have struck many familiar notes for readers of the late-eighteenth-century works of Ann Radcliffe and her followers: Gothic historical romances in which *banditti* carried off beautiful heroines or lonely travellers and took them to castles ruled by cruel, controlling Counts. Now, however, timeless bandits were frequently replaced by *guerrilleros*, characters from a very specific time and place: the Peninsular War, from 1808 to 1814. This was Napoleon Bonaparte's ill-fated invasion and attempted occupation of Spain and Portugal, against which widespread civil resistance took place, including in the form of organised groups of irregular combatants who disrupted communications, intercepted supplies, passed on information, and engaged in skirmishes or occasionally even battles with the French forces. The Spanish word for this type of military action, 'la guerrilla' or 'small war', was mistranslated into French and English at the time and used to mean the irregulars themselves.[2] The modern word 'guerrilla', then, has its genesis in the Peninsular War. As to the concept of what a guerrilla was by 1814, and how they should be perceived and categorised, that was another question entirely.

Composed by both male and female authors over the decades that followed the Peninsular War, literary scenes of guerrilla ambush made villains

[1] Such is, unfortunately, the case in the ambush scene in Jacques Antoine Révéroni de Saint-Cyr, *Le Torrent des passions: ou, Les dangers de la galanterie*, 2 vols (Paris: J. N. Barba, 1818), II, p. 150.

[2] As used in the nineteenth-century literature in this chapter, the French word for an irregular fighter from the Peninsular War was 'guérilla', with the English equivalent being 'guerilla' with only one 'r'.

out of Spanish insurgents in order to shock and thrill, but also to join in a much broader cultural and political discussion about the morality of irregular warfare itself. A key reason why the insurgents of anti-Napoleonic Spain had inspired new terminology was that they were portrayed by the Spanish media, and by many sympathetic liberal commentators internationally, as politically rather than materially motivated: patriots rising up to fight for national freedom.[3] This was a conceptual shift which differentiated the guerrilla from the diverse late-eighteenth-century figures of the mercenary (freelance contractor hired to do the job), partisan (regular detachment commander, leading units of light troops) or member of organised crime.[4] While this shift was visible most readily in anecdotes printed in Spanish papers and Spanish biographies of leading guerrilla commanders, which often told Robin Hood-esque tales of guerrillas liberating prisoners or recapturing stolen goods and redistributing them to the people, a section of literature produced in Britain and France also absorbed this new idea of the irregular to an extent. Guerrillas who were heroic, honourable—and handsome!—appeared in a number of romance novels written by or for women, as well as in military adventure stories set in the Peninsular War.[5] Yet in all these episodes, the figure of the Spanish guerrilla was still held at arm's length as an indisputable 'other'—either exoticized as a mysterious romantic lead, with lavish descriptions of his clothes,

[3] A historiographical debate continues over how much of this was true – Charles Esdaile has influentially argued it was largely a myth, whereas Richard Stites, for example, is more equivocal. Esdaile, *Fighting Napoleon: Guerrillas, Bandits and Adventurers in Spain, 1808–1814* (New Haven and London: Yale University Press, 2004); Stites, *The Four Horsemen: Riding to Liberty in Post-Napoleonic Europe* (New York: Oxford University Press, 2014), pp. 36–39. Studies have also been written in Spanish on how individual guerrilla commanders built this rosy renown, both at home and in Spanish America: e.g. Alberto Ausín Ciruelos, 'Imagen y propaganda de los guerrilleros españoles en la *Gazeta del Gobierno de México* durante la guerra de la Independencia (1810–1814)', *Aportes*, 94.2 (2017), pp. 7–39.

[4] Martin Rink, 'The Partisan's Metamorphosis: From Freelance Military Entrepreneur to German Freedom Fighter, 1740 to 1815', *War in History*, 17.1 (Jan., 2010), pp. 10–14.

[5] See for example: Alicia Le Fanu, *Don Juan de las Sierras, or, El Empecinado. A Romance*, 3 vols (London: A. K. Newman and Co., 1823); Felix M'Donogh, *The heroine of the Peninsula; or, Clara Matilda of Seville*, 2 vols (London: Sherwood, Gilbert and Piper, C. Chapple, A. K. Newman and Co., 1826); Anon. [Francis Glasse], *Ned Clinton; or, The Commissary, comprising adventures and events during the Peninsular War: with Curious and original Anecdotes of military, and other remarkable characters*, 3 vols (London: William Marsh, 1825), I, pp. 118–139.

dark and striking features, and total command over his men, or revealed to have in fact been a British officer in disguise all along.[6]

Another group of novels, by contrast, denied the patriotic motives of the *guerrilleros* and any positive interpretation at all. Reacting with hostility to changing ideas about the waging of war, they wrote scenes that encouraged the sense that an encounter with a guerrilla would most likely be a violent, life-threatening, and traumatic one. Many of these writers were based in France, whose army had suffered directly from guerrilla attacks during the Peninsular War, but many others were from Britain, which had been allied to Spain during the conflict and greatly helped by its irregular auxiliaries. The lines of 'true' enmity in the war were therefore blurred by this literature, aiming to present the irregulars as the real object of hate and fear—while the conventional war of pitched battlefields and sieges seemed to take on a less emotive, background role. Indeed, several studies have noticed a phenomenon in British and French veterans' military memoirs from this period in which troops from these two opposing armies, supposedly enemies, often treated each other with friendliness, admiration and restraint, and openly fraternised between battles.[7] One contemporary Anglo-French lithograph even showed a group of English prisoners nobly choosing to help their French captors to fight off a guerrilla ambush, rather than turn on them and be rescued (see Fig. 1). In these narratives, which form an important background context to the novels discussed here, British and French regular troops shared a set of 'civilised', martial values and were playing an equal game—whereas the *guerrilleros* were subverting it.

The idea that guerrillas or non-state fighters were an unstable 'other' to the regular army, not respecting or living up to the standards of conduct in warfare, suffused attempts to codify the international laws of war in the second half of the nineteenth century. In Francis Lieber's code for the US Civil War and the first Geneva Convention in 1864, irregulars not attached

[6] For example, the heroic guerrilla leader turns out to be an Englishman in disguise in Emma Parker, *The Guerrilla Chief: A Novel*, 3 vols (London: William Lindsell, 1815).

[7] Laurence Montroussier, 'Français et Britanniques dans la Péninsule, 1808–1814: étude de mémoires français et britanniques', *Annales historiques de la Révolution française*, 348 (April-June, 2007), pp. 132–133; Gavin Daly, *Storm and Sack: British Sieges, Violence and the Laws of War in the Napoleonic Era, 1799–1815* (Cambridge: Cambridge University Press, 2022), pp. 122–123, 159; Matilda Greig, *Dead Men Telling Tales: Napoleonic War Veterans and the Military Memoir Industry, 1814–1914* (Oxford: Oxford University Press, 2021), pp. 77–78.

Fig. 1 Lithograph after Giscard, 'English heroism' (London: Lithographie Francaise, ca. 1812). Courtesy of the Council of the National Army Museum, London. Purports to depict an incident in the Peninsular War wherein English soldiers, who had been captured by the French, assisted their captors in repelling an attack by Spanish guerrillas led by Francisco Espoz y Mina, before dutifully returning their borrowed weapons

to a standard national army were exempted from any sort of protection if captured or wounded. In 1907, the Hague Convention extended the definition of a lawful belligerent to include volunteer and militia corps who had clear leadership structures, used a distinctive and recognisable emblem, carried arms openly, and abided by the 'laws and customs of war'. If a rapid popular mobilisation took place in response to an invasion by a foreign country, leaving no time for a clear chain of command and system of identification to be set up, this was also deemed legitimate if arms were still carried openly and 'laws and customs' again obeyed.[8] The paradox was, of course, that non-state groups using irregular tactics could rarely compete in open combat with regular troops and therefore relied on a degree of this forbidden subterfuge to be effective. It was not until a century on, in other words, that the case made by supporters of the guerrilla movement in the Peninsular War for their legitimacy (that the irregular groups had formed to resist the French invasion and were well-organised and patriotic) was included in international legislation. Meanwhile, the opposing contemporary view, that the patriotic guerrilla was a myth and irregulars in Spain and elsewhere were unequal and undeserving participants in conflict, had strongly endured.

This chapter provides important cultural context to the longevity of that marginalising framework. Examining various examples of early nineteenth-century popular literature that deliberately constructed the guerrilla as villainous, I argue that there was a large and complicated literary engagement with the idea of irregular warfare many years before the seismic international accords were signed. Although the Peninsular War is often pinpointed as a significant moment of change in surveys of irregular warfare, it is rare to find studies that engage with the cultural side of this shift, and scholars working on twentieth- and twenty-first-century counterinsurgency doctrine (COIN), international law, or colonial violence often have little in the way of nineteenth-century precedents to consult.[9] Special issues on the origins and global dimensions of 'small wars' organised in the journal *Small Wars and Insurgencies* in the last decade by Beatrice Heuser and Mark Lawrence, for instance, both focus more on the international discussion of the use of irregular tactics than on popular and cultural

[8] *Laws and Customs of War on Land (Hague IV); October 18, 1907: Regulations*, s. I, articles 1–2, <https://avalon.law.yale.edu/20th_century/hague04.asp> [accessed 9 March 2023].

[9] Paul B. Rich, 'Terror and Its Limits: The Historical Understanding of Terrorist Movements, States and Tribes in an Age of Cultural Anxiety', *Studies in Conflict & Terrorism*, 36 (2013), p. 1029.

responses to them.[10] As a whole, survey studies of modern irregular warfare tend to neglect cultural history, with the important exception of Sibylle Scheipers' work, *Unlawful Combatants: A Genealogy of the Irregular Fighter* (2015), which helpfully identifies the underlying narrative strategies used to marginalise and exclude guerrillas and others since the late eighteenth century, and has been influential to the framing of this chapter. Scheipers draws exclusively on legal texts and secondary sources for her study, leaving an opportunity to examine how these narratives were explored, implemented, and diffused by contemporary fiction.[11]

Below, I draw on a survey of the European literature held in the Corvey collection, one of the most important collections of Romantic-era writing in existence. Compiled by the owners of Castle Corvey near Höxter in Germany around the years 1790–1840, re-discovered in the late 1970s and today available as a digital database managed by Gale Nineteenth Century Collections Online, the collection contains more than 9500 works of fiction, prose, drama, and poetry in English, French, and German. Many of these works are little known in scholarship today (Brian Dendle's 1991 survey of the British 'Peninsular War novel' as a genre appears to have only scratched the surface of works touching on the theme).[12] Beginning with a short introduction to the context in which these books were published, the chapter will first show how tension was built in these narratives in order to convince a wide variety of readers that an encounter with a guerrilla, even as a civilian, would be a dangerous and violent event. I then discuss these moments of encounter in more detail, arguing that this section of literature based the illegitimacy or criminality of the *guerrillero* not only on their actions but also on their appearance, attitudes towards women, and ascribed racial identity. Finally, I explore the implication made by these stories—that irregular combatants do not deserve

[10] *The Origins of Small Wars: From Special Operations to Ideological Insurgencies*, ed. by Beatrice Heuser (= *Small Wars & Insurgencies*, 25.4 (2014)); *Insurgency and Counterinsurgency in the Nineteenth century: A Global History*, ed. by Mark Lawrence (= *Small Wars & Insurgencies*, 30.4–5 (2019)).

[11] Sibylle Scheipers, *Unlawful Combatants: A Genealogy of the Irregular Fighter* (Oxford: Oxford University Press, 2015), p. 23.

[12] 'European Literature, the Corvey Collection, 1790–1840', Gale Nineteenth Century Collections Online (NCCO); Brian J. Dendle, '*The Romance of War, or, The Highlanders in Spain:* The Peninsular War and the British Novel', *Anales de Literatura Española*, 7 (1991), pp. 49–64.

equal or humane treatment from the regular army—and the presence, or absence, of a neat resolution to the tale.

Context

The discussion of Spanish guerrillas by northern European authors in the early nineteenth century took place in a context in which the public had already been made quite familiar with the events, themes, and central characters of the Napoleonic conflict in the Iberian Peninsula. For decades prior, information regularly reached people in Britain and France through newspapers, letters from soldiers abroad, and the theatre, which served as a forum for sharing the latest news of the nation's various battles, as well as spreading messages of patriotism and support for each war.[13] Recent research suggests that far from being mostly ignorant of the happenings abroad, the eighteenth- and early-nineteenth-century British public appear to have been keen attendees of wartime-themed plays, even those packed with military and naval jargon, hinting at 'the country's total immersion and detailed knowledge of war'.[14] Indeed, when the Peninsular War itself broke out in 1808, playwrights and theatre managers across England reacted quickly, using it as a setting for reworked stock pieces with an existing Spanish theme.[15] Meanwhile, literature using the Peninsular War as a setting flourished in the early nineteenth century, with the 'Peninsular War novel' forming a significant sub-genre of British fiction and the published memoirs of soldiers returning from the Peninsular campaign achieving lasting popular and commercial success in both Britain and France.[16]

In 1823, events in the Peninsula again made headlines, as another French army crossed the Pyrenees to intervene in Spanish politics, this time to support autocratic King Ferdinand VII against the liberal factions attempting to reform the monarchy—among whom were several famous former guerrilla leaders from the Peninsular War, such as Francisco Espoz y Mina and 'El Empecinado'. The invasion of 1823 strengthened, rather than replaced, literary interest in the Peninsular campaign, bringing it to

[13] See Paula R. Backscheider, *Women in Wartime: Theatrical Representations in the Long Eighteenth Century* (Baltimore: Johns Hopkins University Press, 2022).

[14] Backscheider, p. 263.

[15] Susan Valladares, *Staging the Peninsular War: English Theatres 1807–1815* (Farnham, Surrey/Burlington, VT: Ashgate, 2015), p. 5.

[16] See Dendle; Greig; Neil Ramsey, *The Military Memoir and Romantic Literary Culture, 1780–1835* (London and New York: Routledge, 2016 [2011]).

the attention of a new generation of writers and readers. With it came a surge in writing about Spanish irregular warfare, perhaps reacting to the presence of irregular tactics and fighters on both liberal and royalist sides or hearing warning bells of the continuing risk of insurrection and revolution internationally. As one French author put it in his preface in 1823, 'Spain is now attracting the attention of all Europe. The newspapers of all parties repeat the names of *Guerillas*, of *Guerilleros*, as we have decorated the first *bandoleros* [bandits] who bore the banner of insurrection during the invasion of the French army in 1808'.[17]

In fact, the word 'guerilla' [*sic*] or 'guérilla' features in roughly 97 works of fiction, autobiography, and poetry in the Corvey collection in the 1820s, compared to 12 in the previous decade and 50 in the decade that followed.[18] The vast majority of these mentions refer specifically to the Peninsular War, with five referring to the wars of independence in Latin America and a few being passing references to irregular tactics in general. This was a small but significant development within a much longer literary tradition that cast irregulars in supporting or leading roles: Corvey's data includes thousands of mentions of 'brigands' and 'banditti' between 1734 and 1890, who ran amok in monasteries and castles in Italy, Germany, France, Hungary, Scotland or ancient Britain, as well as Spain. More specifically, the introduction of the term 'guerilla' also overtook the use of the word 'miquelet', a specific term referring to a form of irregular local militia already existing in Catalonia before the Napoleonic Wars.[19]

Building Tension

Before a guerrilla character has even appeared on the pages of a Victorian- or Restoration-era novel or memoir, the groundwork is often laid in the text for the danger they pose to the traveller, both civilian and military.

[17] Auguste Frankoual, *Le partisan des Pyrénées, ou le guerillero, traduit de l'espagnol; par M. Frankoual; suivi des Visions, par Schiller*, 2 vols (Paris: Plancher, 1823), I, p. i. [Although advertised as the translator, Frankoual was most probably the author: see Alexandre Pigoreau, *Petite Bibliographie Biographico-Romancière ou Dictionnaire des Romanciers* (Geneva: Slatkine Reprints, 1968), pp. 35–36.]

[18] Figures compiled from Gale NCCO 'Corvey Collection' NGramData for a text search for 'guerilla' (accented letters included). The figures may be slightly inflated by the fact that Gale includes different volumes of the same work as individual hits.

[19] Gale NCCO NGramData: 'brigand' 2408 volumes; 'banditti' 1449 volumes; 'miquelet' 40 volumes.

There is the sense that there are a great, uncountable number of these criminal gangs preying upon the Spanish countryside—authors writing in both English and French were fond of using the dehumanising word 'infested'—and that they are a serious threat to all but a large and well-armed company of troops, able to blend into the landscape with such ease that their presence is only noticed once the first shots of the ambush ring out.[20] In Peninsular veteran Lt. Col. David Roberts' satirical poem *The Military Adventures of Johnny Newcome* (1815), the naïve Johnny is struck by these fears while travelling to join his regiment in Spain, when his servant's mule stumbles and casts its load on route:

> John now dismounted, and with horror stood;
> They'd told him of Banditti in the Wood.
> "O, Teague! dear Teague! as we are only two,
> If the curst thieves should come, what shall we do?"[21]

Usually, once the 'curst thieves' appear, there is little the vulnerable travellers *can* do, increasing the feeling of high stakes in this type of encounter. Civilian travellers or solitary soldiers are usually immediately taken captive in these stories, while ambushed regiments only manage to escape with heavy casualties. Adding to the tension are rumours of what happens to those who are caught: 'robbed and hung by your feet'; 'sawed in half between two planks'; not to mention simply 'murdered' on the spot.[22]

Women were supposed to doubly fear such an encounter. In Jenny Bastide's *La Belle-Mère* (1828), the female narrator describes her fear of a guerrilla attack:

[20] For example, the verb 'to infest' or 'infester' is used to describe guerrillas in Spain in Etienne Marcel, *Alfred et Coralie: ou, Les Français en Espagne. Par Mlle Emilie M***, auteur de Léopold* (Paris: Armand Aubrée, 1830), II, p. 93; Henri Ducor, *Aventures d'un marin de la garde impériale*, 2nd edn (Paris: Ambrose Dupont, 1833), I, p. 293; Thomas Hamilton, *The Youth and Manhood of Cyril Thornton*, 3 vols (Edinburgh: William Blackwood; London: T. Cadell, 1827), II, pp. 238–239.

[21] David Roberts, *The Military Adventures of Johnny Newcome, with an Account of His Campaign on the Peninsula, and in Pall Mall: With Sketches, by Rowlandson, and Notes, by an Officer*, 2nd edn (London: Patrick Martin, 1816), stanzas 38–39.

[22] All translations from French the author's own. Ducor, 127; Georges Touchard-Lafosse, *Le bosquet de Romainville: confidences du soir*, 2 vols (Paris: Charles Lachapelle, 1833), II, p. 298; Marcel, p. 93.

> Several times on our retreats we heard bands of guerrillas passing by, not far away from us. As such, certain of the fate that awaited us if we were discovered, I armed Henri with an instrument of death, and entreated him to give it to me if we fell into their hands; I did not have the courage to face the ordeal that would follow, and I knew well that after having made me suffer their torments, they would dishonour me again before killing me. But Henri retreated before the cruel duty with which I charged him, and destiny allowed that we were spared that horrible end.[23]

The result was the creation of a shadowy threat that surpassed the formal lines of allegiance and enmity in the war, frightening not only foreign travellers but Spaniards themselves. In Peninsular veteran Thomas Hamilton's novel *The Youth and Manhood of Cyril Thornton* (1827), the hero escapes from French prison in Madrid disguised as a muleteer, with the help of a local guide. They are searched by some French soldiers, but manage to pass undiscovered, yet still do not feel safe: 'Jozé's chief fears, however, were directed not towards the French, but his own countrymen; and had we fallen in with a party of Guerillas [*sic*], even supposing we had the good fortune to escape alive, there is no doubt we should have been left to pursue our journey, in a state of primitive nudity'.[24] Even before the guerrilla ever leaps out from behind a rock, or slides out from behind a tree, therefore, he has been built up as an enemy of humanity, an embodiment of the worst excesses of unregulated warfare: murder, looting, rape.

Moment of Encounter

In crafting, clothing, and scripting a character who could live up to this reputation, early nineteenth-century writers fell upon tropes developed in pre-existing genres of literature and art. From Gothic literature came a fascination with the grotesque, the supernatural, and the terrifying. From a Spanish genre of picturesque landscape painting, stretching back to Salvator Rosa in the seventeenth century, came the idea of 'bandits' as rustic figures in a dark, rocky landscape. Finally, from a rich tapestry of existing preconceptions, fantasies, and ways of writing about Spain, came ideas of a sort of distant, romantic, and exotic land, marred by a despotic

[23] Jenny Bastide [pseud. of Marie-Hélène Dufourquet], *La Belle-Mère*, 3 vols (Paris: A. Boulland, 1828), ii, p. 37.

[24] Hamilton, p. 274.

government, widespread poverty, and superstition.[25] Echoing a quote attributed to Voltaire, 'Africa begins at the Pyrenees', writers and travellers from northern Europe portrayed the Iberian Peninsula as a borderland area more properly considered African than European, encouraging a racist perception of the inhabitants as at best 'exotic' and at worst less human than them.[26]

The resulting villains usually possessed many of the same traits: features described as crooked or unsightly, perhaps obscured by facial hair (one illustrative example has a group of ambushers sporting 'short, but very thick beards, black like the devil's mane, hiding part of their hideous faces'),[27] dark skin colour, ragged or ill-combined clothing, a disparate array of weapons that included a long, keen knife, a cruel disposition, a devious sort of dishonesty wherein evil intent is at first hidden beneath apparently friendly or pious demeanour, a greedy nature, a lascivious approach to women, and a fondness for hideouts in isolated, rugged landscapes. Art could mirror literature: see Fig. 2 for a French artist's combination of many of these tropes in a stereotypical sketch for drawing students. All of these traits served to mark the guerrilla as 'uncivilised' or 'barbaric' and therefore both inferior to and threatening to the audience at home.[28] In addition, many writers used language that dehumanised them entirely: either zoomorphic (comparisons were often made to cats, vultures, spiders, and dogs) or supernatural, with language transforming an ordinary man into 'a regular demon… whose only delight was slaughter and war to the knife! A fellow that could eat fire, as the proverb says, and upon whom lead and steel were alike ineffectual'.[29]

As scholars of modern colonial violence and counter-insurgency will know well, and as Tom Menger demonstrates later in this volume, such

[25] Susan Valladares, 'Romantic Englishwoman and "the Theatre of Glory": The Role of the Peninsular War in Forging British National Identity', *MoveableType*, 4 (2008), p. 112; Stites, p. 5.

[26] Original: 'L'Afrique commence aux Pyrénées', qu. Stites, p. 6.

[27] Alphonse Lorry [pseud. of Pierre Clerjon], *L'Attaque du pont: ou, La fille retrouvée: par Alphonse Lorry, Membre de l'Academie de Robertins, Inspecteur des Eaux thermals de la même ville*, 4 vols (Paris: Th. Berquet, 1830), II, p. 84.

[28] On the long European intellectual tradition of associating primitivism with mountainous or hilly regions, vs. the 'civilized' plains, see Elizabeth Kolsky, 'The Colonial Rule of Law and the Legal Regime of Exception: Frontier "Fanaticism" and State Violence in British India', *American Historical Review*, 120.4 (Oct., 2015), pp. 1225–1226.

[29] James Grant, *The Romance of War: or, The Highlanders in Spain*, 3 vols (London: Henry Colburn, 1846), I, p. 238.

Fig. 2 Nicolas Toussaint Charlet, 'Spaanse guerrillastrijder met dolk in een landschap [Spanish guerrilla with dagger in a landscape]', in *Suite de dessins à la plume à l'usage des élèves des écoles spéciales (…)* (Paris: Gihaut frères, 1839). Courtesy of the Rijksmuseum, Amsterdam. This sketch was part of a book created by a French artist for use in lessons on pen and ink drawing at the Ecole Polytechnique. The students, future officers and engineers, would have copied out these sketches to practice their technique

marginalising and dehumanising rhetorical framing of irregular combatants has two central, politically useful effects.[30] First, it allows all the negative traits of the regulars to be displaced onto them, thereby creating an artificially pure and honourable image of the enlisted soldier or officer.[31] Second, it excuses the regular army if it uses barbaric conduct itself against them, allowing a regiment of British or French troops to ambush and murder guerrillas in these stories with seemingly little guilt or disruption to the narrative. Both strategies are readily noticeable in Peninsular War fiction and memoir, prefiguring the rhetoric around colonial adversaries in the later nineteenth century, and underlining the extent to which the political debate around the changing face of war was also taking place in the type of literature consumed by the ordinary reader.

In these stories, the casting of the Spanish guerrilla as a dark mirror to the ideal early nineteenth-century soldier was often achieved through quite a direct narrative strategy: the repeated side-by-side contrasting of similar characters, who appear as good and bad sides of the same coin. The juxtaposition in these encounters is often communicated through detailed descriptions of the characters' physical appearance, clothing, and weaponry. One striking example comes from the popular novel by Scottish writer James Grant, *The Romance of War: or, The Highlanders in Spain* (1846). On their first night in Spain, the hero Ronald Stuart, an ensign in the 92nd or Gordon Highlanders regiment, and his commanding officer Major Campbell are billeted together in the house of a shopkeeper named Narvaez Cifuentes, who soon turns out to be the book's primary antagonist. Cifuentes is introduced with a slew of unflattering and foreboding physical traits:

> His stature was low, his face was so swarthy as to resemble that of a negro in darkness; his moustaches were thick, fierce, and black, mingling with the matted hair of his huge bullet-head. He wore a long stiletto (openly) in the yellow worsted sash which encircled his waist, and the haft of a knife appeared within the breast of his doublet, or sort of vest with sleeves, which was, like the rest of his attire, in a very dilapidated condition; and altogether, the Senor Narvaez Cifuentes displayed more of the bravo or the bandit, than the saint in his appearance.[32]

[30] Tom Menger, 'Racial Difference and the Limits of Fellow Feeling: British, German, and Dutch Soldiers Encounter the Enemy in Colonial Wars, c. 1900', this volume.

[31] Scheipers, p. 67.

[32] Grant, p. 74.

The portrayal of Cifuentes as very dark-skinned, a racist depiction casually correlating Blackness and badness, is joined by an emphasis on his unkempt appearance that is meant to convey moral failing even more than a lack of personal hygiene.

In comparison, Ronald's companion Major Campbell has been described just beforehand as 'a handsome Highlander, of a most muscular make and herculean form. His dark hair was becoming grizzled, for he was nearly fifty years of age, and his nutbrown cheek had been tanned by the sun and storm in many a varied clime'. His weapon of choice is a sword ('a real Andrea Ferrara, with the maker's name on the blade'), though he does not make use of it unless he has to, often preferring 'a short oak stick furnished with a heavy knob at the end, which he had cut in one of the wild forests of Argyleshire'.[33] While Cifuentes is short and misshapen, therefore, Campbell is tall and built like a Roman demigod; while Cifuentes' dark skin indicates his inferiority, Campbell's tanned skin demonstrates his hardy outdoor life; while Cifuentes wears a thin dagger and a knife, weapons often associated in these stories with cowardice and murder, Campbell favours a prestigious regimental blade that would be used in open battle, and in all other occasions, a stout stick that shows both his restraint (it would not be immediately lethal) and his emotional connection to his homeland.

As was a convention at the time, Cifuentes' physical appearance betrays his true wicked nature before his actions do—just as Campbell's appearance predicts that his reactions will be honourable. When Cifuentes' British guests first arrive, he is jovial and welcoming, sings songs, and entertains them with wine from his store, before showing them to their chamber. Immediately, they notice that the mattress is covered in what looks like dried blood. In the same airy, jolly tone, Cifuentes tells them that when four French officers had been billeted there previously, he had similarly plied them with wine, waited until they fell asleep, and together with 'two other comrades, guerrillas of Don Salvador de Zagala's band, stole softly in upon them, and plunged our stilettoes into their hearts'. They then threw the bodies (described coldly as 'carrion') into the street and hid in the woods until the French army had moved on. An author's footnote to this section adds chillingly that 'This piece of cruelty is no fiction, but actually happened as related here'.[34]

[33] Grant, p. 72.

[34] Grant, pp. 74–76.

Although it seems that Cifuentes expects his Scottish guests to feel reassured by this story (he says 'So now, senors, you see I am a true patriot—a loyal Spaniard, and that you have nothing to suspect me for'), Ronald and Major Campbell are instead disgusted and disbelieving. 'Why, senor', the Spaniard protests, 'the men were only Frenchmen, and you would have taken their lives yourselves'. 'In the open field, when equally armed', responds the passionate hero Ronald, 'but we should not have stolen upon them in the night, and butchered them in their sleep, as you say you did. And you shall die for it, you base Spanish dog!' The more sober and senior Campbell stops Ronald from unsheathing his sword: 'it is beneath a soldier and a gentleman to draw on so vile a scoundrel as this'. He threatens Cifuentes with being reported to the local mayor and hanged, then seizes him by the throat, drags him bodily to the door, and throws him unceremoniously down his own stairs.[35] Cifuentes' mistake, of course, is to assume that the troops sent to his country as allies will applaud any violence against their common enemy. By contrast, the army officers consider that there is an honourable and legitimate way to kill the enemy, and that is only 'in the open field, when equally armed'—anything else is debased, 'beneath a soldier and gentleman'. Unfortunately, their humiliation of Cifuentes is the trigger that changes him from dubious ally into implacable antagonist for the rest of the novel's three volumes. Rather than continuing in the role of vengeful Spanish patriot that might have earned him some very limited legitimacy, Cifuentes is encountered attempting to assassinate a Spanish officer, stealing from local peasants and priests with a gang of bandits, impersonating a priest, and ultimately enacting his personal revenge by raping and murdering Ronald's Spanish sweetheart, a noblewoman called Catalina.

Female characters are often used in these negative portrayals of guerrillas as plot devices, embodying the contested relationship between central male characters rather than having much storyline of their own, and enabling a further contrast between models of 'good' and 'bad' behaviour in terms of idealised femininity as well as masculinity. An example of this can be found in a guerrilla encounter story that takes place in a melodramatic French novel published shortly after the Peninsular War, *Le Torrent des Passions: ou, Les dangers de la galanterie* (1818). Written by a soldier and man of letters, the Baron de Saint-Cyr, the novel is told from the perspective of a French-speaking Austrian officer who takes part in various

[35] Grant, pp. 76–78.

Napoleonic campaigns. In volume two, the narrator is in Spain acting as an escort to a venerable *afrancesado* nobleman, the Count de Casa-Flores, and a young, unmarried British woman, Miss Emma, when they are ambushed by guerrillas with 'sinister faces' in an olive wood.[36]

Both Casa-Flores and Emma are very passive, one-dimensional characters, representing nobility of character and female virtue respectively. Both, of course, are traits that the leader of this gang of bandits, a monstrous monk named Father Ambrosio, longs to destroy:

> The image of Father Ambrosio will never fade from my memory! He was a monk of gigantic height, with a handsome but dark-skinned face. His bulging black eyes seemed to shoot both lightning and desire when they passed by turn from Casa-Flores to the beautiful Emma.[37]

Here again the tropes of dark skin and physical abnormality or deformity are joined to cruelty, lust, and a sense that the guerrilla hides dishonestly behind a peaceful civilian façade: in this story as a supposed man of God, in Grant's novel apparently a simple shopkeeper. Like the Gothic villain archetype he is, Ambrosio ends up murdering Casa-Flores in cold blood, and trying to force Emma to marry him, before the narrator is able to escape the guerrilla camp and return, in the nick of time, with a rescuing force of noble British troops. His essential wickedness is further emphasised by the introduction of a sympathetic, saintly 'good priest' character, Father Dolores, a frail, white-haired, and martyrlike old man who has been trying to hang around the guerrilla camp and intervene in their evil deeds to plead for restraint and mercy, with predictably little success.[38]

Aside from dithering later over whether to resist or submit to Ambrosio's horrible demands, Emma's main act of agency in this episode is to throw herself in front of the Count to protect him from imminent execution, a move described as 'angelic'. Importantly, however, this act of desperation does not lead to any character growth for her. While in a more feminist re-telling it might have led to Emma realising her own strength and bravery and becoming more assertive, in *Le Torrent des passions* she is struck dumb by her own temerity and the trauma of the situation, and cannot

[36] *Afrancesado* refers to Spaniards who were sympathetic to and collaborated with the regime of King Joseph Bonaparte (Napoleon's brother) during the Peninsular War. Saint-Cyr, p. 152.

[37] Saint-Cyr, pp. 154–155.

[38] Saint-Cyr, pp. 162–164.

speak to defend herself, 'ready to faint from horror'.[39] In other words, one self-sacrificing act of defiance is allowed, in a desperate situation, but the woman should not repeat or grow from it. Instead she might find her own actions so unnatural to her that she dies (something that happens in another Peninsular War novel, where a petite and timid young woman's unexpected intervention to save her friend's life, grabbing a weapon and shooting a hostile French soldier in the arm, contributes 'like the undermining worm… to her premature decay'.)[40]

This model of idealised feminine response in wartime is used to emphasise the 'proper' division between gendered civilian and military worlds—women have no place in war, and will only be wounded by any attempt to participate in it. Women taking part in aggressive or violent action, by contrast, is portrayed as a horrific, transgressive act—the type of thing that would only happen, of course, in a group of guerrillas, whose social deviance was underlined further by several authors through an association with aberrant femininity. In *Le Torrent des passions*, for example, Saint-Cyr alleges that the practice of irregular warfare in the Peninsula had resulted in such a degeneration of social norms and military values that 'premeditated murder' had become the 'order of the day, and was cravenly committed by women and children upon the wounded and the dying with the most atrocious barbarity'.[41]

This allegation was fleshed out even more in another melodramatic French novel, *L'Attaque du pont: ou, La fille retrouvée* (1830), this time by Alphonse Lorry, the pseudonym of Pierre Clerjon, a doctor and writer from Lyon. Although the bulk of the novel is set in the 1820s, the second volume contains an account of the Peninsula as told to a fascinated audience by a Napoleonic veteran character named Captain Brandor. Brandor's account of the guerrilla war in Spain again underlines its 'barbaric character' and features himself and his beautiful ward Aurore being captured by predictably villainous-looking bandits.[42] Taken by the guerrillas up into a deep and dark ravine, through which crashes a torrential river, illuminated only by the bright moon, they eventually arrive at a sort of rocky esplanade, shared by an old oak tree. There they witness the arrival of a group

[39] Saint-Cyr, p. 158.
[40] M'Donogh, pp. 167–169.
[41] Saint-Cyr, p. 151.
[42] Lorry, pp. 78, 84–85.

of evil monks who are part of the guerrilla band, led by a terrible, cruel woman whom they refer to as 'la cheftaine' [Captain]:

> The other woman had an impassive face, like that of the most unyielding ruler of hell. That face, that would have been beautiful if crime had not degraded it, produced such an impression on me that I cannot look upon an assembly of low devotees without believing that I see that monster.[43]

The woman addresses the monks, who pull forward a crying woman and her sleeping daughter, and commands them to 'do their duty', upon which one of the monks promptly pulls a long dagger from beneath his habit and plunges it into the heart of the unfortunate mother. (At this instant, the well-behaved Aurore dutifully throws herself into Brandor's arms and cries out in terror.)

> So shall perish all the wives of the courtiers of King Joseph! cried the female Lucifer with a barbaric joy; and with her own hands, she dragged her victim, who was close to death, by the hair... and cast her into a chasm that the torrent had formed not far from us. Then, throwing a purse full of gold into the midst of the assassins: 'Take it, you have earned it.' She disappeared, in the obscurity of the ravine, as Satan set forth into the depths of the abyss when he had lost his humanity. What made this scene of murder more dreadful for us was a black cloud that obscured the brightness of the moon: it heralded the coming of a great storm.[44]

Unnaturally strong, emotionless, her beauty ruined by her evil deeds, comparable to the Devil himself in female form, this chieftainess contrasts spectacularly with the two more traditional models of femininity on display: a mother and a virtuous young woman, both weeping and crying out, one victimised by men, the other leaning on a man for support, while the transgressive woman commands men to do her will.

Ambiguous Resolution

Here, then, is the deviant and devilish irregular, defying gender norms and defiling the institution of modern warfare: as Saint-Cyr put it mournfully in 1818, 'the magical words of glory, military honour and even humanity,

[43] Lorry, p. 87.
[44] Lorry, pp. 88–89.

were erased from the strategic dictionary'.[45] This type of literature explores and approves of the differential treatment of differently categorised enemies during war, condoning a framework in which the irregular can be summarily executed if captured or left to die if wounded, because they deserve nothing better. Interestingly, however, even in the very negative portrayals of guerrilla warfare discussed in this chapter, the villains are not always killed by the heroes. Instead, they often receive vaguer or more indirect just desserts, saving the protagonists from having blood (even villain's blood) on their hands. Narvaez Cifuentes survives all three volumes of Grant's first iteration of *The Romance of War*, like a nightmare character who keeps coming back to commit evil deeds no matter how many times he is thwarted, until eventually he is killed by Catalina's brother in the first volume of the sequel (1847). When his prisoners are rescued by a British detachment in *Le Torrent des passions*, Father Ambrosio and his murderous followers are merely disarmed, and left to go about their business: it is only later that we learn that Ambrosio later met his fate by being killed in a duel by an offended husband in South America, where, true to nature, he was dishonestly passing himself off as a colonel.[46]

Lorry's *L'Attaque du pont* is an exception: at the crucial moment, when the 'thirst for blood' brings fanatical guerrillas to the door of his cell, Captain Brandor manages to kill several with an iron bar and a couple of pistols taken from one of the fallen. He throws a torch onto a bed made of straw, and the twisted monastery goes up in flames, allowing him and the swooning Aurore to escape. As if unleashed by his action, though, a terrible thunderstorm shakes the mountain as they try to descend, striking a sudden tone of uncertainty: 'All of Nature seemed conjured against me or against the monks'.[47] In fact, at the endings of these encounter stories, ambiguity, discomfort, and a lack of closure always seem to sneak in. Even if a few villains have been despatched, more undoubtedly 'infest' the countryside, with Spain still imagined by British and French readers as a dark, dangerous, and perhaps, given the melodramatic style of many of these books, very thrilling place.

Perhaps another reason that the most pejorative portrayals of guerrillas from this period tend to shy away from presenting their moment of death, or from putting responsibility for enacting it onto the protagonists, is that

[45] Saint-Cyr, pp. 150–151.
[46] Saint-Cyr, pp. 199, 267.
[47] Lorry, pp. 106–108.

it is a moment that invariably creates moral ambiguity. In particular, a scene in which French regular soldiers kill guerrilla prisoners recurs frequently across other, more nuanced Peninsular War novels and memoirs, and two striking features are often the same. The guerrilla, who may have been wild, cruel, and vengeful in freedom, is suddenly admirable and stoic in death. In the aftermath of a successful French ambush, a fictional British officer in Colonel Charles White's *The Adventures of a King's Page* (1829) watches an injured Spanish guerrilla be hung from a tree root on the side of a rocky mountain road: 'it was evident he suffered unutterable anguish; still, he neither groaned nor asked for mercy'.[48] And the regular soldier, looking on, experiences a discomfort and a shame he cannot entirely understand or shake off. In W. H. Maxwell's *The Fortunes of Hector O'Halloran* (1853), the eponymous hero, an Irish staff officer, watches two guerrillas meet a French firing squad with calmness, dignity, and defiance. 'A dead silence prevailed, and, at the movement of the muskets, my heart beat wildly as a startled girl. Many of the French soldiery turned pale, and but a very few looked on the scene of death with indifference'. Afterwards, Hector—who was taken prisoner at the same time as the guerrillas, but who has been treated with exaggerated civility by the French because of his regular status—finds he cannot bring himself to eat.[49]

The mention in Maxwell's book of the emotion also experienced by the French soldiers is in keeping with the way many Peninsular War veterans depicted similar scenes in their memoirs.[50] In the memoir of the Marquess of Hautpoul, a French general in the Peninsula and later Governor General of Algeria, there is a striking reflection on the experience of the officer who personally gives the order. Having allegedly been ordered by Napoleon to take no prisoners from among the irregulars, Hautpoul writes that he intended at first to disobey, and to send a captured local guerrilla chief to France instead, but was pressured by Spaniards favourable to King Joseph Bonaparte to execute the man: 'He died with a lot of courage. I have always retained a profound regret for this action; I acted under direct orders, but it was not in my character to have prisoners killed. Waging all-out war against the enemy, weapons in hand, but being gener-

[48] Anon. [Charles White], *The Adventures of a King's Page*, 3 vols (London: Henry Colburn, 1829), III, p. 15.

[49] W. H. Maxwell, *The Fortunes of Hector O'Halloran and his man Mark Antony O'Toole* (London: R. Bentley, 1853), ch. XXIX–XXX.

[50] Greig, p. 47.

ous after victory, such is the mark of the French soldier'.[51] Scheipers notes in her study that the suspension of the 'civilised' rules when it comes to punishing guerrillas can also damage the self-perception of soldiers, for whom the idea that they are operating on a code of martial honour or virtue can be a psychological protection in itself.[52]

Indeed, the conceptualisation of this code of honour by British troops in the nineteenth century plays a key role in the discomfort experienced by soldiers in these stories. Very instructive in this regard is J. B. Potter's study of the 'Breaker' Morant trial from the Second Boer War (1899–1902) and its modern dramatization in a film by Bruce Beresford (1980). In the film, which is based on the real court-martial of three Australian officers accused of killing Boer prisoners of war and civilians in the closing years of the war, the defence conceptualises military honour as a rights-based system: something that could be lost by combatants through perfidious actions (such as lying, not wearing a uniform, playing dead to trick an opponent into letting down his guard). This conceptual framework is as dangerous for irregulars at the end of the nineteenth century as it was in the Peninsular War: 'When Boer commandos did not fight in uniform and waged guerrilla war, they did not conform to British conceptions of lawful combatants. By regarding this behaviour dishonourable – i.e. unworthy of a right to respect – the British could justify the use of harsh tactics that diverged from official policies'.[53]

Crucially, however, Potter notes that this system worked both ways: honour could also be regained through acting in good faith. In Beresford's film, the defence attorney Major J. F. Thomas even cites codes of practice from the Napoleonic Wars to support this idea, mentioning Wellington, whom he says 'influenced nearly all our military law', and claiming that '[t]he performance of a duty of honour and trust after knowledge of military offense ought to convey a pardon'.[54] This, then, explains the stomach-wrenching moment of change that so upsets the regular soldiers in the Napoleonic stories discussed above: by starting to act honourably again just before death, the defeated or imprisoned guerrillas also begin to erode the moral justification for their execution, only, agonisingly, no-one stops

[51] *Mémoires du général marquis Alphonse d'Hautpoul, pair de France (1789–1865), publiés par son arrière petit-fils*, ed. by Etienne Hennet de Goutel (Paris: Perrin, 1906), pp. 63–64.

[52] Scheipers, pp. 18–19.

[53] J. B. Potter, 'Peace suspended by a sword: honor and justifications of violence in *Breaker Morant*', *Small Wars and Insurgencies*, 34.5 (2022), p. 5.

[54] Potter, pp. 6–7.

it from happening. In fact, I would argue that these anecdotes contain an early representation of what could now be considered moral injury, defined by clinical psychology researchers Litz, Stein, Delaney et al in 2009 as 'perpetrating, failing to prevent, bearing witness to, or learning about acts that transgress deeply held moral beliefs and expectations'.[55] When the captive guerrilla suddenly becomes more admirable or respectable or honourable to the watching soldiers at the point of death, it complicates the act of violence against him. It comes dangerously close to being the killing in cold blood, without trial, of a prisoner of war and fellow 'regular' combatant: an uncomfortable ending for these encounter stories to address.

Conclusion

Hostile to contemporary changes in the perception and status of the irregular combatant, a group of early-nineteenth-century writers provided fuel for a developing attitude towards insurgents that damned and demonised them, setting out in clear side-by-side comparisons the miles of distance between this villainous 'other' and the regular soldier or officer who would protect you, protect the country, and uphold the norms of honour, decency, masculinity, and femininity. The fact that the regular army was absolutely *not* a paragon of virtue at this time, with soldiers' looting, raping and murdering being well-documented in primary as well as secondary literature for this period, was conveniently neglected.[56]

Disconnected also from any realistic or nuanced portrayal of the reasons for civilians to take up arms, in this horror-filled and melodramatic strain of literature the guerrilla became a narrative device that acted in much the same way as the figure of the pirate did at the same point in time, deployed to reassure audiences and to shore up certain social mores and political systems during what Alexandra Ganser has called 'crises of legitimacy'.[57] The 'bad guerrilla' demonstrated that war could be legitimately fought by standing armies only; that soldiers were male and that

[55] Brett T. Litz, Nathan Stein, Eileen Delaney et al., 'Moral injury and moral repair in war veterans: A preliminary model and intervention strategy', *Clinical Psychology Review*, 29 (2009), p. 697.

[56] Dendle, p. 54; Philip Dwyer, '"It Still Makes Me Shudder": Memories of Massacres and Atrocities during the Revolutionary and Napoleonic Wars', *War in History*, 16.4 (2009), pp. 381–405; Greig, pp. 46–54.

[57] Alexandra Ganser, *Crisis and Legitimacy in Atlantic American Narratives of Piracy* (Basingstoke: Palgrave Macmillan, 2020).

women's place was in the home; that enemies who had darker skin or came from places seen by northern Europeans as 'African' were dishonest, cruel, and sinister; that civilians who resisted by taking up arms against an invading force were opportunistic, greedy and damaging to the local population; that, more than that, they were often evil, inhuman, and deserved none of the protections that were being given to those wearing uniform and subject to recognised systems of command.

Although the tone of the literature discussed in this chapter does not represent the whole in terms of the vast artistic and literary output sparked by the irregular war in the Peninsula, which could also appropriate, celebrate, and romanticise the *guerrilleros*, it is important to examine these bizarre and upsetting books, because they voiced the same assumptions that would be later applied to military practice in colonial warfare and formalised into international law. Only a few decades later, the idea that regular armies could be justified in using a violent reciprocal approach against irregulars was used in Algeria by Peninsular War veteran Marshal Bugeaud, whose *razzias* or raids deliberately targeted the civilian population. This strategy was approved of by Victorian military theorist C. E. Callwell in his book *Small Wars, Their Principles and Practice* (1896), several editions of which were published by Her Majesty's Stationery Office as 'semi-official manuals'.[58] Besides opining that irregular combatants had a 'natural aptitude for creeping about and… stalking unsuspecting adversaries', Callwell also influentially argued that the key to winning this type of war was to suspend the rules of 'civilised' warfare and create 'moral effect' through extreme, punitive violence.[59] Callwell had non-European opponents in mind in those passages, but this chapter shows that this distinction was flexible and paper-thin. Callwell was not an isolated voice, either: as Tom Menger has shown in his work in this volume and elsewhere, a wide range of colonial service manuals from Britain, France, and Germany at the turn of the nineteenth century used almost exactly the same tropes to describe 'savage' adversaries as Peninsular War

[58] Daniel Whittingham, '"Savage Warfare": C.E. Callwell, the roots of counter-insurgency, and the nineteenth century context', *Small Wars and Insurgencies*, 23.4–5 (Oct.–Dec., 2012), p. 591.

[59] C. E. Callwell, *Small Wars. Their Principles and Practice*, 3rd edn (London: His Majesty's Stationery Office/Harrison and Sons, 1906), pp. 72, 253.

stories had for the Spanish guerrillas.[60] The dichotomies of civilized/uncivilised, white/not-white, masculine/feminine, honest/dishonest, and so on linked this episode from the Napoleonic Wars to the foundations of nineteenth-century imperial ideology.[61]

Yet even in the most exaggerated of these novels, the authors struggle with the implication of their words: that the regular army should act with impunity against these criminal combatants. They tend to avoid portraying scenes of righteous punishment, in which the heroes would have to despatch their demons calmly and face to face, for fear they might transform back to people in the cold light of day. In fact, the potential guilt of the regular army and the damaging psychological effects of war form a subtle but intriguing undercurrent in this nineteenth-century literary debate about the changing way of war. Monsters these fictional guerrillas may be, but whose creation are they? As one novel's version of the real guerrilla leader Julian Sanchez explodes in fury, in response to the criticism of two visiting British officers who call his killing of two hundred wounded French prisoners unacceptable 'butcheries', his atrocities are the products of an atrocious war started by the French Army themselves, and to assert that the conflict can still be somehow limited, restricted and civilised when it is also so personal is to be blind. 'Butcheries! [...] yes, I was once a butcher, but I slaughtered cattle only; until these bloodhounds came amongst us, the thought of spilling human blood would have congealed my own. Butchers! yes, did I not see my grey-headed father and helpless mother murdered before my very eyes?—did I not witness the pollution of my wife and sisters?—was I not plundered of all my worldly possessions, and the very roof under which I was born, reduced to a blackened mass of cinders? Speak not, then, to me of mercy and humanity,—you who have no wrongs to redress!'[62] Published only eleven years after Mary Shelley famously imagined a monster not born from darkness or an aberration of nature but created by scientific, rational man, this speech perhaps bears something of her influence: the monster speaks, rejects reproach, and places the blame on the one who made him.

[60] Menger, 'Racial Difference and the Limits of Fellow Feeling: British, German, and Dutch Soldiers Encounter the Enemy in Colonial Wars, c. 1900', this volume, and Menger, 'The Colonial Way of War: Extreme violence in knowledge and practice of colonial warfare in the British, German and Dutch colonial empires, c. 1890–1914' (Diss., Universität zu Köln, 2021), pp. 150–166.

[61] Elizabeth Kolsky, *Colonial Justice in British India* (New York: Cambridge University Press, 2010), pp. 4–5.

[62] White, pp. 9–10.

Acknowledgements Research for this chapter was made possible by grant funding from the Irish Research Council (2019–2020), the UK Arts & Humanities Research Council (2020–2021), and the Spanish Agencia Estatal de Investigación (2021–2022), while the author was employed as a postdoctoral researcher at University College Dublin, Cardiff University, and Universitat Pompeu Fabra (UPF), respectively. I owe my thanks to Lucy Riall, Robert Gerwarth, and Stephen Jacobson for their feedback on my ideas for this topic in its early stages. Thank you also to Holly Furneaux, John Horne, Sibylle Scheipers, and Margot Tudor for generously agreeing to read and comment on the first draft of this chapter. I am grateful also to Maria Romano at Gale and Jordi Morell at UPF for kindly organising access to the Corvey Collection and to the participants at the Enemy Encounters (July 2021) and War and Peace in the Age of Napoleon (September 2022) conferences for their responses to early versions of this paper.

'I was a stranger and ye took me in': Adopting the Enemy

Holly Furneaux

These photographs capture a startling moment of familial and national refashioning (Figs. 1 and 2). Adopted in the mid-nineteenth century by the British soldiers fighting against their home nations, the boys here undertook journeys to Britain where they entered new familial relationships, were placed in British schools and prepared to become British citizens. Their deeply emotive stories are bound up in wider, politically, and personally fraught questions of the adoption of children in wartime, with motives and effects ranging from the preservation of life to genocide. Current investigations into the Russian removal of Ukrainian children to Russian institutions and families highlight the contravention of international laws of war. Ukraine's Ministry of Foreign Affairs has described these deported children as facing 'illegal adoption' which breaks the UN International Convention on the Rights of the Child and the Convention against genocide, and in March 2023 the International Criminal Court issued a warrant for the arrest of Vladimir Putin and his Commissioner for

H. Furneaux (✉)
Cardiff University, Cardiff, UK
e-mail: furneauxh@cardiff.ac.uk

H. Furneaux, M. Greig (eds.), *Enemy Encounters in Modern Warfare*, https://doi.org/10.1007/978-3-031-56748-3_3

Fig. 1 Roger Fenton, 'Colonel Brownrigg and the two Russian Boys, Alma and Inkerman', 1855 (Royal Collection, RCIN 2500375)

Fig. 2 Jabez Hughes, 'Capt. Speedy and Alameyu', Ryde, Britain, 1868 (British Museum, Af,B7.22)

Children's Rights, Maria Lvova-Belova, 'for the war crime of unlawful deportation of population (children) and that of unlawful transfer of population (children) from occupied areas of Ukraine to the Russian Federation'.[1] In 1948, the UN included 'forcibly transferring children of the group to another group' within its definition of genocide.[2] Depending on national perspective, the adoption of enemy children has been presented either as an act of war violence or as humanitarianism. Often, as in the case of Russia now and, for example, the USA during 'Operation Babylift' at the end of the Vietnam War in which an estimated 3300 children were taken from Vietnam for adoption in the USA, the country removing enemy children presents this as a humanitarian mission aimed at preserving innocent life threatened by war.[3] Now, and in the past, this emphasis on the supposedly philanthropic, life-saving 'rescue' of enemy children eclipses the deaths of children killed in conflict, and endeavours to rewrite the war violence of the 'rescuing' nation. The adoption of the boys in these photographs became popular good news stories in Victorian

[1] Mychael Schnell, 'Ukraine: Deported children facing threat of illegal adoption in Russia', *The Hill*, 14 April 2022, https://thehill.com/policy/international/3267579-ukraine-deported-children-facing-threat-of-illegal-adoption-in-russia/; https://www.icc-cpi.int/news/situation-ukraine-icc-judges-issue-arrest-warrants-against-vladimir-vladimirovich-putin-and See also AP news investigation of 13 October 2022, <https://apnews.com/article/ukrainian-children-russia-7493cb22c9086c6293c1ac7986d85ef6>. Article 7.1 of the UN Convention on the Rights of the Child stipulates that the child shall have 'as far as possible, the right to know and be cared for by his or her parents', and 9.1 states that 'a child shall not be separated from his or her parents against their will', except following due legal process: <https://www.ohchr.org/en/instruments-mechanisms/instruments/convention-rights-child>.

[2] Article 2e, <https://www.un.org/en/genocideprevention/genocide-convention.shtml>. This addition was made at the request of Greece following the relocation of nearly 20,000 Greek children to Eastern European Communist countries by the Greek Communist Party in the Greek Civil War. See 'Forcible Transfer', *Encyclopedia of Genocide and Crimes Against Humanity*, <https://www.encyclopedia.com>; Loring M. Danforth and Riki Van Boeschoten, *Children of the Greek Civil War: Refugees and the Politics of Memory* (Chicago: University of Chicago Press, 2011).

[3] Gloria Emerson critiques the power of the US operation to instill a 'self-congratulatory spirit, a feeling of winning something at last, the need to prove to ourselves what decent people we really are. It is almost forgotten during these excited, evangelical scenes at airports that it is this country that made so many Vietnamese into orphans…. Now we have decided the Vietnamese we will "save" and "love" must be very pliant, very helpless'. 'Operation Babylift,' *The New Republic*, 26 April 1975, pp. 8–10.

Britain, used to exemplify an imperial ideology of benevolent, paternalistic improvement.[4]

Though adoption was not legalised in England until the Adoption Act of 1926, I use this resonant anachronism for the way it points us towards the emotional histories of those involved in what is typically described in modern adoption studies as the adoption triad: birth parents, child, and adoptive parents. As an adoptive parent myself (albeit as a white British parent of white British children), I am acutely aware that the narratives of adoptive parents and the cultures to which they belong are typically the most audible within this triad, potentially causing further harm to children and birth parents who have already suffered substantial loss. This asymmetry is further pronounced where the children are those of an enemy nation, with the voices of children and their birth families subsumed—literally so in the demand that the child, often exclusively, speak the language of their second homeland—within the adoptive nation's narrative of itself and the countries it opposes in war. In her work on modern transnational adoption Laura Briggs directs us to the way that adoption and its cultural representation is 'simultaneously symbolic and material practice, with substantial effects on individual lives and a discursive significance that extended in multiple directions and ultimately took in a whole culture'.[5] Here I attend to the interlinking personal histories of the children in these photographs and the adults attached to them, and to their wider 'discursive significance' in the cultural meanings established around them.

In the twentieth century formalised 'transnational adoption emerged out of war', a response to the suffering caused to children left without parents and resources after conflict, often in former enemy countries.[6] As Briggs argues with a focus on the USA, these post-war adoptions served a variety of personal and political purposes. They could help to assuage

[4] These cases are indicative of wider Victorian practices of international adoption by soldiers, explorers, and missionaries. Notable examples include the Sudanese boy Jimmy Durham adopted by the Durham Light Infantry in 1885, explorer Henry Morton Stanley's guardianship of the formerly enslaved boy Kalulu, and the missionary project rehoming and training African children at Congo House in North Wales. The complex interplay of imperial ideology, personal feeling and highly asymmetric power in these international adoptions is the subject of my next research project.

[5] Laura Briggs, 'Mother, Child, Race, Nation: The Visual Iconography of Rescue and the Politics of Transnational and Transracial Adoption', *Gender & History*, 15.2 (2003), p. 181.

[6] Laura Briggs and Diana Marre, 'The Circulation of Children' in *International Adoption: Global Inequalities and the Circulation of Children*, ed. by Briggs and Marre, (New York: New York University Press, 2009), p. 1.

national guilt about particular conflicts and reidentify the adopting country as rescuer rather than perpetrator of war violence: 'the United States could not even potentially be held accountable for the military, political and economic causes of hunger or poverty – its only role was to rescue the unfortunate victims of such events'.[7] This chapter examines how earlier British removals of enemy children anticipate key debates about the power dynamics and national ideologies bound up in transnational adoption now. Anthropologist Pauline Turner Strong, writing about the adoption of Native children by non-Native parents in North America, registers the political and personal complexity of 'adoption across political and cultural borders [which] may simultaneously be an act of violence and an act of love, an excruciating rupture and a generous incorporation, and an appropriation of valued resources and a constitution of personal ties'.[8] I use the recent and developing field of transnational and transracial adoption studies to think through the extent to which the removal of these children to grow up in Victorian Britain combines acts of violence with acts of love, and how far the rehoming of enemy children reinforced or challenged designations of us/them, and reaffirmed or disrupted national beliefs and prejudices.

I consider the history of transnational adoption from the late eighteenth century, examining the emotional and political ambivalence around the British taking hostage the young sons of Tipu Sultan. Images of enemy children under British guardianship ostensibly work as the inverse to widespread representations—like those discussed in Rio Creech-Nowagiel's chapter in this collection—of the enemy as child killers or as callously deploying children in war regardless of their welfare.[9] Representations of Tipu's sons being separated from their family, however, show especially clearly the possibilities for engagement with the suffering of these children, counter to a narrative of benign imperialism. This concern with the sadness of adopted enemy children continued in relation to the boys in these photographs.

[7] Briggs, 'Mother, Child', p. 198.

[8] Pauline Turner Strong, 'To Forget Their Tongue, Their Name, and Their Whole Relation: Captivity, Extra-Tribal Adoption, and the Indian Child Welfare Act', in *Relative Values: Reconfiguring Kinship Studies*, ed. by Sarah Franklin and Susan McKinnon (Durham: Duke University Press, 2001), p. 471.

[9] See Rio Creech-Nowagiel, '"Children Made to Carry Bombs": Staging the Malayan 'Emergency' in Postwar Britain', this collection.

Their histories of cultural appropriation have made it difficult to correctly name these boys. On the left are the Russian children 'Alma and Inkerman', so named after the campaign's major battles by British soldiers at war in the Crimea who took them prisoner while they were out gathering grapes. They were brought to Britain at the conclusion of the war, where they grew up and made their lives. Though 'Alma's' biography has yet to be discovered, 'Inkerman's' family recognised a reprinting of Roger Fenton's photograph some 40 years later and there was a small media flurry around the discovery that he was still living in the small market town of Wilton in Wiltshire. 'Inkerman', given an opportunity to tell his own life story, confirmed his original name as Simeon Paskiewitch and that he had adopted the surname 'Sinca', a contraction of Inkerman, in Britain. On the right is Prince Alemayehu—also variously called Alamayou (Queen Victoria's preferred spelling), Alamayu and Alamaiou—son of Emperor Tewodros and Queen Tirunesh.[10] Alemayehu's life was radically changed by the punitive Napier expedition to Abyssinia, now Ethiopia. The campaign's formal objective was to free the European hostages Tewodros had imprisoned at his mountain stronghold, Magdala, but was also concerned with securing Britain's international reputation and with acquiring substantial loot along the way. As well as 32,000 soldiers, the expedition included meteorologists, archaeologists, zoologists and a well-funded representative of the British Museum, variously tasked with collecting data, specimens and treasure. Tewodros took his own life rather than capitulate to the British and the prince's mother died of a lung disease, likely tuberculosis, shortly after the British victory. Together with considerable material plunder from the expedition, orphaned Alemayehu was brought to England in the care of the characterful Captain Speedy with whom he is pictured.[11] He survived under British guardianship for 11 years, dying of illness aged 18 years.

[10] I have followed Susanne Gruss's rationale in choosing the first of these variants as 'a self-conscious attempt to use the version that seems to be closest to the Abyssinian original, and mirrors the choice of spelling recent journalistic and political initiatives to recover Alemayehu's story and repatriate his remains have opted for', 'Black, Queer, Victorian? The Precarious Neo-Victorian Afterlives of Prince Alemayehu' in *Black Neo-Victoriana*, ed. by Marlena Tronicke (Leiden: Brill, 2021), p. 61.

[11] Andrew Heavens shows that Alemayehu was not the only child brought from Magdala. At least two others were on the same ship, including Napier's 'little Abyssinian slave girl', and on another vessel a German explorer who accompanied the expedition brought the King of Prussia what the press called 'a rather curious present from Abyssinia – namely an Abyssinian boy'. *The Prince and the Plunder* (Cheltenham: History Press, 2023), p. 105.

A comparison of how these children's narratives were presented is instructive about how Britain preferred to present its way of warfare, with an emphasis on the preservation rather than destruction of life and culture. Racist assumptions of white European superiority, which underpinned the extreme violence of empire, structured the presentation of these adoptions as an unambiguous cultural benefit for the children involved. Narratives of soldiers adopting allied and enemy children orphaned and displaced by war were popular through the Victorian period. As I argue elsewhere, these recurrent plots—in poetry, fiction, opera and narrative painting—helped to secure a benign image of the British soldier as an emotionally sensitive figure more concerned with healing than harm.[12] The military man of feeling was part of the logic that presented the white European as the definition of the human, and the humane, and justified imperial violence as civilising paternalism. As Lisa Lowe puts it, 'as modern liberalism defined the "human" and universalized its attributes to European man, it simultaneously differentiated populations in the colonies as less than human'.[13]

In my current broader project on enemy encounters through the long nineteenth century, I attend to how Britain emphasised its modern liberal position in care for vulnerable European military enemies—including the wounded and prisoners of war—as indicative of national character. The exclusions at the heart of European definitions of humanity, however, also meant that humanitarianism—encoded in new laws designed to protect humans in warfare in the second half of the nineteenth century—was not equally extended to all populations.[14] In this chapter, I place a comparison of the British adoption of Russian and Ethiopian enemy children within a framework that registers extreme imperial violence, adding my voice to those, including Tom Menger in this collection, who reject the exceptionalist narrative of the British empire as unusually restrained and respectful

[12] Holly Furneaux, *Military Men of Feeling: Emotion, Touch, and Masculinity in the Crimean War* (Oxford: Oxford University Press, 2015), ch. 3.

[13] Lisa Lowe, *Intimacies of Four Continents* (Durham: Duke University Press, 2015), p. 6.

[14] Indeed, as Sven Lindquist has noted, ideas of 'philanthropic' extermination – especially in speeding up the eradication of races already deemed doomed to extinction – could be in intimate proximity to 'the heart of humanism', '*Exterminate all the Brutes*', first published 1990, trans. Joan Tate (London: Granta, 2018), p. 9.

of life.[15] While recognising the violence inherent in and enabled by fictive binaries on which empire rested—civilised/savage, humane/brutish, adult/child—I suggest that these could also be called into question by the powerful affects that circulated around the figure of the vulnerable child.

As Mary Louise Pratt documents, from the late eighteenth century as 'contradictions between egalitarian, democratic ideologies at home and ruthless structures of domination and extermination abroad became more acute', there was an effort to rehabilitate Euroimperialism through sentimentality.[16] Colonists were presented as exemplary men of feeling, suffering in their efforts to bring enlightenment to a 'contact zone' 'rehabilitated by European sentimentality'.[17] The European military man's self-sacrificing care for the vulnerable, including children caught up in war, was a part of this sentimental narrative. Similar emotive imagery is endemic today, with, as Briggs demonstrates, the image of 'Madonna and waif' having become 'the grammar of hunger or need in contemporary US culture', an image that 'directs attention away from structural explanations for poverty, famine and other disasters, including international, political, military and economic causes. It mobilises ideologies of "rescue", while pointing away from addressing causes'.[18] The images of Captain Speedy and Colonel Brownrigg with their young charges suggest that a differently gendered configuration of elective 'fathers' and waifs was the preferred iconography through which attention was redirected through a substitution of gentle for violent soldiering. The range of emotional responses generated by the suffering of those children, however, were less easily contained within a narrative of benign, beneficial imperialism.

[15] See Tom Menger, 'Racial Difference and the Limits of Fellow Feeling: British, German, and Dutch Soldiers Encounter the Enemy in Colonial Wars, c. 1900', this collection. Michelle Gordon counters continuing belief in the 'beneficial agenda' of the British empire with evidence that 'violence, or the threat thereof was endemic across the British empire'; 'colonial administrators, politicians in London British military men accepted the need for exemplary violence and the utilization of extreme tactics'. *Extreme Violence and the 'British Way': Colonial Warfare in Perak, Sierra Leone and Sudan* (London: Bloomsbury, 2021), p. 31. See also: Elizabeth Kolsky, *Colonial Justice in British India: White Violence and the Rule of Law* (Cambridge: Cambridge University Press, 2010); Kim Wagner, 'Savage Warfare: Violence and the Rule of Colonial Difference in Early British Counterinsurgency', *History Workshop Journal* 85.1 (2018), pp. 217–237.

[16] Mary Louise Pratt, *Travel Writing and Transculturation* (London: Routledge, 1992), p. 74.

[17] Pratt, p. 75.

[18] Briggs, 'Mother, Child', pp. 193, 180.

The affective and political ambivalence around enemy children taken by the British is apparent in depictions of the taking hostage of two young sons of Tipu Sultan at the conclusion of the 3rd Mysore War (1790–1792). Lieutenant-General Lord Cornwallis, Governor-General of India, required that the treaty Tipu signed ceding half his territories and thirty million rupees, be guaranteed by his children.[19] They were held in Madras by the East India Company and returned to their family just over two years later.[20] The delivery of the boys did not lack drama; Princes Abdul Khaliq (aged eight) and Muiz-ud-din (aged five) were accompanied by a British gun salute as they proceeded on elephants to Corwallis's tent, where he presented them each with a gold watch. In his 1793 account of the campaign, Alexander Dirom emphasised the emotional satisfaction of the children and onlookers as Tipu's vakeel (ambassador) transferred them to Cornwallis:

> 'These children were this morning the sons of the Sultan my master; their situation is now changed and they must look up to your lordship as their father.'
>
> Lord Cornwallis, who had received the boys as if they had been his own sons, anxiously assured the vakeel and the young Princes themselves that every possible attention would be shewn to them, and the greatest care taken of their persons. Their little faces brightened up; the scene became highly interesting; and not only their attendants but all the spectators were delighted to see that any fears they might have harboured were removed, and that they would soon be reconciled to their change of situation and their new friends.[21]

[19] For details on Tipu's opposition to the East India Company, Cornwallis's laying siege to Tipu's capital, Tipu's capitulation and the treaty of Srirangapattana, see Partha Chatterjee, *The Black Hole of Empire: History of a Global Practice of Power* (Princeton: Princeton University Press, 2012), p. 87.

[20] Penderel Moon, *The British Conquest and Dominion of India* (London: Duckworth, 1989). Moon describes the boys being taken to plays and performances of Handel's *Messiah* and notes the popularity of the boys with Madras society 'which was surprised by the charm and excellent manners of these children of a "mad barbarian and bloodthirsty tyrant"', p. 259. Chatterjee details the further captivity and displacement of Tipu's family by the British administration after he was killed in 1799 again defending Srirangapatna in the fourth Anglo-Mysore War. Fearing the family's potential as a focus for hostility to British rule, his sons and other relatives were imprisoned first at Vellore and then in Calcutta, so that 'there was no question of allowing a son of Tipu to claim even a ceremonial right to succession', pp. 97, 102.

[21] Alexander Dirom, *A Narrative of the Campaign in India, which Terminated the War with Tippoo Sultan in 1792* (1793, repr. New Delhi: Asian Education Services, 1985), pp. 228–229.

Robert Home, an artist with the army in India, similarly infused his depiction of the transfer of the Princes to Cornwallis with this 'benign' paternalism. As the National Army Museum in London, which holds the painting, acknowledges: 'Cornwallis is depicted as a benevolent English gentleman, exercising direct paternal authority over a supposedly despotic oriental kingdom. After the excesses of the corrupt and greedy East India Company "nabobs" (Britons who had become rich while living in India) in the 1760s and 1770s, this was a very rosy propaganda image indeed'.[22] As Catherine Anderson and Sean Willcock point out, following the trial of Warren Hastings, former Governor of the East India Company for alleged crimes including extortion and embezzlement in India, British paternalism—and the morality of just rule it supposedly indicated—was in need of rehabilitation.[23] The calm, trusting expressions Home gives the Princes suggest they can place their faith in Cornwallis; the younger boy takes his hand while the older boy gazes across at a Mysorean noble, who turns away unable to bear witnessing the boys' removal. As Franklin and Mary Wickwire interpret it in their 1980 history of Cornwallis's imperial career, 'the memory of the event has endured so well because it symbolised the way the British conceived their role in India. The natives were children who needed the fatherly care and guidance of the British in order to grow to maturity. They should be loved but watched for without British care they could go astray'.[24] In depictions of Britain taking on the care of enemy children, in this case forcibly, the racist imperial ideology of a 'civilising' 'white man's burden' was accompanied by a literal enactment of the logic of supposed humane paternalism played out by replacement British father to displaced enemy children. However, these emotive scenarios could also evoke a range of feelings, in which the benevolence of

[22] Robert Home, 'The Reception of the Mysorean Hostage Princes by Marquis Cornwallis', c.1793, National Army Museum <https://collection.nam.ac.uk/detail.php?acc=1976-11-86-1>.

[23] Catherine Anderson, 'Tipu Sultan's Sons and Images of Paternalism in Late Eighteenth-Century British Art', in *Romanticism and Parenting: Image, Instruction and Ideology*, ed. by Carolyn A. Weber (Newcastle upon Tyne: Cambridge Scholars Publishing, 2007), p. 20; Sean Willcock, 'A Neutered Beast? Representations of the Sons of Tipu Sultan – 'The Tiger of Mysore' – as Hostages in the 1790s', *Journal for Eighteenth-Century Studies,* 36 (2013), p. 122.

[24] Franklin and Mary Wickwire, *Cornwallis: The Imperial Years* (Chapel Hill: University of North Carolina Press, 1980), p. 171.

British imperialism could come under scrutiny, and the supposedly secure limits between civilised and uncivilised could be destablised.

The taking of these boys hostage was a highly popular artistic subject in Britain, represented in additional large scale history paintings including by Henry Singleton, Arthur William Devis and Mather Brown at the National Gallery, in a host of engravings typically sold for a shilling, and on domestic objects like tea trays.[25] As Brian Allen summarises, 'these images are consistent in their portrayal of the vulnerability of the young princes in the company of the corpulently avuncular figure of Cornwallis'.[26] While the pain of familial separation was only subtly registered in Hone's painting in the turning away of the Indian figure in the foreground, many others more explicitly drew out disproportionate vulnerability within the encounter. James Gillray's engraving after a painting by James Northcote, for instance, reverses the ascription of benign paternalism, depicting an elderly Indian hiding his face in his arms as he delivers the reluctant boys to an unsmiling Cornwallis who bears down on them, punitive rod in hand. The boys' resistance to their separation from home and family is made literal in their holding on to the weeping white-bearded man.[27] Most emotive are those prints that move the focus to the boys' parting from their mother, such as Robert Sayer's plate. As Willcock notes, the scene is a Europeanisation of this family. The Princes had separate mothers, and Muiz-ud-din's mother had been killed in the British bombardment shortly before Tipu's capitulation. The single maternal figure depicted here is in line with British ideals of supposedly monogamous nuclear family; imaginatively displacing the boys' birth mothers, she is also presented as familiar to British audiences with pale skin and a cross around her neck (Fig. 3).[28]

The individuality of the princes is also sacrificed in favour of an emphasis on their diminutive size and their sorrow, an infantilisation which recurs in representations of Alemayehu as frightened, frail child rather than

[25] Brian Allen, 'From Plassey to Seringapatam. India and British History Painting c. 1760–1800' in *The Raj: India and the British 1600–1947*, ed. by C.A. Bayly (London: National Portrait Gallery, 1990), pp. 26–128, p. 35. Queen Charlotte was so delighted by Mather Brown's treatment of the subject that she ordered it to decorate her apartments at her new estate. Derinda Evans, *Mather Brown* (Middletown: Wesleyan University Press, 1982), p. 116.

[26] Allen, p. 35.

[27] James Gillray, after James Northcote, 'Marquis Cornwallis receiving the royal hostages at Seringapatam', 1794. British Museum, 1851,0901.1354.

[28] Willcock, pp. 134–135.

Fig. 3 Robert Sayer, 'Tippoo Saib's Two Sons taking leave of Their Mother', 1792, (British Museum, 2010,7081.1367)

potential king in waiting. In Sayer's depiction, a tiny Muiz-ud-din stretches up towards the kneeling mother's face, encircled by her arms, while his older brother holds his hands, in a familiar melodramatic stage posture of distress, crossed over his chest. A sorrowing white-bearded man, similar to Gillray's gentle Indian patriarch, stands behind this grouping, his hand also on his heart. Similarly sentimental plates celebrated the joyful family reunion on the boys' return in 1794.[29] In these prints the sensibility widely claimed in Britain at the end of the eighteenth century as a marker of social superiority is clearly attributed to Tipu's suffering family. The sympathetic engagement with the distress of the princes that these prints encourage might create uncertainty about the humanity of Cornwallis's insistence in taking the boys as surety of their father's word and, by extension, open a space for disquiet about wider acts of imperial violence, which are domesticated in the symbolism of this hostage exchange as hurting families and threatening the most intimate relationships between parents and children. The violence of empire is indicated here, though, through an iconography which operates through its own violent reimagining of the princes and their birth mothers in line with European ideals of family.

These competing threads of interpretation—both shoring up and undercutting Britain's preferred self-representation as exceptionally benign in its international expansion and humane in its ways of warfare—structure representations of the adoptions of Alma, Inkerman and Alemayehu. As with the visual iconography of Tipu's young sons, presentations of these children as they adapted to their new guardians also emphasise their difficult feelings in doing so. In a letter home to his wife from the Crimea photographer Fenton gives a very brief history of how the Russian boys became British prisoners: 'They went out nutting last autumn & were taken, cried sadly but now would cry to go back'.[30] Though his reassurance that 'both would like to come to England', the distress briefly registered at capture—at which they 'cried sadly'—and the recalibration and continuation of this distress at the prospect of return, suggests a particularly difficult emotional position for these children in which they inhabit an in-between identity over which they can exert no control, their only reported action being tears.

[29] See for example, Haines and Son, 'Return of Tippoo Saib's Two Sons to their Mother', 1794, BM 2010,7081.3180.

[30] Roger Fenton to Grace Fenton, 29 April 1855, collected in <http://rogerfenton.dmu.ac.uk/showLetter.php?letterNo=11>.

Philip Boys has traced 'Inkerman's' experiences after his removal to England where he was educated at St Mark's school in Windsor, apprenticed as a mid-shipman and settled as a carpenter in Wilton, Wiltshire, where he married a British nurse-maid and had several children.[31] These details are known as there was a resurgence of interest at the end of the nineteenth century in this narrative of British humane treatment of enemy children amidst the controversy about the suffering of civilians within British concentration camps in the South African War. Fenton's photograph was republished in a memorial retrospect on Queen Victoria by the Marquis of Lorne in 1901 with a caption celebrating the adjustment of these 'tiny prisoners of war' to their new situation, airily dehumanising the children as faithful pets: 'They attached themselves so much to the Colonel as to follow him everywhere with the fidelity of spaniels'.[32] Recognising his father in the boy standing in the image, 'Inkerman's' son contacted the *Daily Mail* who then interviewed his father.[33] The paper presented the feature as a 'romance of adventure and lost identity' and assigns itself a philanthropic role in the effort to discover Simeon's birth relatives: 'He has little expectation of ever tracing his Russian relatives, but marked copies of the "Daily Mail" will be sent to officers in the neighbourhood of the Crimea in the hope of finding them out, if possible'. Though presented as benevolent involvement in the 'adventure' of restoring 'lost identity', this reference to Simeon's Russian relatives might have given readers pause over the motivation for his initial removal. The very brief biography of the boys given by Fenton after photographing them stated that 'one is an orphan and one has or had relatives in the town', which similarly opens the question of the legitimacy of the British claim to them and highlights the damage to civilians caused by the war. Further details of Simeon's recollections were printed in the *Dundee Evening Courier*:

> After the battle of Alma, when the English, French and Turkish soldiers got into Balaclava, the Russian farmers became frightened, and ran inside the walls of Sebastopol, leaving the grape crops behind them. We boys got out and began picking the grapes, but one day we saw some English soldiers in front of us. We all ran away, and I and the other little one in the picture got

[31] 'Whatever happened to "Alma" & "Inkermann"?', 1 December 2018, <https://livesofthelightbrigade.wordpress.com/2018/12/01/whatever-happened-to-alma-inkermann/>.

[32] Maquis of Lorne, *V.R.I. Her Life and Empire* (London: 1901), p. 231.

[33] 'A Romance of the Crimea', *Daily Mail*, 13 August 1901.

> under a big tub. Here we had to stay in fright all night and part of the next day. In the afternoon one of the soldiers came across our poor old tub and knocked it over, and there was a surprise for him to see us two frightened little nippers.
>
> The English soldiers were more than kind to us. I don't know how Englishmen, whether pro-Boer or not, can bring charges of brutality against English soldiers. They gave us blankets and food, and looked after us.[34]

Simeon favourably contrasted the treatment he received from the English with having been 'chased and ill-treated by Turks', with whom Britain were allied in the Crimean War. Similarly comforting for British audiences was his extrapolation that the 'more than kind[ness]' he met with from English soldiers might cast doubt on current claims of English brutality in South Africa. This immediate context is apparent from the surrounding material in the papers carrying this uplifting enemy adoption story. In the *Daily Mail*, this piece sits between a dispatch from Kitchener, Commander-in-chief of the British Army from November 1900, recording Boer deaths in the column to its right and a letter from a soldier serving in South Africa headed 'Lord Kitchener's Generalship' in the column to the left, which concludes: 'In spite of the clap-trap now being talked in Parliament by ignorant and vain knights of the shires, I can assure you that a more humane and pacific General never directed a force on active service and this testimony is from Boer and British alike'.[35] Testimonies to the humanity of British command and soldiery endeavoured to counter-balance reports of the high mortality and suffering, particularly of Boer women and children, in the British concentration camps instituted under Kitchener's scorched earth policy; Emily Hobhouse had reported on this to parliament and in the *Guardian* two months before the coverage of Simeon's story.[36]

Simeon Paskiewitch's survival in Britain, which met the key requirements of a respectably productive life of industrious education,

[34] 'A Romance of the Crimea', *Dundee Evening Telegraph*, 14 August 1901.

[35] A piece on the same page of the *Dundee Evening Telegraph* moves the focus of atrocity to the enemy. In 'The Murders by Boers at Vlakfontein', a young Scottish soldier reports on the Boers shooting British soldiers 'in cold blood whilst standing with hands up, and I saw a corporal in the artillery have his brains beaten in for refusing to show the Boers how to explode the shells'.

[36] 'The Concentration Camps: Miss Hobhouse's Report', *Guardian*, 19 June 1901. On Kitchener's policies for civilian internment see Elizabeth Van Heyningen, *The Concentration Camps of the Anglo-Boer War: A Social History* (Sunnyside: Jacana Media, 2013).

apprenticeship, work, marriage and fatherhood, allowed his adoption to be revisited as a good news story. Prince Alemayehu's narrative is more unsettling, marked by his early death and continuing calls for the repatriation of his remains, and characterised throughout by the trauma of his removal from his home. Encounters with Alemeyehu, then and now, are melancholic. As Ellani Centime Zeleke puts it in her 2011 meditation on Ethiopia, love, war and belonging, 'whenever I see a photograph of Alemayehu I can stare at it for hours. There is always something so tragic in his pose, the delicacy of his finger's grip and the way he is always decorated with African paraphernalia'.[37] For Zeleke, the young prince's story points to the broader imperial processes by which those 'born in the land of Alemayehu's birth were orphaned by the violence that is associated with the scramble for Africa (especially the Battle of Adwa and the Battle of Magdella)'; 'Another poet warned me, you do not have to teach an orphan how to cry'.[38] Though taking a very different perspective, Queen Victoria who took a keen interest in Alemayehu, similarly registered the orphan's isolation and distress in her various journal entries about him, particularly her account of his death:

> Very grieved & shocked to hear by telegram, that good Alamayou passed away this morning. It is too sad! All alone, in a strange country, without a single person or relative, belonging to him, so young, & so good, but for him one cannot repine. His, was no happy life, full of difficulties of every kind, & he was so sensitive, thinking people stared at him on account of his colour, that I fear he would never have been happy. Everyone is sorry.[39]

The Queen's harrowing description of the boy as 'all alone', her acknowledgement that 'his was no happy life' and fear that 'he would never have been happy' in Britain casts doubt on the decision to remove him from Ethiopia. Her emphasis on Alemayehu's otherness in a 'strange country' and her suggestion of the 'difficulties' of racism he experienced, resonates with the only detailed account of Simeon 'Inkerman's' life which presents him as physically and temperamentally set apart from British characteristics. Novelist and Wilton Mayor Edith Olivier recalled Simeon's involvement in the community in highly colourful anecdotes of his liveliness. She

[37] Elleni Centime Zeleke, 'My People', *Callaloo*, 34.3 (2011), p. 824.

[38] Zeleke, pp. 824, 825.

[39] *Queen Victoria's Journals*, 14 November 1879, <http://www.queenvictoriasjournals.org>.

recalled 'he never walked, but he ran everywhere, on the level, or up and down the ladders which seemed to be his native place' and his love of making and setting off fireworks: 'he himself was not unlike a cracker, mischievous and erratic'.[40] Olivier's marking of Simeon's differences are also explicitly national and racial as she notes his local moniker 'the Russian sailor', and that 'he was obviously a Slav, with his short square figure, and his flat face marked with smallpox. He spoke very quickly, in a deep hollow voice, which sounded as if he had his head in a bucket'. She goes on to suggest that Simeon was characterised by his otherness as an immigrant: 'The Wilton Mummers had a phrase, "a foreign-off man", and this describes Sinca, who was always "off" somewhere with his foreignness'.[41] This being 'off' through 'foreignness' is also manifested in what she sees as the physical difference of Simeon's offspring, 'a family extremely unlike the Wiltshire children among whom they played'.[42] Olivier's emphasis on the combination of 'off' physical and behavioural characteristics at variance with the Wiltshire people illustrates the designation of racial difference even between white Europeans, with Slavic Simeon presented, albeit affectionately, as less civilised in his 'mischievous and eccentric' habits.

Though British willingness to adopt enemy children was presented as an indicator of magnanimous humanity, the full renationalisation of these children was limited by an emphasis on cultural and racial difference which persisted even in assessments of Simeon who lived the majority of his life in Britain, where he married, raised his 'extremely unlike' children and died. Alemayehu's race was repeatedly scrutinised, the Queen being amongst those who 'stared at him on account of his colour'. She writes in detail about his physical appearance after their first meeting, finding him to be 'a very pretty, slight graceful boy of 7 with beautiful eyes & a nice nose & mouth, though the lips are slightly thick. His skin is a dark bronze. His hair, which has been shaved, is crisp & curly. There is nothing whatever of the negro about him'.[43] The fascination with adjudicating how far Alemayehu's appearance accorded with 'negro' or stereotypically Caucasian features recurs throughout many private and public descriptions, as Andrew Heavens has documented.[44] An account of Alemayhu's arrival in

[40] Edith Olivier, *Without Knowing Mr Walkley* (London: Faber, 1938), p. 159.
[41] Olivier, p. 158.
[42] Olivier, p. 160.
[43] *Queen Victoria's Journals*, 16 July 1868.
[44] Heavens, pp. 110–114.

Britain for the *Exeter and Plymouth Gazette*, for instance, is similarly interested in his skin tone, hair and lip shape. This report goes on to invoke the pseudo-science of phrenology, enthusiastically used through the nineteenth century to support theories of white racial superiority, and concludes that 'his ears are larger than those of an average English child, but not large enough for a personal defect, if his hair were grown'.[45] The photographic archive of Alemayehu shows an ongoing tension between the desire to present him as exotic and other—pictured with the 'African paraphernalia' that haunts Zeleke—or to show his enculturation to British costume and customs, wearing black suits and boots.[46]

The details of how the British came to take guardianship of Alemayehu are sketchy and disputed. The explanation widely circulated in Britain was that given by Captain Tristram Charles Sawyer Speedy, who had been recruited to the Napier expedition for his previous knowledge of Abyssinia and of its language Amharic. Speedy presented his guardianship as the dying wish of Queen Tirunesh, who had him swear on the Bible that he would raise her son in the Christian faith. Speedy quickly became Alamayehu's only guardian and the only person around him with any knowledge of his own language as his Ethiopian tutor, Zeneb, who was appointed by his mother, was dismissed part way through the voyage to Britain. In a letter Zeneb recalls repeatedly asking 'why do you separate us?'.[47] In another rare counter-narrative to Speedy's, Hormuzd Rassam—a pre-expedition envoy of the British, liked and then imprisoned by Tewodros—described Speedy as pestering the Queen in her final illness for custody of her son.[48]

[45] *Exeter and Plymouth Gazette*, July 1868.

[46] This is particularly stark in the pair of photographs taken by Leandro Preziosi in Malta on a stop on the voyage before Alemayehu arrived in Britain. Set against an identical background, the companion pieces are 'Portrait of Prince Alemayehu in Abyssinian costume, 27 May 1868' and 'Portrait of Prince Alemayehu in western clothes, with the sailor's cap of HMS Urgent'.

[47] Zeneb to J. L. Krapf, 16 December 1868, in *Tewodros and his Contemporaries 1855–1868*, ed. by Sven Rubenson (Addis Ababa: Addis Ababa University Press, 1994), p. 370.

[48] Hormuzd Rassam, 'Letter to the Editor', *The Times*, 3 November 1872. Having detailed the Queen's complaints against Speedy he concludes: 'All the late European captives knew with what hatred Captain Speedy was regarded by Theodore, nor can anyone say that he was much liked by the Abyssinians generally; so it is not likely that the Queen, to whom he was a stranger, can have meant him to take charge of her son'. He also comments on Tewodros's feeling for Speedy, 'whom he always called an enemy', in *Narrative of the British Mission to Theodore, King of Abyssinia*, 2 vols (London: John Murray, 1869), II, p. 330.

Accounts by those involved in Alemayehu's life on his arrival in Britain emphasise his emotional dependence on Speedy. Cornelia Cotton—in a kind of grandmother role to Alemayehu as Speedy married her daughter that winter—described 'the attachment which soon sprang up between him and his ward' as 'of no ordinary nature':

> The child clinging to his powerful protector as if he felt only safe when hand was clasped in hand; and it was often very touching to see the figure 6 ft 6 inches tall, bending with tenderest sympathy over the little black child at his side, or when seated on his knee, telling him some Abyssinian story, or trying to win a smile from the grave features by some merry passing incident.[49]

These contrasts of size, strength, age and race were similarly emphasised in photographs of the pair together, harnessing the affective appeal, for British audiences, of the white saviour narrative. Such narratives remain potent in national self-imagining in the USA and Europe, reinforcing old colonial binaries, as Briggs shows, as 'discourses of rescue extend in only one direction'.[50] As, though, with the counter-readings suggested by the distress depicted in iconography of the removal of the sons of Tipu, the repeated recognition of Alemayehu's 'grave' featured, rarely smiling melancholy, simultaneously undercuts the messaging of British benevolence.

Cotton is voluble on the work of enemy adoption as signalling British Christian magnanimity. In a pean to Queen Victoria she exclaims, 'How gracious, how beautiful this act! God Bless our noble Queen'; 'perhaps in no single instance has her generous, disinterested sympathy displayed itself more remarkably than in her kindness to this poor boy, the child of *that* man who caused her own subjects such untold, needless suffering, and involved her country in such an expensive war'.[51] In an extension from figurehead to whole nation, Cotton suggests that 'notwithstanding the character of the father, every heart seemed ready to feel compassion and pity for his orphan child, and to welcome him to the shores of this free and happy country with no ordinary degree of warmth'.[52] Lemn Sissay—who draws parallels between Alemayehu's experience and his own difficult history of being separated from his Ethiopian mother for adoption in a white

[49] C[ornelia] C[otton], '*Anecdotes of Alamayu: The Late King Theodore's Son* (London: Willian Hunt, no date), p. 19.

[50] Briggs, 'Mother, Child', p. 197.

[51] Cotton, p. 30.

[52] Cotton, p. 32.

British family—notes the British saw Alemayehu as a 'route for them to fulfil who they were, forgetting that nobody was truly fulfilling who he was'.[53] Cotton's rhetoric that asserts British 'compassion and pity' for Alemayehu as indicative of superior British liberal democracy—'this free and happy country'—resonates with the recent claims, critiqued by Lee Edelman, that US national policy and character is guided by its exceptional care for children's futurity. This emotional appeal to the care of the child operates to obfuscate the curtailment of freedoms and nixes opposition, as the rhetoric of 'fighting for the children' is compulsorily 'one sided', 'a value so unquestioned, because so obviously unquestionable'.[54] In Alemayehu's case, mandatory compassion for the child was complicated as neither the Queen nor the many involved in his education and development could fully ignore his suffering. As repeatedly recorded in personal and newspaper accounts, the distressed child would often be unable to sleep alone and experienced day and night terrors. Speedy, though, found an explanation which could support the British saviour narrative by rooting the boy's trauma not in his displacement and isolation from family, home and nation but in the barbarity of his early life. The Queen recalled that 'Capt. Speedy said that the poor child had a recollection of the horrible massacre of the captives by his father's hand & orders, having heard the shrieks'.[55]

While suspicious reading is necessary to register the violence done by culturally legitimated mis-explanations of empire as benign, a different, more reparative, approach might be appropriate to recognise the feelings and motivations of the adoptive parents involved in enemy adoptions, as well as the significantly broader category of transnational adopters. There is a gap, as Briggs notes between the 'symbolic and material practice' of adoption, with the personal experience not typically accounted within the broader cultural representation. While the actions of these adoptive parents clearly take place within unequal structures of power, including purchasing power, their own individual investments in the well-being of the children in their care might both parallel and go beyond or around national

[53] 'Prince Alamayu', Lemn Sissay's choice on BBC Radio 4, *Great Lives*, 18 December 2012.

[54] Lee Edelman, *No Future: Queer Theory and the Death Drive* (Durham and London: Duke University Press, 2004), p. 2. Though Edelman's insistence that queer must be asocial and counter to the 'familiar familial narrativity of reproductive futurism' is overly proscriptive (p. 17), his arguments about the cultural force of an appeal to the child to forestall debate apply very clearly to war propaganda.

[55] *Queen Victoria's Journals*, 17 July 1868.

narratives. A reading able to hold competing possibilities is needed to understand Speedy's emotional investments in both Abyssinia and in Alemayehu himself. Speedy's learning of Amharic, delight in his version of Abyssian warrior dress, apparent in Hughes's photograph and others, and in the moniker 'Basha Felika' (roughly Capt. Speedy in Amharic) which he claimed Tewodros called him, could be seen both as an outward-looking cosmopolitan fascination with another culture, and as part of colonial appropriation.[56] Similarly his attachment to Alemayehu might combine elements of cultural appropriation with a longing for adventure and advancement, as well as genuine affection and concern for the boy himself.

From available records, Captain Speedy, his wife (also named Cornelia) and her wider family became passionate advocates of Alemayehu's well-being. Weighing up contestations of Speedy's account of Tirunesh's dying wish that he take care of her son, Heavens suggests that Speedy recognised the prestige and potential advancement that could come with guardianship of the prince, a role that could 'open the door to further opportunities and adventures'.[57] But Heavens also registers the depth of feeling Speedy expressed for the boy over the two years he was allowed to retain his guardianship by the British establishment, and his anguished lobbying against the Chancellor of the Exchequer's decision that he be removed from the Speedy family at the end of that period. As Speedy wrote to Sir John Clayton Cowell, master of the household to Queen Victoria: 'The child has somehow entwined himself round our hearts, and as we have no family it seems as if Providence has given himself to us'.[58] Though such Providential emotional entwining of the adopted child of especial value to the adoptive father was widely celebrated in Victorian fiction, most notably in George Eliot's 1861 novel *Silas Marner* which was published at the

[56] Bradley Dean argues that by the end of the nineteenth century adaptability, shown in the wearing of local costumes by figures like T. A Lawrence and Richard Burton, had become a 'strategy to transform the British from outsiders to insiders and thus consolidate their power'. He also discusses the delight that Robert Baden-Powell and imperial agents took in their bestowed names, which 'represented to British audiences an influence over the foreign men who were moved to offer these signs of recognition and respect'. Bradley Deane, *Masculinity and the New Imperialism* (Cambridge: Cambridge University Press, 2014), p. 73.

[57] Heavens, p. 92.

[58] Quoted by Heavens, p. 141. Heavens documents the extensive efforts of both Captain and Mrs Speedy to prevent Alemayehu's removal from them, including the other testimonials they gathered to their parental love of the boy and their willingness to take full financial responsibility for him, rejecting the continuation of the annual guardian's salary of £700.

start of the decade in which Speedy became family to Alemayehu, the Speedys' 'gift' was removed. The government, unmoved by their evidence of attachment and the Queen's support for Alemayehu remaining with Speedy, took over Alemayehu's guardianship and boarded him at school.

While Eliot's story is able to explore the emotions of both adoptive father and adopted child—she has Eppie articulate what Silas's elective care means to her—the child's perspective is too often lost in accounts of adoptive experience. For Sissay that loss casts doubt on the truth of Speedy's account of Alemayehu as 'he brought the story back and he wrote the story for the young boy'.[59] The problematic silencing of the adoptee's own history and that of their birth parents, particularly in transnational adoptions, continues. Adoptee, activist and graphic artist Lisa Wool-Rim Sjöblom's recent exploration of her experience resonates with Alemayehu's, as her own narrative is subsumed within one of the benevolence of her adoptive country and her space for complex feeling compressed by the expectation of gratitude.[60] British perspectives eclipsed the experiences of Tipu's sons and their mothers, with imaginative engagement with their suffering enabled by a Europeanising of their family. Similarly, the promise of hearing Simeon Sinca's own account was sidelined by the British press investment in becoming part of the adventure of his origins story and co-opting his words as a reassurance of British wartime humanitarianism. Alemayehu's voice, however, does occasionally break through, notably in a desire, near his death, to return to Ethiopia. Debilitated and ill after bruising experiences of the British education system at Rugby and Sandhurst he expressed a 'hankering after his own country where he said he had two aunts and a half-brother living'.[61] His family in Ethiopia also yearned for him, with his grandmother writing an unanswered letter to Queen Victoria about his care and a letter which was never given to Alemayehu envisaging his return: 'All the people of

[59] Lemn Sissay, opening talk at 2018 V&A exhibition *Magdala 1868–2018*, <http://www.rastafari-in-motion.org/new-chapter-vanda.html>.

[60] In her graphic novel *Palimpsest* Sjöblom explores the overwriting of the adoptee's story by the adopting county: 'The narrative of the good country of Sweden, with its noble people, who through adoption are saving vulnerable children of the world. Regardless of what I have to say about my own experience of being adopted, I am repeatedly told – I was lucky I was saved by adoption'. Lisa Wool-Rim Sjöblom, 'A Tool for Activism', in *Superheroes: Orphans and Origins* (Lewes: Unicorn, 2022), p. 67.

[61] Quoted by Heavens, p. 181.

Abyssinia are awaiting you, are desirous of you'.[62] Ongoing campaigns for the repatriation of Prince Alemayehu's remains are infused with familial feeling. Sissay, for example, casts the call for his return as family restitution of a child 'stolen from his mother and his father and his sisters and his brothers and everything he was ever relative to'.[63]

The emotive invocation of family bonds in the repatriation campaign contests the British saviour narrative distilled in the Biblical inscription Queen Victoria chose for the memorial marking Alemayehu's grave at Windsor Chapel: 'I was a stranger and ye took me in'.[64] This parable of the behaviour of 'the righteous' (Matthew 25.37) to the distressed, is contested by the Ethiopian memory of Alemayehu, not as 'a stranger' but a son. Though commentators, like the Wickwires, suggest that 'the memory' of the British taking on the care of enemy children has 'endured so well' as symbolic of British ideologies of paternalist imperialism, the remembering of Alemayehu is of a different order. In the campaign for the repatriation of the prince's remains, familial feeling for a displaced child is linked to an emergent, in Britain startlingly belated, remembering of extreme British imperial violence.

Acknowledgements I am grateful to Andrew Heavens for generously sharing his extensive knowledge of Alemayehu with me and to Catherine Butler and Matilda Greig for their insightful comments on drafts of the chapter.

[62] Woizero Laqaya to Alemayehu, 12 January 1872, reproduced in *Letters from Ethiopian Rulers early and mid-nineteenth century* (Oxford: Oxford University Press, 1985), p. 155.

[63] Sissay, V&A *Magdala*. Ethiopian politician Mulugeta Asarate described the Ethiopian President's 2007 call for the return of the Prince's remains as 'reflecting the wishes and desires of the people of Ethiopia to see the remains of a one-time prisoner of war return and unite with his father in burial', <https://richardpankhurst.wordpress.com/2007/09/03/ethiopia-asks-for-return-of-stolen-child-prince/>. See also Heavens, p. xii.

[64] Holy Bible, King James Version, Matthew 25.35.

'He said he was an Irishman also': Irish-American Union and Confederate Encounter Stories in the American Civil War

Catherine Bateson

'I had a hard fight myself as I was either a prisoner or a dead man', Daniel Crowley informed his friend Cornelius Flynn as he wrote from his regiment's entrenched camp 'Before Petersburg, Va.' on 30 June 1864. Crowley explained how 'my Irish Blood would not yield' before his Confederate opponent, and that he took his fighting chances by 'Bayonetting one Reb and shooting another. It was a hand-to-hand affair and believe you me the Johnnies are afraid of the steel while it is in an Irishman's hand'.[1] This Cork-born Union Army soldier emphasised how his ethnicity aided the adrenaline behind his despatching of two enemies determined to do the same to him.

[1] Daniel Crowley letter to Cornelius Flynn, 30 June 1864, Crowley-Flynn Letters, Boston Athenaeum Special Collections. Union soldiers and civilians referred to Confederates as variations of 'Johnny Reb' (rebel), while Confederates described the Union as 'Yankees'.

C. Bateson (✉)
University of Kent, Canterbury, UK

H. Furneaux, M. Greig (eds.), *Enemy Encounters in Modern Warfare*, https://doi.org/10.1007/978-3-031-56748-3_4

In so doing, Crowley added another layer to the 'fighting Irish' concept, long-present in histories and attitudes about Irish and Irish diaspora foreign military service—the American Civil War (1861–1865) was yet another addition to a well-established trope.[2] Crowley was one of over 180,000 Irish-born, and an even greater number of second-/third-/fourth-generation soldiers born in North America to Irish parents and grandparents, who fought for the Union (Federal) Army during the Civil War. Around 20,000 Irish-born Confederate Army soldiers could be found on the opposing side along with thousands more Irish-descended southern Americans. In addition, thousands of ethnic-Irish filled the Union and Confederacy's respective naval ranks. The experiences of these military and naval servicemen have been documented in broader Irish-American Civil War studies by David Gleeson, Damian Shiels, and Susannah Ural.[3] Given the sizeable levels of Irish migrant populations across the United States by the time of the conflict (numbering approximately 1.6 million Irish-born in 1860), the Irish diaspora's sons could be found throughout Union and Confederate regiments, brigades, militias, and smaller units. Some, like the Union Army's official Irish Brigade, were designated official Irish regiments. Others, such as many of the units from Illinois, Indiana, Louisiana, Massachusetts, New York, and Pennsylvania, were not Irish-dominated but did contain a significant proportion of Irish-born and descended soldiers in their ranks. The composition of all of these military groups reflected the Irish diaspora's vast American spread by the mid-nineteenth century.

Thus the likelihood of Irish soldiers meeting across battlefields and picket lines was high. Consequently, Ireland's sons crossed paths at nearly every major battle and minor skirmish. They appeared in reports from the conflict's opening engagement at the First Battle of Bull Run in July 1861

[2] There is debate about whether the 'fighting Irish' was uniquely specific to Irish foreign military service (both domestic and international) but the impression is distinctive and pervasive in American popular memory, public heritage, and sporting culture. For more on the construction and internationalisation of warrior ethnicity and 'martial races', see Heather Streets, *Martial Races: The Military, Race and Masculinity in British Imperial Culture, 1857–1914* (Manchester: Manchester University Press, 2010).

[3] See David T. Gleeson, *The Green and the Gray: The Irish in the Confederate States of America* (Chapel Hill: The University of North Carolina Press, 2013); Damian Shiels, *The Irish in the American Civil War* (Dublin: The History Press, 2013); Susannah J. Ural, *The Harp and the Eagle: Irish-American Volunteers and the Union Army, 1861–1865* (New York: New York University Press, 2006).

through to its last days in 1865. Indeed, at the First Bull Run, the Irish 69th New York State Militia fought Louisianan regiments comprising southern Irishmen. Wherever you went in Civil War America, soldiers and sailors from and connected to Ireland fought each other as committed members of the American Union and Confederate forces.

This chapter assesses the way encounter stories between Union and Confederate Irish soldiers were articulated in first-hand accounts and memoirs and in fictionalised postbellum stories that recalled both sides' wartime experiences. These encounters—real and imagined—reveal much about the memory of Irish diaspora Civil War service, how they differed from other non-Irish encounters, and how, at times, they reflected desires for a common purpose connected to Irish nationalism. Reference to such meetings indicated a sense of fellow-feeling Irish-Americans had towards each other while on opposite sides, as well as to their non-Irish fellow Americans, especially when fighting was not occurring. Exploring the ways in which national connections and depictions were articulated through various encounter scenarios during moments of combat and during lulls in fighting, this chapter will investigate how wartime actors interacted during tumultuous civil conflict periods.

Furthermore, this chapter will explore how, in the case of Irish-born and descended soldiers in the American Civil War at least, friendly fraternisation between Union and Confederate enemies seemingly occurred in 'quieter' moments when bayonets and guns were lowered. In particular, sharing some food, a fire, a smoke, news of the old Irish home country, and connections over trenches, and in field hospitals, appear to have engendered connections not present when fighting occurred. Addressing aspects of the 'memory versus history' of Irish-American encounters, this chapter will also assess the 'image versus reality' of these moments, using case studies from the battlefield and from contemporary literature. Such tales reveal how preconceptions were reinforced and altered, and how experiences reshaped the meanings behind 'the enemy' they encountered. This chapter considers whether—beyond some moments of real and fictional possible connections—shared 'Irishness' across sides affected fighting morale to any significant extent.

Irish-American Civil War Encounters: Memory versus History

Enemy encounter and interaction stories appear regularly in contemporary soldiers' writings, officer reports, newspaper accounts, and post-war memoirs but have received little in-depth study of their importance to the overall conflict experiences of combatants, prisoners-of-war, and those in the more immediate occupied home-front. This is surprising given the vast field of American Civil War studies covering a plethora of wartime topics. When they are discussed, analysis can be as fleeting as the original encounters themselves. As a result, they can be exaggerated, misremembered, and turned into moments far from what happened according to surviving records.

Lauren K. Thompson's recent research on soldier fraternization, discourse, trade, and moments and memory of interactions is a much welcome effort to help Civil War historians and wider military culture researchers understand how modern warfare encounters with the enemy played out and were discussed by soldiers involved. Thompson observes that we need to move beyond the Civil War battlefield and shift focus to encounters on the front-lines' peripheries: the picket duties, campfires, prisons, and home-front communities that were turned into war zones. 'When men were in small groups and close to their enemy, soldiers identified with one another and formed fraternal bonds', Thompson argues, connecting over shared experiences, rumoured reconnaissance, wider political and military policies, national news, warm drinks, and food.[4] For Irish-born and descended soldiers there was greater chance of this fraternal bond forming through their shared ethnicity. Crucially, as will be shown in this chapter's case studies, this was only in cases where fighting was *not* taking place. The fog of war intervened once guns and bayonets were in their hands. This is in keeping with Thompson's broader assessment of soldier fraternization across the whole of the American Civil War; this chapter uses similar methodological framing about common combatant fraternization to apply it to specific ethnic-Irish encounter moments.[5]

[4] Lauren K. Thompson, *Friendly Enemies: Soldier Fraternization Throughout the American Civil War* (Lincoln: University of Nebraska Press, 2020), p. 3.

[5] Ethnic aspects of wartime service and soldier fraternization are only briefly covered in Thompson's work. In addition, this chapter considers wider encounters with civilians and in prisoner settings, which Thompson argues 'resulted in a different set of circumstances' that she does not discuss (*Friendly Enemies*, p. 6).

When it comes to the memory of Irish-Americans' Civil War battlefield encounters, the reality is less romantic than twentieth-century popular culture depictions would have audiences believe. Scholars refute the idea that the sons of the diaspora were aggrieved to fight each other just because they had Irish roots. David Gleeson, for instance, stresses that while 'Irish units on both sides did face each other on the battlefield…there is very little record that these meetings left much of an impression on Irish Confederates'. If anything, he argues, 'despite the strong ethnic connections between themselves and Irish Union soldiers, Irish Confederates seem to have felt no compunction about fighting their compatriots'.[6] The same is true for Irish-American Union fighters. Crowley's 'do-or-die' depiction of facing the enemy at close quarters was a shared combat sentiment.

This scholarly intervention has been generated, in part, by Hollywood's attention to one particular Irish battlefield encounter. In mid-December 1862, at the Battle of Fredericksburg (Virginia), the Irish Brigade made several futile charges up an embankment called Marye's Heights. They were almost annihilated by Confederate guns and cannon in front of them, resulting in the loss of forty-five per cent of the Brigade's fighting strength.[7] Yet in the 2003 Civil War film *Gods and Generals*, Hollywood's military revisionism shapes events to bend reality for the story to have more emotive drama.[8] The Confederate counterparts behind the defensive stonewall are Irishmen too, who shoot and cry simultaneously at the fact they are killing their fellow countrymen and possible Union Irish family members. Such scenes appeared designed to appeal to 'modern portrayals of soldier fraternization [that] perpetuate the brother versus brother narrative', Thompson posits, making the historical reality more palatable to audiences' imaginations.[9] It is an evocative image, made all the more bittersweet by the notion of immigrant brethren being forced to fight and kill each other. To increase the film's inaccuracy further, the Confederate

[6] Gleeson, p. 104.

[7] The Union Army's Irish Brigade comprised the 63rd, 69th, and 88th New York, 28th Massachusetts, and 116th Pennsylvania regiments.

[8] *Gods and Generals*, dir. by Ronald F. Maxwell (Warner Bros. Pictures, 2003). The film is notorious amongst Civil War scholars for its inaccuracies and sympathetic portrayals of Confederate leaders such as General Stonewall Jackson. This is in sharp contrast to Maxwell's 1993 prequel *Gettysburg* (New Line Cinema), known for its dryly accurate military movements in re-enactment of the July 1863 Battle of Gettysburg, Pennsylvania.

[9] Thompson, p. 165.

attackers applaud and cheer the Irish Brigade's bravery when the latter finally retreat.[10]

Contemporary and post-battle accounts from Fredericksburg dispute this rose-tinted encounter memory. True, the Union Irish Brigade faced the Confederate 24th Georgia Infantry but this was amongst other southern units. The infantry contained only a handful of Irish Georgians but was not an exclusively Irish regiment. They were commanded by County Antrim-born Robert Emmet McMillan, who ordered his men to hold fire until the Irish Brigade were closest. In subsequent popular military histories, this delay is perceived falsely to be because McMillan sensed his men's unease at firing upon their fellow countrymen. In reality, McMillan held fire to the last moment to make the most impact, inflicting maximum damage by mowing down the advancing Union line.

The commander was even recorded yelling to his men that because it was the Irish Brigade coming at his Confederate line, they were to 'Give it to them!' with as much indiscriminate firepower and remorseless speed as could be mustered. McMillan saw the enemy—not the ethnicity—in this encounter. His official post-battle report did not even mention the Irish Brigade. One observer later noted that the 'Southern Irishmen…have no sort of scruple in killing as many of their northern [Irish-American] brethren as they possibly can' at Fredericksburg specifically, and in the wider conflict.[11]

Despite the prevalence of ethnic-Irish soldiers throughout Union and Confederate ranks, the majority of encounter references by Irish-American Civil War participants did not involve their fellow brethren. That is not to say that their ethnicity was never alluded to; at times this even became a specific detail in encounter tales. In his postbellum memoir about serving in the Irish-dominated 116th Pennsylvania, Derry-born Union Private William McCarter recalled one October 1862 occasion when he was around the evening campfire in West Virginia 'with other members of our company, talking and chattering'. This typical Civil War scene is interrupted suddenly:

[10] There are conflicting contemporary reports about whether any cheering occurred in admiration of the Irish Brigade's constant advance into Confederate fire, or simple Confederate victory yells at another failed Union wave of attack.

[11] Ryan Quint, 'Civil War Myth Busting: The Fictional Confederate Irish Brigade at Fredericksburg', *Emerging Civil War* (7 January 2021), https://emergingcivilwar.com/2021/01/07/civil-war-myth-busting-the-fictional-confederate-irish-brigade-at-fredericksburg/ [accessed 2 August 2022].

> I was somewhat surprised by the appearance of a Rebel soldier. He came boldly forward and sat down in our midst. The Reb took out his pipe, lit it and commenced to smoke. By way of introduction, he stated that he was one of five left in charge of the Rebel wounded then occupying a little church as a hospital. He had just come out to enjoy like ourselves a good quiet smoke...He was a young man, quite communicative.[12]

Perceiving no danger, this smoking non-Irish 3rd South Carolina regiment soldier relaxed in the company of McCarter's fellow-Irish Pennsylvanians. The interloper was made to feel welcome, and conversation turned to the conflict's current state, the composition of the opposing armies in the region, and whether the Confederate soldier wanted the war to end. Thompson has observed how Union and Confederate soldiers 'could quickly shift their perception of the enemy from one of fear and hatred to one of empathy and commonality'.[13] McCarter's story is indicative of this.

The account then recalls the 'voice' of this enemy soldier, quoting his response in a passage of 'Federal'/'Rebel' dialogue like a play: 'Not as long as one of her men is left' would the young Confederate give up the secessionist fight. 'You [the Federal forces] may whip us, but it will only be by unequal and overpowering numbers'. The South Carolinian soldier also reveals his awareness that he is surrounded by Irishman when asked why he and his fellow Confederates had not attempted to break through the Union line. 'Well, I'll just tell you', he replies:

> Because we saw that damned infernal Green Flag of yours a coming, and knew well who was behind it. It gave us a sickening dose at Antietam...the bayonet charges you fellows made there which broke our centre...We can lick you out of your Yankee shoes every day in the year with the musket or rifle, given equal numbers. But your 'cold steel' and the way you wild Irishmen use it, we cannot stand.

In other words, he and his remaining wounded regiment were not prepared to get into another engagement with this particular fighting Irish-American opposition regardless of the fact that passionate Confederate

[12] William McCarter, *My Life in the Irish Brigade: The Civil War Memoirs of Private William McCarter, 116th Pennsylvania Infantry*, ed. by Kevin E. O'Brien (Cambridge: Da Capo Press, 1996), pp. 39–41.

[13] Thompson, p. 3.

'rebel yells' and battle commitment were as commonplace as among their Union opponents. Instead, smoking a pipe was preferable to shedding blood.

Expressing such praise for Irish battlefield bravery was also not an attempt at flattery. The Union Army Irish gained praiseworthy reputations for their commitment on the field through their bayonet charges, willingness to face attack head-on, and forming fighting rearguards for retreating forces. These actions were undertaken from the war's very first clashes, and earned the Irish a notable bravery status that only grew for the remainder of the conflict. Daniel Crowley's depiction of his own fighting in June 1864 presents an idea of this mentality. The fighting at the Battle of Antietam on 17 September 1862 referenced in McCarter's story was no different, as the Irish Brigade made fierce charges throughout what is still widely regarded as the single bloodiest day in United States military history.

McCarter clearly enjoyed his campfire encounter. Not only did he deem it an important enough incident to write about in his 1870s memoir, but he also recollected that when he left the scene, he 'very reluctantly bid good night to Johnny [Reb], leaving him and the rest of our boys to continue the dialogue and enjoy themselves in the best way they could'.[14] Such disclosures signify that there *could* be moments of connection between combatants. Non-Irish specific Civil War encounter references suggest that soldiers enjoyed these calmer instances of being in their enemy's presence, even if it was more for reassurance about their opponent's location. Union Corporal Zenas Haines, who served alongside Irishmen in North Carolina during the conflict, recalled one morning when 'the music of a rebel band floated to our ears from over the river'. By that afternoon, however, 'we have had the thorough base of their big guns on the same side'.[15]

McCarter was not the only Irish soldier who liked gentler enemy encounter moments. Following capture at the First Bull Run in July 1861, County Sligo-born Michael Corcoran, then-colonel of the Irish 69th New York State Militia, spent the subsequent thirteen months in various southern prisons. His post-release memoir, published during the war itself,

[14] McCarter, pp. 39–41.

[15] Zenas T. Haines, *Letters from the Forty-Fourth Regiment M.V.M.: A Record of the Experience of a Nine Months' Regiment in the Department of North Carolina in 1862–3* (Boston: The Herald Job Office, 1863), p. 99.

expressed frequent dislike for Confederate secession and strong declarations of Corcoran's deep-seated American Union loyalty. It also recorded differences with how southerners treated him during various interactions. On one move between prisons, Corcoran was allowed to stretch his legs after the commanding Confederate guard granted the Irish officer's request to walk alongside the transfer column. 'Though, of course, my enemy, I have remembered, and shall ever remember, this courteous soldier for his noble, generous behaviour toward the captives under his charge', Corcoran later reminisced.[16] The same sense of respite that McCarter exhibits in his own tale is presented here; a brief mutual gratefulness pervades both stories.

Additionally, Corcoran was appreciative of encounters with southern Confederate civilians, particularly when they came bearing gifts. In October 1861 while at Castle Pinckney in Charleston Harbor (South Carolina), the Union prisoners' 'dinner to-day was improved by the addition of a large glass of jelly, the gift of a kind-hearted lady of Charleston'. This delighted Corcoran, by then tired of three months' worth of prisoner rations. Several days later, the same generous southern woman sent 'a half pound of excellent coffee'. Though 'we never found out the name of our benefactress', Corcoran was clearly touched by this gift-giving provisioning. He would 'never cease to be thankful for the numberless attentions she paid us while we were confined'.[17] This was a far better form of encountering the enemy for the Irishman.

'Fighting Like Tigers': Irish Union and Confederate Battlefield Interactions

It is understandable that any encounter between family and brethren on opposing sides would generate interest from a Civil War home-front society eager to hear stories that were not always about destruction and loss. Most individual encounters did not make it into the newspapers, though it is telling that soldiers and officers' memoirs like William McCarter's and Michael Corcoran's included encounter tales. Along with considering them significant enough to recollect in and of themselves, to serve 'as closure for some, and for others a way to cope', there were additional

[16] Michael Corcoran, *Captivity of General Corcoran* (Philadelphia: Barclay & Co., 1864), p. 43.

[17] Corcoran, p. 62; p. 65.

publishing considerations to take into account.[18] Encounter stories between enemies were public interest topics that sparked fascination for an avid mid-nineteenth-century readership wanting to hear about how fathers, sons, brothers, uncles, and cousins were faring.

That explains the appearance of one specific Irish wartime encounter in the press in early 1863 when Ohio's *Dayton Daily Empire* published a front-page article headed: 'The Irish Boys'. The majority of the piece was a letter received by the newspaper from First Lieutenant Edward G. Breene, 35th Indiana Infantry Regiment, a unit nicknamed the '1st Irish Regiment' due to significant numbers of Irish in its ranks.[19] As part of the Union's Army of the Cumberland, they had been based around Murfreesboro, Tennessee, in the Civil War's western theatre. Since October 1862, they had seen much action in the region, especially as the nation welcomed in the New Year at the Battle of Stones River.[20] Breene's letter was written not long after the engagement, with the *Dayton Daily Empire* publishing his account on 27 January 1863.

With delight, the newspaper expressed that the correspondence 'gives us some interesting items about the Irish Boys, which we have not seen in the papers', and quoted much of Breene's letter in full. It related immediately to Irish Union versus Irish Confederate battlefield encounters, as Breene stated how his regiment 'were pitted against Claiborn's famous Irish Brigade'.[21] This is a misnomer; the lieutenant meant the recently-promoted Confederate Major General Patrick R. Cleburne, born in Ireland and once a member of the British Army before migrating and settling in Arkansas in the 1850s and serving in the Civil War as the Confederacy's foremost Irish-born general.[22] His division was part of the

[18] Thompson, p. 15.

[19] Edward G. Breene service record (M540 Roll 8), National Park Service Soldiers and Sailors Database, https://www.nps.gov/civilwar/search-soldiers-detail.htm?soldierId=2636B585-DC7A-DF11-BF36-B8AC6F5D926A/ [accessed 10 August 2022]. The newspaper called the unit 'the 35th Indiana (1st Irish) Regiment'.

[20] The Stones River Campaign, Tennessee, lasted late fall 1862–January 1863, with the battle discussed taking place between 31 December 1862–2 January 1863. The Confederates referred to it as the Second Battle of Murfreesboro.

[21] All quotes hereafter – Lieutenant Edward G. Breene, 'The Irish Boys', *Dayton Daily Empire* (Dayton, Ohio), 27 January 1863, Library of Congress Chronicling America: Historic American Newspapers, https://chroniclingamerica.loc.gov/lccn/sn85026002/1863-01-27/ed-1/seq-1/ [accessed 10 August 2022].

[22] Breene possibly confuses Cleburne's name with Claiborne, a Tennessee county whose Confederate sons likewise served at Stones River. The misspelling may also be the newspaper's typographical error.

Confederate's 2nd Corps at Stones River, and was a well-honed, effective fighting force in a comparable reputable manner to the Union's official Irish Brigade. During the battle, Cleburne and his men helped repel Union wings during several days of attack and counter-attack, earning praise from superior commanders for commitment on the field and skill under fire even within a larger Confederate battle defeat.

Cleburne's divisional success is made clear in Breene's letter, which records how his 'brigade outnumbered our men, and we were being driven back' by the Irish Confederate leader. They were only 'finally forced to fall back before the dreadful storm' of Union artillery reinforcements, but Breene praised his Confederate opponents as 'fighting like tigers every step'. They had caused 'the 35th [to] suffer so severely', and Breene acknowledged a lucky escape for both his 35th Indiana and for the 19th Illinois Infantry Regiment who fought alongside them. This latter Union regiment likewise contained many ethnic-Irish soldiers, who mostly came from Chicago's predominate Irish-American diaspora. Thus the combination of two Irish-heavy Union regiments and Cleburne's Irish command—along with Irish southerners in Cleburne's ranks—is behind Breene's suggestion that 'Greek meet Greek' during the desperate fighting.

This classical history reference is an interesting one, revealing Breene's knowledge of early modern literature proverbs; the phrase 'When Greeks join'd [to fight] Greeks, then was the tug of war' first appeared in Nathaniel Lee's play *The Rival Queens, or the Death of Alexander the Great* in the mid-1670s. The saying has more than war-related meaning, however, as it can additionally refer to a 'battle-of-the-wills' between two equally matched sides. In other words, it can be applied to a metaphorical battle as well as a physical one. In Breene's contextual use and understanding of the phrase, he was stressing that when matched together in a warring context at Stones River, equal characteristics on each side met. Here, it was applicable in both terms of perceived fighting strength and in terms of mutual national origins.

'After this severe and bloody battle encounter between the Irish Boys of the North and South was ended', Breene continued by describing the bittersweet impact of this reunion between the Irish Indianans, Irish Illinois, and Cleburne's Irish soldiers:

> The scene presented on the field was one of sad and unusual interest. Old neighbours and friends from the Green Isle, who had accidentally cast their lots with the northern and southern portions of this country, met for the first time for years as – *enemies.* There many who had been friends in boyhood fired upon and killed each other, and lay side by side in death.

Breene reported further how one wounded 35th Indiana sergeant 'was deposited alongside a rebel' in a nearby temporary hospital, with the two wounded combatants encountering each other while receiving treatment. 'In this man', the Union sergeant 'recognised an only brother, whom he had not seen for years!' The *Dayton Daily Empire* editor added a few lines after this to say that Breene recalled how 'a number of recognitions took place on the field between the Irish belligerents, and those between the wounded men', with familiar encounters between long-lost friends and possible relatives being 'especially affecting'.[23]

Breene's whole tale is emotionally affecting, but is it too much of a reunion coincidence to be true? The connections in the battlefield's aftermath could well be, particularly as the story shares similarities with other comparable accounts (including the stories discussed across this chapter). The battle depiction itself is, however, an exaggeration; Irish did not meet Irish at Stones River in quite the way Breene described events. All three regimental combatants discussed from Indiana, Illinois and Cleburne's men *were* there at the engagement, and military records do reflect some of Breene's account. However, Patrick Cleburne, while in command of a division that included some fellow Irish-born and descended soldiers, did *not* face Breene's companions. He was on the other end of the battlefield. That is not to say that the letter is entirely inaccurate, as the Irish-American Indiana and Illinois Union regiments did face Louisianan Confederate units that contained members of the southern state's Irish diaspora. What is certain is that no official 'Irish Brigade' served in the Confederate Army (unlike in the Union). The romantic vision of 'Irish versus Irish' at Stones River, and the inclusion of the by-then familiar wartime figure Cleburne, is one of poetic licence akin to the Battle of Fredericksburg depiction in *Gods and Generals*.[24]

Apart from just how truthful Breene's tale was, the description of wounded Irish Union and Confederate kin being reunited in the aftermath of just one battle brought home very real concerns about the possible damage the war could do to familial and friendship ties. Like many civil

[23] The *Dayton Daily Empire* concludes Breene's letter by highlighting another Irish post-battle encounter: rather than meeting more enemies, wounded 35th Indiana soldiers were 'being well cared for [by] several charitable Irish ladies' from their home state who had travelled 'down from Indianapolis to care of them' in Tennessee.

[24] My thanks to Brendan Hamilton for our conversations and his expertise on this aspect of the Stones River battle, the placing of Cleburne on the battlefield, and information on the missing Louisiana Irish in Breene's account. For more on this and Breene's wartime history, see *Irish in the American Civil War*, https://irishamericancivilwar.com/.

conflicts, the American Civil War is often seen as a 'brother against brother' contest ripping families apart at the same time as tearing the country asunder. Irish-Americans exemplified this particular problem in their wartime cultural outputs, with contemporary songs occasionally exhibiting remorse about how fighting encounters brought harm to their brethren. Lyrics in the 1863 ballad *Pat Murphy of Meagher's Brigade* voiced momentary apprehension that: 'It's a shame for to see, / Brother's fighting in such a quare [queer] manner'.[25]

Other Irish-American Civil War songs echoed *Pat Murphy*'s concern as they detailed battle outcomes to those across the Atlantic in Ireland when they were printed and circulated there in the middle of the conflict. *A New Song on the Last Battle Fought in America* sang about another Tennessean battle, this time the brutal Battle of Shiloh in April 1862, where again Irish-born and descended Union and Confederate soldiers fought each other.[26] Evocative lyrics emphasised the devastating effects this had for their families across the ocean:

> The loss on both sides it was great
> The dead lay in heaps, like shoren wheat
> Belonging to each Army;
> Fully thirty thousand men or more,
> Left wives and orphans to deplore,
> And friends behind on Erin's shore
> Lamenting that dreadful morning.[27]

The following year, *A Lamentation on the American War* expressed similar mournful scenes with its depiction of the dead at the 'Awful Battle of Vicksburg', which culminated on 4 July 1863 after two months of Union besiegement on the Mississippian town. The song noted how 'Old Erin's flag on both sides' had been 'hoised [hoisted] by Irish hearts'. However, this fighting reality came at considerable personal and familial cost. *Lamentation on the American War* highlighted how the events at

[25] Unknown lyricist, *Pat Murphy of Meagher's Brigade* (Boston: Horace Partridge, 1863). For more on Irish-American Civil War song culture, see Catherine V. Bateson, *Irish-Americans Civil War Songs: Identity, Loyalty, and Nationhood* (Baton Rouge: Louisiana State University Press, 2022).

[26] Also known as the Battle of Pittsburgh Landing.

[27] Unknown lyricist, *A New Song on the Last Battle Fought in America* (1862), quoted by Robert L. Wright, *Irish Emigrant Ballads and Songs* (Bowling Green: Bowling Green Popular Press, 1975), p. 463. 'Last Battle' here means 'most recent'.

Vicksburg witnessed: 'the brother fight the brother and father fight the son'. The broadside ballad continued by hammering home news of this deadly encounter, the impact it had on the diaspora's Union and Confederate sons, and how it dislocated them from those at home elsewhere in America and back on the island of Ireland:

> Heaps of Irish heroes brave on the plains there lay,
> That was both killed and wounded there all in America...
> Some thousands here do lie far from their native clay...
> Their bones lie mouldering with the dust all on Columbia's shore.[28]

Nonetheless, it must be stressed that there is no indication any of these lyrical sentiments caused hesitation or altered Irish-American military service. Nor did it stop any of the combatants in Breene's story from firing upon each other in the first place. Killing indiscriminately regardless of *who* was in the opposition and *where* they came from was an attitude actually encouraged in some Civil War balladry, including those produced by and about the Irish. *Song for the Irish Brigade*, an early Confederate wartime ballad written in New Orleans in late 1861, includes lyrics aimed directly at the Union Army's 'Irish ranks' who were 'knaves' and 'snakes' for stopping Confederate secession. It believed the North's Irish Brigade needed trouncing as 'Dastard Yankees, base and hollow'. The true Irish heroes were, according to the song, the stalwart 'Celtic race' of Irish-American southerners, who would 'from their battle place charge at dastard Yankees'. Amongst Confederate ranks 'the Irish green shall again been seen', billowing with 'A burning wind from the South behind', which would cause 'the Yankee rout before it!'[29] It is hard not to imagine such a tactical view being in Irish minds when opposed against each other—an extension of mutual animosity exhibited by both Union and Confederate soldiers and society in the Civil War.

Union Irish-American songs were equally as ferocious in their anger towards their Confederate enemy as they readied themselves to meet on the battlefield. One of many songs about the official Irish Brigade viewed disruptive secessionists as 'Traitors, unholy' who had 'conspired / To pull down the flag of the free'.[30] For that, the Irish in the Confederate Army,

[28] P. J. Fitzpatrick, *A Lamentation of the American War—Awful Battle at Vicksburg*, quoted by Wright, pp. 459–460.

[29] Unknown lyricist, *Song for the Irish Brigade*, in *Hopkins 5 Cent Song Book* (New Orleans: John Hopkins, c. 1861), pp. 14–15.

[30] Unknown lyricist, *The Irish Brigade* (Boston: Horace Partridge, 1862).

along with all eleven states of the Confederacy, needed to be punished. *The New York Volunteer*, 'Dedicated to the Brave Sixty-Ninth' New York—whose militia and later regiment were the Irish Brigade's founding members—went as far to express what Irish soldiers in the Union Army would do if they captured the Confederate President Jefferson Davis and his senior military commanders: 'The rebels soon must yield... / And Davis, Wade and Beauregard will in the air be hanging', executed for their treachery against the nation.[31]

These lyrical sentiments are in keeping with broader enmity felt by American Union defenders and secessionist Confederates when opportunities to face each other arose. Such feelings of hatred towards one another and the greater causes they fought for would erupt, as Michael Corcoran observed when he was moved from Confederate prison holdings in South Carolina to North Carolina. Southern civilians observing the transfer were 'violent in their treatment of the Federal captives'. By comparison to South Carolina, where only one civilian had yelled at the Union prisoners to be executed, in North Carolina the Irish-born officer and his fellow cellmates 'were jeered and hooted at by the ignorant mobs, who sometimes were rather violent in their manifestations of hatred and anger'.[32]

As Corcoran's account highlights, desires to put down the opposition by whatever means seeped into public home-front consciousness and culture on both sides. They were not the sole preserve of soldiers. Yet, for those fighting at least, that feeling lessened when the firing of guns and clash of bayonets stopped. It dissipated when dealing with the wounded and in quieter moments, including the sharing of food and familiar connections as both Corcoran and Breene noted. It was in these latter, more gentle, encounters where fellow ethnic-Irish Union and Confederate connections came fully to the fore.

'Laughter and Cheers': Non-Fighting Irish Encounter Tales

Conversely, when Irish soldiers encountered each other in more peaceful moments, the fact they were on different sides became less significant on the surface at least. One reason for this was because it offered an opportunity to talk about something of greater significance to a section of the

[31] Thomas Donnelly, *The New York Volunteer* (Unknown publisher, 1862).

[32] Corcoran, p. 91; p. 94.

diaspora. They could discuss another future, potential civil conflict over Ireland's desire for independence. Writing to Cornelius Flynn three months after his hand-to-hand combat ordeal, Daniel Crowley informed his friend about a second enemy encounter he had 'a few nights ago' while he was out 'placing videttes between our lines and theirs' at Petersburg (Virginia).[33] When he was 'not more than 100 yards distance' from the Confederate line, he came across an enemy soldier who 'said he was an Irishman' just like Crowley. The two 'had a chat' and exchanged details about what corps and regiments they were in, with Crowley stating he was part of the Irish Brigade's 28th Massachusetts.

This led to enquiries about 'when I [Crowley] left the old land and if I heard from there and how things were looking at present'. Much like Breene's experience at Stones River, encounters between migrant Irish-Americans on opposing sides could facilitate important information network exchanges regarding mutual connections, reports from relatives about the general state of post-Great Famine Ireland in the early 1860s, and specific local news that could alleviate homesickness via a sense of familiarity. This could then be passed on through letter writing to friends and family, and to newspapers on occasion as Edward Breene did. Beyond this, Crowley and his new Confederate associate talked about bigger issues, such as whether former Union commanding General George B. McClellan would be elected to presidential office in the November 1864 election. Most importantly, they discussed when 'there will be peace and we will have a chance for trying our mettel [mettle] for the old sod'.

By turning conversation attention towards Ireland and hopes about fighting for Irish independence from British rule, Crowley and the Confederate Irishman he met reflect the diaspora's underlying sympathetic desires for such an eventuality, even if it did not immediately materialise. Beyond doubt, Irish nationalist groups in the United States like the Fenian Brotherhood saw the Civil War as a chance to gain military experience for Ireland's cause. Crowley expressed, 'I hope the day is not far distance [*sic*]' when the fight for the old home nation would begin, 'as I would rather spill my blood in that cause than in the present one'. The remainder of the letter does not reveal if the Irish Confederate agreed.[34]

[33] Videttes are mounted sentinels before picket defences.

[34] Daniel Crowley to Cornelius Flynn, 29 September 1864. Crowley's original spelling of 'distance' (when he means 'not far distant') is maintained.

Breene's witnessing of wounded Irish Union and Confederate soldiers coming together, and Crowley's own short encounter with his fellow-Irish brethren, indicate much about how the diaspora's sons reacted towards their opposite side when meetings were in close surroundings. In addition, Crowley's story hints at postbellum reconciliation and the wish to fight for Irish independence via the uniting of the Union blue and Confederate grey under the green flag of Erin. However, the mournful caution of wartime ballads counters this idea. Along with enlistment and war service came great sacrifice for all those involved, including the Irish. The lyrical 'shame' *Pat Murphy of Meagher's Brigade* referred to, and the rhetoric heard in the two war-reporting ballads which detailed graphic accounts of the Irish Union and Confederate war dead at Shiloh and Vicksburg, were all produced purposely in relation to battlefield encounters where the cost was eventually too high.

Even after one battle alone, and especially after four years of war, the impact of fighting, killing, and wounding each other was considerable. One Irish nationalist leader declared this a concerning 'impediment' to potential Irish-American fighting hopes; there were not many experienced men left who wanted to, or physically could, go back to Ireland and participate in another conflict.[35] A poem submitted to the diaspora's leading newspaper *The Irish-American* stressed this aspect about soldier loss within Irish wartime service in July 1864, revealing sentiments about combatant loss. Due to the American conflict, Irish soldiers' 'hopes, bright dreams, alas! / Themselves, with the mournful past, have fled'.[36] The loss was a transnational tragedy, and those harbouring independence plans would shed 'many a tear' and give 'tender thought' to 'the lonely graves' of the non-returning dead. This explains why the thought of military encounters between Irishmen on either side was so sorrowful, and reunions so bittersweet. Not only were they killing fellow kin without compunction, they were destroying potential hopes for their old homeland country in the process.

For the vast majority of Irish-born and descended soldiers though, serving in the war was not for any Irish nationalist cause first and foremost. Their biggest concern was more immediate, as it was for many American

[35] Christian G. Samito, *Becoming American Under Fire: Irish Americans, African Americans, and the Politics of Citizenship During the Civil War Era* (Ithaca, NY: Cornell University Press, 2009), pp. 121–122.

[36] 'Bessie', *The Irish Brigade*, quoted in *The Irish-American* (New York), 30 July 1864.

Civil War participants. They fought for what they needed: military salary and subsequent pensions to bring some financial stability to themselves and family members. This monetary reality is at the heart of an enemy encounter story that circulated in 'tales-from-the-war'-type compendiums in the later nineteenth and twentieth centuries called 'Pat and Johnny'. It originated from the memoir recollections of Major John L. Cunningham of the 118th New York Infantry Regiment. His unit was also part of the Union siege at Petersburg alongside Crowley's Irish Brigade. Cunningham's storied memory begins on 'one moonlight evening, when both sides were top of the trenches, [and] the other [Confederate] side were being particularly noisy'.

The reason for 'vociferously noisy' catcalling across the lines was due to 'hilarity' caused by recent Confederate capture of Union cattle.[37] Not content with this bovine kidnap, the Confederates responsible 'loudly and boastingly invited' the Union soldiers opposite them 'to roast beef and beefsteak feasts', knowing full well that they had depleted their meat supplies. Amidst the goading, one Union soldier shouted back 'with a distinct "old sod" accent'. In other words, the provoked soldier had a thick Irish accent, hardly surprising given that the 118th New York contained many from New York's sizeable Irish-American diaspora. Cunningham's tale recounted the ensuing conversation:

> 'Say, you Johnies, stop yer hollerin' so soldiers can slape [sleep]!' Back came, 'Hello, Pat! How long you been over? Suppose you are soldiering for the Yankees'—beautiful greenbacks'. 'Well', said the Irishman, '"yees is fightin" for money that has nayther [neither] beauty nor value'. Laughter followed from both sides with cheers on our side for the Irishman's rejoinder. Then from the other side: 'Come over with us, Pat; We are fighting for honor and you are fighting for money'. 'Then we is both fightin' for what we most nade [need].' This second witty 'hand-out' was followed by more laughter and cheers. As those on our right and left inquired for the cause of our hilar-

[37] Corporal Haines likewise refers to enemy livestock encounters, suggesting the Union Army were 'equally indebted to the pigs and potatoes of the country, for soldiers certainly never could march ten days upon hard tack and coffee alone', Haines, p. 59. Thompson has observed that such were the constraints on Confederate resources during the Petersburg siege 'Union men would even exchange whiskey if they had it', Thompson, p. 12.

ity and being told, they laughed and cheered—and so on until this incident probably traveled our line by repetition from the Appomattox [River] to the south and west of Petersburg, for we could hear the laughter passing on.[38]

Unlike Crowley's account, it is not clear if the Confederates shouting back at the Irish Union soldier are themselves Irish. Yet, as with William McCarter's recollection discussed earlier in this chapter, here the Irishness and identity of at least one of the participants in question is important for the story. This is emphasised more by the fact Cunningham writes in an Irish brogue to help the reader 'hear' the conversation. Instead of fighting with bayonets, the tale is one of mocking wordplay barbs, particularly about the dire state of worthless Confederate money. The idea that fighting for the Union lacked any honour mirrors the sentiments heard in *Song for the Irish Brigade*, with the added caveat that the Confederate soldier in Cunningham's tale thinks that the Irish Union soldier is *only* fighting for money rather than patriotic duty.

The Union 'win' this jovial fight, with Cunningham highlighting how the encounter provided light relief and laughter that dissipated tension in a comparable manner to the quieter enemy encounters McCarter and Crowley described. It was certainly one way to break up eleven months of battles, skirmishes, attacks, and siege warfare at Petersburg. Indeed, Thompson has pointed to the fact that siege circumstances at Petersburg in 1864 led to 'rampant fraternization due to the lack of material supplies, particularly for the Confederates'; the cattle rustling in Cunningham's story is symptomatic of this reality.[39] This encounter incident's message was that both sides could maintain rivalry via less aggressive verbal battles. They could also turn an Irish Union soldier into a local trench hero for putting a Confederate soldier in his moral place.

[38] All quotes hereafter, John Lovell Cunningham, 'Pat and Johnny' (19 June 1864), in *A Civil War Treasury of Tales, Legends, and Folklore'*, ed. B.A. Botkin (New York: Promontory Press, 1981), pp. 440–441. This account was first published in Cunningham's memoir, *Three Years with the Adirondack Regiment: 118th New York Volunteers Infantry* (Norwood: The Plimpton Press, 1920), pp. 137–138.

[39] Thompson, p. 74.

Conclusion: Image versus Reality

Regardless of enjoying a laugh or a chat over the front-line as Crowley and Cunningham did at Petersburg, picket duty along reinforced battlefield defences was not always the safest place for soldiers to be. One of the Civil War's most popular cultural outputs in both the Union and Confederacy was a poem by Ethel Lynn Beers, first written in 1861 and subsequently set to music in 1863. *All Quiet Along the Potomac To-night* depicted a scene familiar to experienced soldiers, describing the quiet over the battlefield and encampments as combatants slept. Those on watch were peaceful too, though danger waited in the shadows:

> All quiet along the Potomac to-night,
> Except now and then a stray picket
> Is shot, as he walks on his beat to and fro,
> By a rifleman hid in the thicket.[40]

Even if two belligerents were able to talk to each other, trade news, information and resources, the point here is clear: death lurked in all aspects of the war. For every non-battlefield enemy encounter tale about a shared joke, story, cup of coffee, or cigar, there was always a counter tale depicting the cruel reality of warfare.

In the case of ethnic-Irish connections, romanticised encounter moments focused on intimate meetings rather than battlefield combat. Most of Breene's tale about Stones River is a case in point—the majority of the story details events *after* the battle when the wounded Irish Union Indiana and Illinois soldiers met Irish Confederate soldiers instead of their aggressive clashes. The bittersweet feeling of loss that their combat caused, while not enough to stop them actually fighting, does explain why encounters appeared more positive when they removed any mention of face-to-face engagement. This also meant they did not have to confront the truth that the diaspora lost much-needed men in the Civil War, arguably delaying Irish independence for other generations to obtain.

The image of soldiers at peace is far better than the reality of them being constantly shot, as *All Quiet Along the Potomac To-night* captures. This attitude was furthered after the Civil War in postbellum memoirs and

[40] Ethel Lynn Beers, *All Quiet Along the Potomac To-night* (New York: *Harper's Weekly*, 30 November 1861). Confederate Private Lamar Fontaine was claimed erroneously to be its author in the southern states.

fictional stories that blurred the truth of encounter meetings with a dash of popular rose-tinting. Bernard Wayde painted one such friendly non-combatant encounter image in his December 1882 dime-store novella *Fighting Pat; Or, the Boys of the Irish Brigade*. A fictional tale from the Civil War, it centres on a New York Union regimental Irish soldier called Pat Mooney who serves in General Michael Corcoran's Irish Legion.[41] Chapter Ten—dramatically subtitled 'What Pat Hears in the Forest'—details how Mooney wandered into a forest one night and hears someone 'singing an unmistakably patriotic Irish song' about late eighteenth and early nineteenth-century nationalist uprisings.[42] This pricks Mooney's ears, as he too often sang as he 'enjoyed the beauty' of a wartime evening.[43] He goes to meet the singer 'to thank him for the spirit of patriotism the song breaths', and finds a man whose 'countenance was characteristically Celtic'. He is an older Confederate Irish soldier on picket duty, guarding a group of nearby Confederate guerrillas. As he keeps watch, he sings 'in a low, plaintive voice' about 'the sufferings of too many of [Ireland's] noble but unfortunate sons'.

Pat Mooney is undeterred to meet this singing opponent, with Wayde breaking the narrative to give readers the protagonist's own internal voice: 'Whether he be for the South or North the man who sung that hasn't the bad drop in him'. Their following conversation describes how the Confederate soldier came from Ireland, and both express their mutual wish for Irish independence. As with Crowley's real encounter at Petersburg, this fictional tale presents shared Irish nationalist desires as the common subject that unites these enemies. Moreover, Mooney refuses to believe that his new associate is part of the Confederate guerrilla force planning to attack Corcoran's Legion in the morning. 'The true Celt is incapable of treachery...The true Celt is never a betrayer', Wayde posits.[44] Armed with information from the Confederate (who technically does betray his own side to benefit Corcoran's Irish-born and descended soldiers), the Legion attack first with Mooney leading the way in a scene illustratively depicted on the novella's front-cover.

[41] Corcoran's Irish Legion, formed in late 1862 after Corcoran's return from Confederate captivity, comprised of four New York regiments filled with Irish-American soldiers.

[42] Bernard Wayde, *Fighting Pat; Or, The Boys of the Irish Brigade* (1882), p. 8.

[43] Wayde, p. 3.

[44] Wayde, pp. 8–9.

Despite this brief moment of combat, most of Wayde's Pat Mooney tale focuses on serene interactions between the Irish soldiers in the Union ranks, and between Mooney and his Irish Confederate acquaintance. Instead of killing and lamenting on wartime aspects more broadly, this fictional encounter ends with 'song and story claim[ing] a fair share of the night's entertainment'.[45] It is an evening of music, mirth, and peace that Daniel Crowley, William McCarter, Edward Breene, and John L. Cunningham would surely love to have been a part of alongside their fellow diasporic countryman on both sides of the American Civil War.

* * *

The truth to these sentiments and enemy encounters is somewhere between; when it came to fighting there was little compunction about Irish-Americans killing each other. They remained enemies who just happened to also be Irish. Regardless of this, Irish-Americans could still lament the overall loss of their opposing sons on a grander scale, especially in relation to the impact this had for Ireland's future independence. On these occasions, a sense of a double grief was above any anti-Union and anti-Confederate sentiment each side felt towards the other. Despite their shared ethnic familiarity, divisions remained in place even beyond the battlefield. There was no eager call in the examples quoted in this chapter to stop hostilities for the sake of their old mother country. Instead, they kept on fighting for the defence of their new American home.

Additionally, the particular real and fictional enemy encounter stories discussed here demonstrate that when it came to military aspects, both sides had embraced soldier life fully. These are very much soldiers' perspectives of encounters regardless of *who* they came across, be that fellow soldiers, prisoners, guards, or civilians from the opposing side. In this, Irish-American combatants shared both a 'brothers-in-arms' mentality and a reality, especially in the ways stories loved to mention enemy kin encountering each other in the aftermath of fighting. Comparable themes appear in Irish Union and Irish Confederate encounters—and in stories involving at least one Irish soldier on one side—from recognising family and friends to swapping news about Ireland, talking about the war, and sharing reputation-enhancing tales about fighting honours gained on American battlefields. Yet again, though, divisions remained amongst the

[45] Wayde, pp. 10.

familiarity. While is it often easier to think about enemy encounters in modern warfare as being moments where combative elements dissipate or take on new characteristics, as this chapter and this wider collection stresses, we must also remember the 'enemy' is still embedded in the narrative. Even when amity *was* recognised, the violence of war was not necessarily immediately redirected or discarded outright.

Even with moments of connection, jokes, a quick talk, a swapping of supplies, and shared grander ideas for Ireland's future, these former fellow countrymen remained in opposition to each other at the end of every encounter tale referenced and analysed in this chapter. Ultimately, their overall outlooks remained Union American and Confederate American foremost, regardless of any other national or familial ties. One postbellum dime-store tale summarised this mindset well: 'Brothers Divided', recalled a meeting 'along the banks of the Rappahannock River after the Battle of Fredericksburg' where the Union Irish Brigade had suffered so severely.

In the story, a Confederate 8th Alabama soldier asks the opposing Union guard to send for his brother Harry to come from his Pennsylvania regiment camp. Harry goes to this fraternal meeting, they have a discussion, and then returns 'after a while, to his own side' to take up guard duty opposite. The siblings then watch each other 'as closely as if he were some stranger rebel'. Whether reality or not, such 'truly, queer events grew out of this war'.[46] Enemies from Irish-American backgrounds who encountered each other on the banks of the American Civil War's rivers, trenches, battlefields, and in every conflict scenario would no doubt agree. They continued to see the enemy beyond their mutual ethnic connections. As William McCarter stated sorrowfully in his memoir: 'Alas, such is civil war'.[47]

[46] 'Brothers Divided', quoted in *Civil War Treasury*, p. 90.
[47] McCarter, p. 39.

Racial Difference and the Limits of Fellow Feeling: British, German, and Dutch Soldiers Encounter the Enemy in Colonial Wars, c. 1900

Tom Menger

In 1893, H. Lynn Stevens was a young Englishman serving in the Bechuanaland Border Police, participating in the invasion and conquest of the Kingdom of the Ndebele by the British South Africa Company, in what is current-day Zimbabwe. In his later autobiography, Stevens described one incident of that war that he said remained very vividly in his memory. Stevens wrote that, having come upon a party of Ndebele raiders and driven them off, he was riding in the rear when he noticed the following:

> (…) suddenly I saw a black man lying hidden cunningly in the grass under a tree, where he must have dropped unnoticed during our advance.

T. Menger (✉)
Ludwig Maximilian University Munich, Munich, Germany
e-mail: t.menger@lmu.de

H. Furneaux, M. Greig (eds.), *Enemy Encounters in Modern Warfare*, https://doi.org/10.1007/978-3-031-56748-3_5

> His gun, when I spotted him, was pointing straight at me; and not realising that this effect was due to his attitude, and involuntary on his part, I reined in sharply and covered him.
>
> Even then he made no move, but lay still among the tree roots watching me; his starting eyes alone betraying life or movement (…)
>
> And it was a piece of foolishness on his part which was to cost him dear. For now, seeing me pull up and point my rifle, Jeffries, who was with me, rode up and saw him.
>
> "By God," he cried, "a Matabele! Hurry up, man; what the h—l are you waiting for? Shoot the —, and come along….".
>
> But I hesitated. Killing a man in fight never gave me a qualm; nor (…) was I over sensitive about taking a sporting shot (in the way of duty) at a running one. But a man on the ground… at our mercy… I could not do it.
>
> Nor did the circumstances require it. So I said no (…). But Jeffries was troubled by no such squeamish fancies as bothered me.
>
> With a curse he shoved past me, and without further parley, or troubling to get down from his horse, he levelled his rifle across his knee, and pulled trigger.
>
> Then we rode on after the others; but I did not like it, and told him so; saying that, much as I loved him, if he ever did it again I'd shoot him for a murdering villain!
>
> But he only laughed good-humouredly, and told me not to talk a lot of d—d Exeter Hall rubbish; that a dead Matabele was the only sort he had any use for; and that I was not cut out for a job of this sort evidently, if I made such a fuss about nothing.[1]

Steven's exact motivations remain unclear here. Was it simply abhorrence at 'unsporting' conduct, some amount of mercy, or even the experience of fellow feeling? What is clear is that such 'squeamishness' in the encounter with the enemy in *fin-de-siècle* colonial wars, was, unfortunately, rather rare. As Stevens' fellow soldier made clear, that was not supposed to be part of the 'job'; shooting a man in such circumstances was 'nothing'. This chapter asks why that was so often the case. A frequent argument here is that it was the structural circumstances of colonial wars that made the violence and cruelty of such encounters nearly inevitable. In most cases, cultural differences between coloniser and colonised were so large that they generally precluded understanding, while a protective legal

[1] H. Lynn Stevens, *The Autobiography of a Border Policeman* (London: Witherby, 1927), pp. 281–282.

framework was absent. The form of most colonial wars, which generally took the shape of guerrilla and counter-guerrilla warfare, meant that colonial soldiers were regularly under high pressure and fear, while the frequent lack of a clearly targetable enemy army meant that soldiers would rather indiscriminately attack or assault any human groups they encountered, whether genuinely hostile or not.[2] While all these factors certainly influenced the enemy encounter in colonial settings, it is my argument here that there was something else at play too, something that went deeper than just situational factors. Rather, a racialised body of thought on how to conduct colonial war generally set the parameters for such encounters. It is these ideological parameters, and the room they left (or not) for encounters that were not physically violent, that will be the focus of this piece.

In doing so, I draw here on my research on extreme violence in *fin-de-siècle* colonial wars in the British, German, and Dutch Empires.[3] From that research, three major colonial conflicts figure here as case studies: the wars fought against the Ndebele and Shona in the early years (1891–1897) of what was to become Rhodesia (current-day Zimbabwe), the German-Herero and German-Nama War in German South West Africa (current-day Namibia), and the Maji Maji War in German East Africa (today Tanzania). These empirical cases were complemented by the investigation of a corpus of British, German, and Dutch manuals of colonial warfare published between 1860 and 1920, though mostly in the 1890s and 1900s. Such manuals constituted a new genre in the nineteenth century. Virtually all were written by men who had a lengthy experience in colonial campaigns themselves, and the authors generally indicated they hoped to

[2] Dierk Walter, *Colonial Violence: European Empires and the Use of Force* (New York: Oxford University Press, 2017), pp. 156–157, 76–81; Michael Hochgeschwender, 'Kolonialkriege als Experimentierstätten des Vernichtungskrieges?', in *Formen des Krieges: von der Antike bis zur Gegenwart*, ed. by Dietrich Beyrau, Michael Hochgeschwender, and Dieter Langewiesche (Paderborn: Schöningh, 2007), pp. 269–290 (pp. 276–277); Daniel Karch, *Entgrenzte Gewalt in Der Kolonialen Peripherie. Die Kolonialkriege in "Deutsch-Südwestafrika" Und Die "Sioux Wars" in Den Nordamerikanischen Plains*, (Stuttgart: Franz Steiner Verlag, 2019), pp. 314–317.

[3] The publication of a book on this research project is forthcoming. A project description can be found here: Tom Menger, 'The Colonial Way of War: Extreme Violence in Knowledge and Practice of Colonial Warfare in the British, German, and Dutch Colonial Empires, C. 1890–1914 (Project Description)', in *Colonial Paradigms of Violence: Comparative Analysis of the Holocaust, Genocide, and Mass Killing*, ed. by Michelle Gordon and Rachel O´Sullivan (Göttingen: Wallstein Verlag, 2022), pp. 273–278.

convey this knowledge to future practitioners of colonial war. Most were published without official imprimatur; only at the very end of the fin de siècle do we find manuals being issued with official military approval.[4] As a historical source, they should be treated with caution; they are mainly *prescriptive* and certainly do not adequately describe the *reality* of colonial war. Nevertheless, being written by soldiers with longer colonial experience and directed clearly towards future colonial soldiers, they offer a window onto what the former individuals considered important and, as some of the empirical examples will also show, such notions were certainly influential on the ground. More particularly for our purposes, they reflect a general conception of the 'native enemy' that influenced how colonial actors encountered that enemy in war. It is, however, important to remember that these publications as well as the examples reproduced below represent the views of the European practitioners, and not of the non-Europeans who, at least outside of settler colonies, frequently constituted the majority of the rank-and-file of colonial armies.[5]

Colonial Warfare and Racial Othering

Colonial military violence, as Kim A. Wagner has written, was 'based on deeply encoded assumptions concerning the inherent difference of local opponents'.[6] Long before the actual encounter, the 'native enemy' had generally already been imagined in highly stereotypical and racist ways. To an extent, such conceptions were already being imparted to Westerners from an early age by education and popular culture. Schoolbooks, popular print media, exhibitions and fairs, youth and adventure literature: virtually

[4] See General Staff, War Office, *Field Service Regulations. Part I. Operations* (London: H.M. Stationery Office, 1909), Chapter X; N.a., *Anleitung zum Felddienst in Deutsch-Ostafrika (Entwurf!)* (Daressalam: Druck "Deutsch-Ostafrikanische Rundschau", 1911); M.J.E. Bos, *Aanhangsel op het leerboek der tactiek Hoogeboom en Pop: de strijd tegen den Inlandschen Vijand* (Breda: KMA, 1913). In general on these manuals: Tom Menger, 'Of "Golden Bridges" and "Big Bags": Thinking the Colonial Massacre in British, German and Dutch Manuals of Colonial Warfare, c. 1860–1910', in *Europe across Boundaries*, ed. by Noëmie Duhaut and Johannes Paulmann (Berlin: De Gruyter Oldenbourg, 2022), pp. 79–98 (pp. 83–86).

[5] On colonial armies, see for instance *Guardians of Empire: The Armed Forces of the Colonial Powers c. 1700–1964*, ed. by David Killingray and David E. Omissi (Manchester: Manchester University Press, 1999).

[6] Kim A. Wagner, 'Savage Warfare: Violence and the Rule of Colonial Difference in Early British Counterinsurgency', *History Workshop Journal*, 85 (2018), 217–237 (p. 220).

all conveyed certain images of the non-European as Other.[7] While all generally stressed the inferiority of the latter and frequently painted them in highly negative and racist ways, occasionally such media left some space for a more positive image and consequently also a friendlier encounter. Oftentimes, such images drew on the trope of the 'noble savage' that had been known since the eighteenth century.[8]

The European practitioners of colonial war, however, would rarely allow for such a conception. Their manuals are revealing in this aspect. One, written by the German officer Kurd Schwabe, explicitly placed a rejection of such 'noble' images in his general introduction. 'Shaking the head, one reader or another might think of the noble, brave, magnanimous native characters, as our German youth literature prefers to paint them', Schwabe told his readers, 'but nothing has been more distorting of a just assessment of savage peoples as exactly this fantastical literature, which in most cases is only missing one thing: the truth'.[9] Instead, Schwabe propagated an essentialist and generalised derogatory image of the 'native'. Supposedly based on his own experiences in Africa and China, the author went on to generalise: 'yet in the Chinese too we find the big-headed, brutish and deceitful character of the negro. They are all same and equal in this, these natives, be they Malays or Zulu, Chinese or [American] Indian'. Bloodthirstiness, brutality, and bestiality were then other characterisations that Schwabe was to add in the pages that followed.[10] While Schwabe was exceptionally strident, the extent of racial denigration one can find in virtually all manuals is startling, particularly when one keeps in mind that these were conceived as handbooks and thus supposedly bound to an expectation of 'sober' and 'objective' analysis. Degradation was already inherent in naming: classifications such as 'uncivilised', as of a 'lower' or 'inferior' race, as 'barbaric', as 'savages' (sometimes even more dismissively as 'mere savages'), 'fanatics' or 'cut-throats' were all in

[7] The literature on 'empire' in popular culture is now extensive. For a recent, multi-volume work, see: J. Griffiths, *Empire and Popular Culture*, 4 vols (London: Routledge, Taylor & Francis Group, 2022).

[8] Stelio Cro, *The Noble Savage: Allegory of Freedom* (Waterloo, Ontario: Wilfrid Laurier University Press, 1990).

[9] Kurd Schwabe, *Dienst und Kriegsführung in den Kolonien und überseeischen Expeditionen, dargestellt und an Beispielen aus der kolonialen Kriegsgeschichte erläutert* (Berlin: Ernst Siegfried Mittler und Sohn, 1903), p. 3.

[10] Ibid., pp. 3–4, 6–7.

routine use to refer to the enemy of colonial conflict.[11] However, it was particularly the characteristics generally ascribed to the colonial opponent that made the latter the object of an intense, and mostly highly degrading process of racialisation, making him or her appear in most cases as an absolute Other, a not 'fully human species'.[12]

Looking at the handbooks, we can discern that three recurring characterisations stood at the centre of the process of dehumanisation and degradation of what was referred to as the 'native enemy'. These display some parallels to the motifs described in this volume by Matilda Greig for the Spanish irregulars in the Peninsular War, though in our case they were more clearly linked to race and appear even more strident in their degradation of the opponent. While the irregular on the European theatre had to be painted as 'dark-skinned' in order to appear uncivilised, this was here already accepted as a given. Consequently, the room for viewing the enemy in a way that was not essentialist and negative was further narrowed here.

The first recurring characteristic attributed to the 'native' enemy was an inborn treacherousness. This was a ubiquitous motif in the manuals, one often dwelt on at length. The *Soldier's pocket-book* of General Viscount Wolseley even included a separate entry in the index for what it called 'savage treachery'.[13] Jochen Arndt has traced the emergence of this image to the British Army's experiences of the Frontier Wars against the Xhosa in South African between 1834 and 1853, when in the knowledge system of the army the 'savage' became the 'treacherous savage'.[14] We have no such detailed accounts of the image's development in other empires, but there it was certainly equally prominent by the *fin de siècle*. For instance, a Dutch manual, written 1896, dedicated no less than sixteen pages to describing 35 cases of supposedly 'treacherous action' against the Dutch colonial army.[15]

[11] Menger, p. 88.

[12] See also: V.G. Kiernan, *Colonial Empires and Armies, 1815–1960* (Phoenix Mill: Sutton Publishing, 1998), p. 158.

[13] Garnet Joseph Wolseley, *The Soldier's Pocket-Book for Field Service*, 5th edn (London: Macmillan, 1886), p. 544.

[14] Jochen S. Arndt, 'Treacherous Savages & Merciless Barbarians: Knowledge, Discourse and Violence During the Cape Frontier Wars, 1834–1853', *Journal of Military History*, 74 (2010).

[15] Klaas van der Maaten, *De Indische Oorlogen: een boek ten dienste van den jongen officier en het militair onderwijs*, Vol. I (Haarlem: Loosjes, 1896), pp. 105–121.

Secondly, savagery and cruelty were generally perceived as common characteristics of the colonial enemy, something which was considered to flow from their 'uncivilised' state; they were supposedly cruel by nature.[16] There was something distinct about colonial readings of violent dissent, as Ann Laura Stoler has argued: it had to be shown as being outside of rationality.[17] And thus, without any apparent trace of exaggeration, tracts could speak of the 'ferocity and cunning of wild beasts' of 'Asiatics and savages' (who allegedly both revelled in bloodshed) or the 'bloodthirsty bestiality' of Africans and Chinese.[18] Hair-raising stories were reproduced to further underline the irrationality of 'native' violence. For instance, a Dutch publication told on its first pages how the enemy had exhumed fallen Dutch soldiers after their burial for the purpose of the mutilation of the bodies.[19] The oft-repeated warning not to let any of one's fellow soldiers, dead or alive, fall into the hands of the opponent, who would invariably cruelly mutilate, torture, kill, or even cannibalise these Europeans, conveyed the supposed savagery of the other side in a way that must have been especially resonant with soldiers, playing on the latter's own fears.[20] Certainly, the killing of the captured or wounded was a regular occurrence in colonial wars (on both sides). Unlike the claims, it was however not universal on the non-European side, nor were torture and mutilation; cannibalism furthermore was rare or non-existent.[21]

The horror stories about the 'savage' enemy often sprouted from the fantasy of insecure and disoriented European soldiers, who drew on well-known preconceived images of savagery. One manual even recognised this, stating '[these facts] indeed are frequently exaggerated in their [our

[16] See for instance: Schwabe; N.a., p. 2. See also Arndt, pp. 723–731; Walter, p. 176.

[17] A.L. Stoler, *Capitalism and Confrontation in Sumatra's Plantation Belt, 1870–1979* (New Haven: Yale University Press, 1985), p. 48.

[18] C.E. Callwell, 'Lessons to Be Learnt from the Campaigns in Which British Forces Have Been Employed since the Year 1865', *Journal of the Royal United Services Institution,* 31 (1887), 357–412 (p. 390); Schwabe, p. 7.

[19] Bos, pp. 2–3.

[20] Dierk Walter, 'Kein Pardon. Zum Problem der Kapitulation im Imperialkrieg', *Mittelweg 36,* 21 (2012), 90–111 (p. 92).

[21] Ibid. See also: Gesine Krüger, *Kriegsbewältigung und Geschichtsbewusstsein: Realität, Deutung und Verarbeitung des deutschen Kolonialkriegs in Namibia, 1904 bis 1907* (Göttingen: Vandenhoeck & Ruprecht, 1999), pp. 104–115; J.P. Daughton, *In the Forest of No Joy: The Congo-Océan Railroad and the Tragedy of French Colonialism* (New York: W.W. Norton & Company, 2021), p. 279.

men's] imagination'.[22] In the context of war, they served two objectives: they dehumanised the opponent and made it easier to justify one's own cruelty.[23] Outside of that context, however, such imaginations had simply become 'generalised truth' in colonial manuals.

Thirdly, the enemy was often also imagined as profoundly uncanny and threatening. While viewed as inferior, the 'native enemy' could also be attributed virtually supernatural powers that appeared inexplicable to the European side. Charles Callwell, author of the famous *Small Wars* textbook, remarked that 'news spreads in a most mysterious fashion' in the lands of those he considered uncivilised.[24] As many others, he believed that the enemy's powers of observation, during the day and especially at night, were far superior to Europeans'.[25] A 'natural aptitude for creeping about' and an 'instinctive capacity for stalking' were further, rather sinister, attributes ascribed to the opponent. In addition, the latter was regularly believed to possess an almost superhuman quick-footedness that was incomprehensible to most colonial soldiers, and which induced fear of being suddenly overwhelmed. Referring to an incident of the Mahdist War, one essay noted: 'At McNeil's zareba they crept up stealthily on all sides, there were no words of command and no confusion, the enemy appeared to spring out of the earth as if by magic (…)'.[26] In this way, othering the enemy of colonial wars also operated on the level of bodily characteristics. Sometimes, this could directly serve to legitimate brutality: the widespread claim that the nerve system of the non-white opponent was less sensitive to pain served as an argument for the employment of

[22] R. da Costa Porter, 'Warfare against Uncivilised Races; or, How to Fight Greatly Superior Forces of an Uncivilised and Badly-Armed Enemy (Prize Essay, 1881)', *Professional Papers of the Corps of Royal Engineers,* 6 (1881), 305–360 (p. 342).

[23] Walter, p. 172; Larissa Förster, *Postkoloniale Erinnerungslandschaften: wie Deutsche und Herero in Namibia des Kriegs von 1904 gedenken* (Frankfurt: Campus, 2010), pp. 161–162.

[24] C.E. Callwell, *Small Wars. Their Principles and Practice* (London: Printed for H.M. Stationery Office, by Harrison and Sons, 1896), pp. 47–48. For another case, see: Kim A. Wagner, "Treading Upon Fires': The 'Mutiny'-Motif and Colonial Anxieties in British India', *Past & Present,* 218 (2013), 159–197 (pp. 166–167).

[25] Callwell, p. 48; General Staff, p. 185; W.A. van Rees, *Handleiding tot de kennis der velddienst en vechtwijze van het Nederlandsch Oost-Indisch Leger tegen Inlandsche Vijanden*, 3rd edn (Breda: Broese, 1869), p. 98; Klaas van der Maaten, *De Indische Oorlogen*, Vol. II (Haarlem: Loosjes, 1896), p. 78. In the secondary literature see: James Hevia, *The Imperial Security State: British Colonial Knowledge and Empire-Building in Asia* (Cambridge: Cambridge University Press, 2012), p. 220.

[26] Callwell, p. 384.

exploding ('dum-dum') bullets against such enemies.[27] In general, an uncanny and threatening alterity of the racialised adversary served to inject further distance and fear into the enemy encounter (Fig. 1).

Vilification or dehumanisation of the opponent was thus not something which arose only on the spot. It had already become codified and become formative to defining colonial campaigning for the practitioners themselves. Frequently, these practitioners were not reticent in also conveying such attitudes to newcomers to the colonial realm, who generally tended to internalise them quickly. We can find one example in William Harvey Brown, an East Coast American who went to South Africa in 1890. Attaching himself to the 'Pioneer Column' moving into MaShonaland, he was out one day scouting with one Montague, a veteran of several colonial wars, who introduced him to the trope of 'native treachery' when they were offered to spend the night in a local village: 'They invited us to their village to spend the night, but Montague objected, saying that although they were ostensibly amiable, they were not to be trusted, and that in all probability we should be murdered by them during the night for the few things which we possessed. As he had been born and reared in the Cape Colony in the neighborhood of the Kafir tribes, I naturally heeded his judgement, although it did not seem to me that such innocent-looking creatures could be capable of treachery'.[28]

The Performance of Superiority and the Denial of Equality

Apart from this a priori denigration and dehumanisation, there was a second important barrier to understanding among the two sides in colonial war. As recent scholarship on colonial violence and war has emphasised,

[27] Wagner, *Savage Warfare*, pp. 223–232; Florence Bernault, 'The Shadow of Rule: Colonial Power and Modern Punishment in Africa', in *Cultures of Confinement: A History of the Prison in Africa, Asia, and Latin America*, ed. by Frank Dikötter and Ian Brown (London: Hurst & Company, 2019), pp. 55–94 (p. 81).

[28] William Harvey Brown, *On the South African Frontier: The Adventures and Observations of an American in Mashonaland and Matabeleland* (New York: Charles Scribner's Sons, 1899), pp. 92–94.

Fig. 1 This cover image of H.D. Hutchinson's *The Campaign in Tirah, 1897–1898* (London 1898) portrays the colonial enemy in typical fashion: wicked, treacherous, and threatening, lurking in the shadows

colonial warfare always had important performative aspects.[29] First, it was meant to perform the supposed European superiority—a superiority understood as racial, cultural, civilisational, and masculine and

[29] See, for example, Kim A. Wagner, "'Calculated to Strike Terror': The Amritsar Massacre and the Spectacle of Colonial Violence', *Past & Present,* 233 (2016); Gavin Rand, 'From the Black Mountain to Waziristan: Culture and Combat on the North-West Frontier', in *Culture, Conflict and the Military in Colonial South Asia,* ed. by Kaushik Roy and Gavin Rand (New Delhi: Routledge India, 2017), pp. 131–156; William Gallois, 'Dahra and the History of Violence in Early Colonial Algeria', in *The French Colonial Mind,* ed. by Martin Thomas (Lincoln, NE, 2012), pp. 3–25; William Gallois, *A History of Violence in the Early Algerian Colony,* (Basingstoke: Palgrave Macmillan, 2013); James Hevia, *English Lessons: The Pedagogy of Imperialism in Nineteenth-Century China,* (Durham: Duke University Press, 2003), pp. 182–206, 220–229; Elizabeth Kolsky, 'The Colonial Rule of Law and the Legal Regime of Exception: Frontier "Fanaticism" and State Violence in British India', *The American Historical Review,* 120 (2015), 1218–1246.

underpinning the claim to colonial domination—to those who were to be colonised or had dared to challenge colonisation. Not rarely, the colonisers even sought to give their violence forms that they believed would be particularly deterrent or frightening in the cultural framework of the Other.[30] Second, however, performativity was equally meant to showcase this assumed superiority to the colonisers themselves. Notwithstanding the common association of empires with power and domination and the fact that empires were in most cases able to muster a powerfully violent, if not to say devastating response when (locally) challenged in their existence, the actors of empire often felt weak and threatened, and were regularly influenced by feelings of anxiety, panic, and fear.[31] The performance of colonial violence was therefore also intended to reassure the perpetrators themselves of their supremacy.

Such thinking interposed a consequential precondition for the possibility of a non-violent encounter with the enemy. While elsewhere the amicable encounter with the opponent rested on the fact that the two sides largely recognised each other as equals, for the practitioners of colonial warfare there could be no question of such equality—warfare itself first had to establish *inequality* in status. Heavy and unsparing violence was generally thought of as conducive to conveying that impression of superiority. This was a frequently repeated line of thought in colonial war. Three examples must suffice here. Callwell's handbook stated that '[uncivilised races] must be thoroughly brought to book and cowed or they will rise again', while his fellow British writer Heneker held that 'No leniency or half measures are of any use until the savage has felt the power of force'.[32] In a Dutch publication, one could read the following: 'Generally a restoration of peace is not possible before the warring party has severely

[30] Wagner, *Savage Warfare*, p. 221.

[31] Ranajit Guha, 'Not at Home in Empire', *Critical Inquiry,* 23 (1997); *Helpless Imperialists: Imperial Failure, Fear and Radicalization*, ed. by Maurus Reinkowski and Gregor Thum (Göttingen: Vandenhoeck & Ruprecht, 2013); Robert Peckham, *Empires of Panic: Epidemics and Colonial Anxieties* (Hong Kong: Hong Kong University Press, 2015); *Anxieties, Fear and Panic in Colonial Settings: Empires on the Verge of a Nervous Breakdown,* ed. by Harald Fischer-Tiné (New York: Palgrave Macmillan, 2016); Mark Condos, *The Insecurity State: Punjab and the Making of Colonial Power in British India* (Cambridge: Cambridge University Press, 2017); Kim A. Wagner, *Amritsar 1919: An Empire of Fear and the Making of a Massacre* (New Haven: Yale University Press, 2019).

[32] Callwell, p. 115; W.C.G. Heneker, *Bush Warfare* (London: H. Rees, 1907), p. 162.

experienced its inferiority'.[33] As is already hinted at in Heneker's quote, there was widespread agreement that showing leniency, in contrast, would only be misunderstood as weakness; this is one of the main tropes of thinking on colonial warfare.[34] With many colonial actors anxious that colonial power was precarious, the last thing one wanted was to openly display weakness (whether the other side actually perceived it as such is questionable in many cases).

We can find many instances where such an attitude was actively conveyed by colonial 'old hands' to those they considered newcomers, thus discouraging fellow feeling or clemency from the outset. The German naval officer Hans Paasche, who operated in the hinterland of German East Africa during the Maji Maji War, received several letters from experienced colonial officers or administrators urging him to be as harsh as possible against the enemy.[35] Another German officer in that same war, Philipp Correck, noted in his diary how he returned to camp to find a senior officer having ordered the execution of a captured African chief and making clear to Correck that so far he had been too 'tame'.[36] Although personal attitudes of soldiers and officers could certainly vary, it is clear that the whole body of colonial thought pushed towards harshness and discouraged clemency.

The degree of violence in the colonial encounter with the enemy—the frequent killing of prisoners or the refusal to take any, as well as the violent abuse of those who surrendered—also depended on the emotional involvement of the actors, as for instance in Rhodesia 1896, where the murder of

[33] Koninklijke Militaire Academie, *Aanteekeningen betreffende het oorlogsrecht der Inlandsche Volken in Nederlandsch-Indië en de tegenover die volken in acht te nemen oorlogsgebruiken* (Breda: KMA, 1904), p. 20.

[34] Huw Bennett, *Fighting the Mau Mau: The British Army and Counter-Insurgency in the Kenya Emergency* (Cambridge: Cambridge University Press, 2012), p. 265; Hevia, *Imperial* Security, p. 221, note 45. Typical for the primary sources is Carl Zimmermann, 'Gefecht', in *Deutsches Kolonial-Lexikon,* ed. by Heinrich Schnee (Leipzig: Quelle & Meyer, 1920), pp. 683–686 (p. 683).

[35] Aus Briefen vom Bezirksamtmann Keudel, Mohorro, letters dated 17 and 21 August and 6 September [probably 16 September] 1905, Bundesarchiv-Militärarchiv Freiburg (BA-MA), RM 121-I/452, 1, 2, 8; P. Werner Lange, 'Die Toten im Maisfeld: Hans Paasches Erkentnisse aus dem Maji-Maji-Krieg', in *Der Maji-Maji-Krieg in Deutsch-Ostafrika, 1905–1907,* ed. by Felicitas Becker and Jigal Beez (Berlin: Christoph Links, 2005), pp. 154–167 (p. 162).

[36] Philipp Correck, Ausreise nach Deutsch-Ostafrika, entries 14 and 15 May 1906, Bayerisches Hauptstaatsarchiv – Kriegsarchiv, Abteilung IV, Munich (BayHStA – IV), HS 908.

several hundred settlers at the beginning of the Ndebele and Shona risings meant that white soldiers also waged war with the intent of wreaking a bloody revenge.[37] However, as shown above, such violence was already preconditioned by the ways colonial warfare was generally thought and performed. The refusal to recognise the other side as belligerents and their classification instead as 'insurgents' or 'rebels' then in many cases offered an 'easy' legalistic excuse for executions or punishment (a surprising number of those captured were still subjected to on-the-spot court-martials, however, farcical these procedures might be).[38] The mistreatment or killing of those who surrendered frequently even occurred where it was easy to see how counterproductive it was. At Gwelo in Rhodesia, a Native Commissioner noted how Ndebele coming in to surrender had been 'stripped of all they had' and unmercifully flogged; unsurprisingly, this then kept many others from surrendering.[39] Similarly, Paasche in German East Africa noted that the 'brutal treatment' of prisoners did hinder submission there.[40]

It is hard to find cases where the killing of prisoners caused moral injury in the perpetrators, as Matilda Greig notes in this volume for some of the perpetrators of executions of captured guerrillas in the Peninsular War.[41] We are still lacking studies on *fin-de-siècle* perpetrators of colonial violence and the question of trauma, but generally I would agree with William Gallois, who remarks of the French nineteenth-century perpetrators in Algeria that their 'senses of self were stronger and harder to break' (in comparison with those of the 1950s and 1960s).[42] While the diaries of some colonial soldiers certainly reveal cases where the initial eyewitness confrontation with colonial violence was met with unease or shock,

[37] Arthur Keppel-Jones, *Rhodes and Rhodesia: The White Conquest of Zimbabwe, 1884–1902* (Kingston: McGill-Queen's University Press, 1983), pp. 442–443, 462–464.

[38] Walter, p. 101. In Rhodesia, a later investigation of the execution of an African woman and her child found that the proceedings amounted to a 'parody of justice', but the officers involved were in the end only 'severely reprimanded', see below, note 61.

[39] William Impey Stanford Driver: Diary as Native Commissioner, Gwelo, Rhodesia, Vol. I: 6 Oct. 1896–17 May 1897, British Library, Add. Ms. 42174, 4.

[40] Telegram Paasche to Kommando 17 January 1906, Der Aufstand in Deutsch-Ostafrika. Zusammenstellung von Kriegstagebuch-Notizen, Briefen und anderen Beobachtungen, BA-MA, RM 121-I/452, 40.

[41] Matilda Greig, 'Counterinsurgency for Readers at Home: Chilling Encounters with Guerrillas in Post-Napoleonic Novels', this volume.

[42] Gallois, *History of Violence*, p. 7.

longer-term traumatisation is hard to spot.[43] Rather, most appear to have grown accustomed to such violence relatively quickly. The long-time colonial adventurer Hamilton-Browne put it most bluntly in his later recollections: 'a scout must have no feelings and years of savage warfare had blunted any I might have ever possessed'.[44] The acknowledgement that one was only fighting 'savages' probably still provided a stronger protection against moral injury than if one was acting against European irregulars, however, much 'othered' those might have been.

Despite all this, contemporary thinking on colonial warfare still offered a small window of opportunity for another sort of encounter. If the practitioners of colonial warfare felt that superiority had indeed been established, clemency could be shown. Strikingly, a 1907 British manual even advocated 'extreme leniency' after the power of force had been felt:

> No leniency or half measures are of any use until the savage has felt the power of force. Leniency is treated as a sign of weakness, and half measures as an undecided and wavering policy. When once the savage comes in with the white flag, and it is evident that he desires to submit, the opposite extreme, almost, is the best plan to follow. He has felt and knows the strength and determination of civilization properly handled in war should he be again at his old practices, therefore extreme leniency should be meted out with absolute justice.[45]

The same manual also featured one of the few instances of a friendlier encounter when it described how the enemy who had surrendered was told to bring in its wounded for medical treatment:

> This practice of getting the enemy to bring in their wounded for treatment was often resorted to in Southern Nigeria. It meant hard work for the medical officer and his assistants, these extra cases, but it was interesting; it brought the troops into touch with the late enemy in a most satisfactory and

[43] Among the many diaries, letter collections and memoirs consulted, I only found two items in which some of the symptoms noted by the writers might indicate trauma: Rudolf von Hirsch, Deutsch-Ostafrika. Tagebuch des Hptm. von Hirsch Oktober 1905 bis März 1906, BayHStA – IV, NL Hirsch 9; Memoirs of M.T. Velsing, Deel IV, pp. 38–40, and Deel VII, pp. 140–141, Nederlands Instituut voor Militaire Historie, The Hague, Toegang 543, Inv. nr. 35.

[44] G. Hamilton-Browne, *A Lost Legionary in South Africa* (London: T. Werner Laurie, 1912[?]), p. 171.

[45] Heneker, p. 162.

friendly way, and tended to establish at once the reputation of the white man and his soldiers for kindness.[46]

As is hinted at here, such behaviour was also viewed from a utilitarian standpoint, as it might induce others to submit and have a favourable effect on any post-war settlement (even if it can reasonably be doubted whether such a reputation for kindness could possibly be established after the violence of such a war). The quote also points to the assumed non-application of the Geneva Conventions, under which application it would actually have been self-evident to treat the wounded of both sides.[47]

One of the main questions for the practitioners of colonial war, therefore, was to decide when 'enough force had been felt' and clemency could indeed be shown—an entirely subjective measure. Consequently, a German manual classified the point when utmost severity was to be substituted for by leniency as a complex balancing act, even as an 'art', only to be learned through long years of experience and even then often resting on 'a matter of luck and [gut] feeling'.[48] Needless to say, this made the duration and extent of 'severity' very much an arbitrary matter and it will not be a surprise that in reality a heavy emphasis was often on severity rather than on leniency.

There were a few cases where a certain sense of equality towards the colonial foe, generally elusive in such wars, could indeed make an appearance. Two instances can be briefly mentioned here. In the first, a classification of the opponent as a noble and courageous warrior (characteristics one unsurprisingly also claimed for oneself) could prompt a certain identification across racial boundaries. Obviously, by recognising the other as a worthy opponent, one also upgraded one's own feats in war: the adversary had been a 'good match', and the best fighter had carried the day (a motif that could also serve to alleviate the humiliation of European defeat).[49] Occasionally, a certain amount of identification could also occur when upper-class white officers recognised equivalent class status in their enemy counterparts, or in those cases where the enemy wore

[46] Ibid., p. 123.

[47] I thank Holly Furneaux for this last point.

[48] Zimmermann, p. 683.

[49] Robert H. MacDonald, *The Language of Empire: Myths and Metaphors of Popular Imperialism, 1880–1918* (Manchester: Manchester University Press, 1994), p. 30.

Western-style uniforms and fought according to Western ways.[50] And yet, there were generally clear limits to this as well. For instance, concerning the motif of the worthy opponent, the 1879 British war against the Zulu, later often eulogised as a war of open, 'fair' field battles against a noble warrior race, in fact saw colonial troops engage in the systematic burning of villages, the confiscation of cattle, and the killing of wounded Zulu, as Michael Lieven has shown.[51] One should thus read such idealisations with caution. In part, these views were later constructions, when the colonisers, having quashed all opposition and established colonial rule, could afford such narrative embellishments.[52]

Encountering the Enemy? Encountering the Civilian?

As a majority of colonial wars in this age were fought as guerrilla wars with the typical hit-and-run attacks, the enemy encounter was oftentimes a quite rare and fleeting moment.[53] Therefore, a close encounter with the other side occurred predominantly when prisoners were brought in. Frequently, these prisoners were actually civilians who had either been unable to flee in advance or been caught by surprise—though regularly they were immediately accused of being spies. Interrogation, generally mediated by a translator, was a common way of encounter. On a grander scale, such encounters with civilians took place when colonial armies abducted women and children *en masse* in order to pressure their male relatives to give up fighting (a common course of action in colonial campaigns) or when communities as a whole came to surrender.[54]

[50] For some examples from the Boxer War: Dietlind Wünsche, *Feldpostbriefe aus China: Wahrnehmungs- und Deutungsmuster deutscher Soldaten zur Zeit des Boxeraufstandes 1900/1901* (Berlin: Ch. Links Verlag, 2008), pp. 280–281, 328–332.

[51] Michael Lieven, 'Butchering the Brutes All over the Place: Total War and Massacre in Zululand, 1879', *History*, 84 (1999), 614–632.

[52] See for instance James O. Gump, *The Dust Rose Like Smoke: The Subjugation of the Zulu and the Sioux*, 2nd edn (Lincoln: University of Nebraska Press, 2016), pp. 146–152, 168.

[53] Walter, pp. 175–176.

[54] Walter, pp. 94–95; Susanne Kuß, *German Colonial Wars and the Context of Military Violence* (Cambridge, MA: Harvard University Press, 2017), p. 72. For on-the-ground, relatively detailed chronicles of this routine during the Maji Maji War, see the military diaries (*Kriegstagebücher*) in BA-MA, RM 121-I/448-460.

Asking about the enemy encounter in colonial wars, therefore, inevitably means also asking about the civilian encounter. This encounter was differently conditioned, and also more paradoxical. In principle, codes of military and male honour for Western soldiers at the time forbade harming women, children, and the defenceless, and urged their protection. They were not generally considered legitimate targets of military violence. The manuals of colonial warfare therefore barely mentioned civilians at all. Nevertheless, we know that violence against civilians, physical abuse, plunder, and sexual violence, were common occurrences in colonial wars. Leaving aside here the different dynamics of those military campaigns that were intended to exterminate local populations,[55] generally several factors account for this. First, where othering and dehumanisation was so clearly racialised, contempt and hatred could easily turn against the racial group as a whole, and make those civilians appear as humans (or even subhumans) not worth protecting. Particularly where revenge was an important motivating force for European colonial soldiers, civilians could easily become legitimate targets in the eyes of the former. The specific forms that colonial wars often took tended to reinforce these tendencies. Such conflicts generally mobilised all able men of the local communities for an elusive guerrilla of hit-and-run attacks, most of the time leaving women, children, and the elderly (if they had not equally been taken into hiding) as the only targets for the frustration and anger of colonial forces. What is more, the differentiation between combatants and non-combatants in such a war was generally blurred: fighters rarely wore uniforms, and they frequently were forced to live on active support of the non-fighting population, turning that population as a whole, in the eyes of many of the counter-guerrilla force, into the enemy.[56] Nevertheless, it is clear that violence against civilians, particularly women, often still needed more explicit legitimation for the perpetrators. In many colonial wars, this appears to have been provided by the circulation of horror stories about women participating in war and committing atrocities, thus violating both moral and gender norms by European standards.[57]

[55] On this, see, for example, Lyndall Ryan, 'Settler Massacres on the Australian Colonial Frontier, 1836–1851', in *Theatres of Violence: Massacre, Mass Killing, and Atrocity Throughout History*, ed. by Philip Dwyer and Lyndall Ryan (New York: Berghahn, 2012), pp. 94–109.

[56] Walter, pp. 163–165, 175–180.

[57] For the circulation of such stories in German South West Africa, see for instance: Krüger, pp. 105–106. The motif is also identified by Matilda Greig in some of the literature on the guerrilla war in Spain, see Greig, this volume.

The contribution by Todd, Bailey, and Bennett in this volume presents several examples of the circulation of such stories during the war in German South West Africa.[58] In Rhodesia, a trooper writing a letter also felt the need to attenuate feelings of guilt about the (apparently accidental) killing of a woman at night by his addition that 'Of course we should not shoot a woman intentionally, although it has been proved [sic] that the women are the chief torturers'.[59] Alleged brutality against, and murder of, whites even gave rise to cases where soldiers on their own initiative killed prisoners even after these had been slated for release by their commanding officers. Such cases, documented both for Rhodesia and for GSWA (in the latter case even involving women), testify to the extraordinary lengths to which soldiers could go to exact revenge even on civilians (though obviously they would not regard them as such anymore).[60]

Oftentimes, several motives appear to have come together in civilian killings, such as in the case of the execution of an African woman and her young child at Makalaka Kop, Matabeleland, in May 1896, after being captured and questioned by a military detachment under Captain Gibbs that was building a fort there. When the execution later came to the attention of the British High Commissioner for South Africa, a court of inquiry was held, in which the accused claimed that the woman had been a spy, and that she could have endangered them by telling nearby Ndebele forces about the numerical weakness of the detachment. Supposedly, the necessary guards as well as rope to keep the woman as a prisoner had been lacking, leaving no other option but execution. The killing of the child was explained as an accident, as they had supposedly lost sight of it and later found that the woman was carrying it under her kaross at the moment she

[58] Todd, Bailey, and Bennett, 'Global Sexual Encounters with Racialized 'Others' in War, Genocide, and Captivity, 1905–1945', this volume.

[59] Letter by Arthur Smith to his parents, reproduced in: 'Stroud man's experiences in South Africa', *The Echo* 1 December 1897.

[60] See for such cases: Matthias Häussler and Trutz von Trotha, 'Brutalisierung >von unten<. Kleiner Krieg, Entgrenzung der Gewalt und Genozid im Kolonialen Deutsch-Südwestafrika', *Mittelweg 36*, 21 (2012), 57–89 (p. 79); Richard Hodder-Williams, 'Marandellas and the Mashona Rebellion', *Rhodesiana*, 16 (1967), 27–54 (p. 42); Keppel-Jones, p. 463. It must be noted that in the last case the actual status of the prisoner is unclear. However, had he in any way been recognisable as a combatant, it seems unlikely that he would have been slated for release. For another case, see: Letter John Archer to his mother 28 August 1896, GB 0099, King's College London, Liddell Hart Military Archives, KCLMA Archer, box 1.

was shot.[61] Many of these claims seem highly doubtful (the detachment having ventured out despite knowing they were too weak to withstand a larger assault, the complete lack of any material to constrain one woman, the losing sight of the child despite the simultaneous fear that the woman might escape and tell others about the force), and so the case rather seems to point towards a paranoia about spies, combined with at best a manifest indifference towards the lives of women and children, or worse towards a killing inspired by active hatred and vengefulness.

Violence was thus widespread in the encounter with the (enemy) civilian. Still, even if it was not always clear who was a combatant and who a civilian, the encounter with the latter was generally of a different sort, and violence needed different justifications. That is not to say that one cannot also find examples of friendly encounters with civilians in colonial war: of handing out food, attending to them medically, or playing with children, among others. Naturally, whether and to what extent such actions occurred varied with the personal dispositions of the individuals involved. However, it was also exactly through such actions that colonial soldiers were able to reassure themselves that they still held fast to the codes of honourable warfare. On a personal level, it equally served perpetrators in preserving their self-image as an 'integer personality'. Studying Nazi perpetrators, the sociologist Harald Welzer argues that many of these individuals could think of themselves as being 'forced' by the circumstances to engage in terrible violence, but that it was through occasional and isolated interpersonal deeds of mercy, charity, or justice that they proved to themselves that they still possessed morally 'integer personalities'.[62] It should therefore come as no surprise that such acts, also in colonial war, could appear in combination with a context of gruesome violence. The German officer Hirsch, for instance, oversaw during the Maji Maji War how his askari troops and Hehe auxiliaries burned and destroyed villages and crops on a massive scale, murdered prisoners, delivered to him the severed heads of 'rebel' chiefs who had had a price put on their head, and 'traded' with captured women as if they were slaves. At the same time, however, Hirsch could at one point write with affection about some 'lovely small children' whom he petted and fed with chocolate. These were the children (and

[61] Letter Lord Rosmead to Chamberlain 3 October 1896 and enclosures, TNA, CO 879/47/3, 590–599.

[62] Harald Welzer, *Täter: Wie aus ganz normalen Menschen Massenmörder werden*, Kindle edn (Frankfurt: Fischer e-books, 2007), chapter Tötungsmoral.

their mothers) who he had just captured and who were now to be used to exert pressure on their father, a 'rebel' leader still at large.[63]

Conclusion

How European soldiers encountered the enemy in colonial wars is a complex question that cannot be addressed in its entirety in the scope of a chapter. In the above, I have nevertheless attempted to highlight some more general and distinctive characteristics of these encounters. While situational and personal factors were always influential, I have argued that we should also pay close attention to the ideational determinants of this encounter.

It is undeniable that these encounters were frequently marked by high levels of violence. One reason for this was that the 'native enemy' had already been conceived a priori in starkly racialised and derogatory terms. Treacherousness, cruelty, and uncanniness were some of the main characteristics ascribed to this enemy, all of which made him/her appear as clearly inferior, less human, and generally not worthy of compassion. This represented a first major barrier to understanding among belligerents. The sheer amount of racial denigration that went into knowledge formation on colonial warfare is something we have to take into account if we ask about the possibilities of fraternisation in such war. This preconception of the enemy is reflected in manuals of colonial warfare, but it was more generally conveyed by (popular) literature and media at the time, and also reproduced by many experienced colonial soldiers themselves (as was the supposed need for severity).

This necessity of harshness was related to a second constitutive element of colonial warfare: its practitioners conceived of it as acting out, performatively, the supposed racial, civilisational, and masculine superiority of the Europeans. The sense of a certain equality between the foes, generally a precondition for understanding and fraternisation with the enemy, was therefore excluded by them from the outset. With leniency generally considered a sign of weakness when dealing with 'uncivilised races', there was little room for a fraternal encounter between the belligerents. Only when the European side felt that their superiority had been definitively established by the violence of war, could leniency be shown. When exactly that point had been reached was obviously a very subjective matter and

[63] Hirsch, Tagebuch, diary entry 27 February 1906, BayHStA – IV, NL Hirsch 9.

considered a delicate decision to make. The result was that in most cases the emphasis was plainly on harshness rather than on leniency.

The forms of colonial armed conflict meant that the encounter with the enemy was often an encounter with civilians—even if in such wars the distinctions were often blurred. Nevertheless, soldiers generally remained conscious of the distinctive status of civilians, women, and children in particular, and the codes of military honour that forbade harming them. Encounters with civilians could be cordial and protective. Violence was, however, often a characteristic of them. The racial othering that was used to demonise the enemy generally did not differentiate, and thus could make racial groups as a whole appear unworthy of protection, and an object of hatred and contempt. This was particularly the case when colonial soldiers were seeking to take revenge. As most wars took the form of a guerrilla and counter-guerrilla, civilians could also easily be suspected of supporting the guerrillas and thus lose their civilian 'innocence'. Despite all these factors, however, it is noticeable that European soldiers still felt the need for special justifications if they harmed women. In that case, horror stories about the cruelty of women in mutilating or torturing European prisoners often served to provide such legitimisation.

The above-described dynamics of violence and othering in the enemy encounter could also make their appearance in wars against European irregular fighters, as shown for instance in this volume by Matilda Greig. Still, there appears to be a difference in degree to the colonial wars of the *fin de siècle*. Not only did the violence as well as the violent othering of the enemy appear to come even more naturally when the opponent was non-European and non-white; both seemed to take more intense forms as well. Many writers on colonial war thought the same: while they sometimes recognised the likeliness between colonial wars and the guerrilla wars of the Napoleonic and French Revolutionary Era, they also insisted on distinctions. If greater leniency could work against European foes (as it had for instance in the later phase of the war in the French Vendée), it was inadvisable against 'uncivilised' races. The following quote from Callwell's manual sums up the point: 'Uncivilized races attribute leniency to timidity. A system adapted to La Vendée is out of place among fanatics and savages'.[64]

[64] Callwell, p. 115.

Global Sexual Encounters with Racialized 'Others' in War, Genocide, and Captivity, 1905–1945

Lisa M. Todd, Jordyn Bailey, and Allison L. Bennett

In December 1942, the *Ottawa Journal* published an article entitled, 'Canadian Soldiers Rated Physically Below Hun Prisoners.' As part of ongoing media coverage of enemy prisoner of war camps across the country, the article compared the physical appearance of Canadian and German soldiers and found the former lacking. Indeed, the journalist maintained the 'European' appearance of the captured German men, with their blue eyes, blonde hair, light skin, and 'thighs like hams,' led to many admiring looks from men and women alike, and as the headline suggests, even led to serious conversations about the physical superiority

L. M. Todd (✉) • J. Bailey
University of New Brunswick, Fredericton, NB, Canada
e-mail: ltodd@unb.ca; j.bailey@unb.ca

A. L. Bennett
Memorial University of Newfoundland, St. John's, NL, Canada
e-mail: allison.bennett@mun.ca

H. Furneaux, M. Greig (eds.), *Enemy Encounters in Modern Warfare*, https://doi.org/10.1007/978-3-031-56748-3_6

of the enemy prisoners.[1] In other words, that an Allied newspaper was paying compliments to the appearance of Nazi soldiers at the height of the Second World War is indicative of the complicated nature of constructed gender and sexual identities, the power of racial hierarchies favoring 'whiteness,' and reminds us of the often uneven application of enemy 'Other' stereotypes.

In this chapter, we use three case studies to investigate the multifaceted ways 'whiteness' worked to facilitate violence, coercion, and consent during soldier-civilian heterosexual encounters in war, genocide, and captivity. The first section uses military records, missionary reports, and eyewitness accounts to illustrate the widespread rape and sexual coercion perpetrated by German soldiers on women and girls in battle zones, occupied territories, and in concentration camps in German-occupied Namibia. We show how the exigencies of colonial warfare, coupled with the anti-Black racism of settler colonialism, resulted in extreme military violence against girls and women. In First World War Egypt, Australian servicemen were cautioned to avoid sexual encounters with local women in brothel areas when they arrived in this 'Oriental' region for fear of venereal disease (V.D.), but such warnings often went unheeded. Our second case study uses first-person writings to consider the racialized differences between Australian serviceman Percy Deane's relationships with an Egyptian woman in Cairo and a French woman in Port Said, both in the context of an imperial military occupation. In 1940s Canada, the confluence of racial and sexual hierarchies operated differently. Our final example examines public reactions to the 35,000 German military personnel imprisoned in Canadian Prisoner of War (P.O.W.) camps. We show how wartime ideals of 'whiteness' and sexualized masculinity led many civilians to treat the prisoners, even those engaged in illegal sexual relationships with Canadian women, with hospitality and affection, despite the men's status as 'enemies' and 'Nazis.' Each case study, from a distinct twentieth-century global conflict, demonstrates how intersectional identities framed wartime enemy encounters, and especially how they were understood, accepted, and rejected. These examples also show how lines between friend and foe, combatant and non-combatant, were often blurred in complicated ways.

[1] 'Canadian Soldiers Rated Physically Below Hun Prisoners,' *Ottawa Journal*, 9 December 1942.

For over three decades, scholars have been working to establish the complex connections between gendered identities, sexual behaviors, and people at war. Historians have recently turned their attention to important links between rape, forced prostitution, attacks on reproductive bodies, and genocidal violence.[2] In a landmark volume, Dagmar Herzog recommends historians examine the connections between 'intimacy and massacre' by considering the impacts of sexual violence but also how conflict opens new opportunities for 'consensual pleasures.'[3] In this chapter, we extend that scholarship to new theaters of war. Literature on war and sexuality in the First World War has disproportionately focused on European theaters; here, we shift the focus to Egypt, one of the war's peripheral, often forgotten, fronts.[4] Likewise, the growing scholarship on relationships between P.O.W.s and local women has largely neglected the study of Canada, so we investigate encounters between prisoners and local

[2] Recent works include Joanna Bourke, *Disgrace: Global Reflections on Sexual Violence* (Chicago: Chicago University Press, 2022); Linda Connolly, 'Sexual Violence in the Irish Civil War; A Forgotten War Crime,' *Women's History Review* 30, no. 1 (Feb 2021): 126–143; Regina Mühlhäuser, *Sex and the Nazi Soldier: Violent, Commercial and Consensual Encounters during the War in the Soviet Union, 1941–5* (Edinburgh: Edinburgh University Press, Translation Edition, 2020); Sara Meger, 'Rape of the Congo: Understanding Sexual Violence in the Conflict in the Democratic Republic of Congo,' *Journal of Contemporary African Studies* 28, no. 2 (Apr 2010): 119–135; *Sexual Violence against Jewish Women during the Holocaust*, ed. by Sonja M. Hedgepeth and Rochelle G. Saidel (Waltham, MA: Brandeis University Press, 2010).

[3] Dagmar Herzog, 'Introduction: War and Sexuality in Europe's Twentieth Century,' in *Brutality and Desire: War and Sexuality in Europe's Twentieth Century* (London: Palgrave Macmillan, 2009), pp. 1–5 (pp. 2, 4–5).

[4] See Stefan Hock, 'To Bring About a "Moral of Renewal": The Deportation of Sex Workers in the Ottoman Empire during the First World War,' *Journal of the History of Sexuality* 28, no. 3 (Sept 2019): pp. 457–482; Joanna Bourke. 'Love and Limblessness: Male Heterosexuality, Disability, and the Great War,' *Journal of War & Culture Studies* 9, no. 1 (2016): pp. 3–19; Martina Steer, 'Nation, Religion, Gender: The Triple Challenge of Middle-Class German-Jewish Women in World War I,' *Central European History* 48, no. 2 (June 2015): pp. 176–198; Clare Makepeace, 'Male Heterosexuality and Prostitution during the Great War: British Soldiers' Encounters with Maisons Tolérées,' *Cultural and Social History* 9, no. 1 (2012): pp 65–83; Michelle K. Rhoades, 'Renegotiating French Masculinity: Medicine and Venereal Disease during the Great War,' *French Historical Studies* 29, no. 2 (2006): pp. 293–327.

citizens in Ontario.[5] Finally, studies of genocide in Namibia have usually ignored the use of sexual encounters as part of the German military campaign.[6] Here, we place them front and center to highlight the everyday workings of racialist violence. By putting diverse examples of global sexual encounters with racialized 'Others' into conversation, we examine how cross- and counter-cultural experiences, rooted in racialized and military hierarchies between soldiers and civilians, impacted enemy encounters across the twentieth century.

'The soldiers interfere with our women': Sexual Violence in German-Occupied Namibia, 1904–1908

Germans had been traveling to, and living in, southern Africa since at least the late eighteenth century, but after 1880 the metropole government began to establish a settler colony on the ancestral homelands of about 200,000 people, including communities of Ovambo, Herero, Nama, Himba, and San. The ensuing land expropriation and dual legal system enforced white supremacy, eroded African political and economic independence, and resulted in increased tensions in daily life. These tensions

[5] *See Prisoners of War and Local Women in Europe and the United States, 1914–1956: Consorting with the Enemy*, ed. by Matthias Reiss and Brian K. Feltman (London: Palgrave Macmillan, 2022); Reiss, *Controlling Sex in Captivity: P.O.W.s and Sexual Desire in the United States during the Second World War* (London: Bloomsbury, 2019); Raffael Scheck, *Love between Enemies: Western Prisoners of War and German Women in World War II* (Cambridge: Cambridge University Press, 2021); Fabien Théoflakis, '"Rodolph—How Nice He Is!": Contact Between German Prisoners of War and French Civilians, 1944–1948,' in *Wartime Captivity in the Twentieth Century: Archives, Stories, Memories*, ed. by Anne-Marie Pathé and Fabien Théoflakis (Oxford: Berghahn, 2016), pp. 163–176; Lisa Todd, *Sexual Treason in Germany during the First World War* (London: Palgrave Macmillan, 2017); Cornelie Usborne, 'Female Sexual Desire and Male Honor: German Women's Illicit Love Affairs with Prisoners of War during the Second World War,' *Journal of the History of Sexuality* 26, no. 3 (2017): pp. 454–488.

[6] Elizabeth R. Baer, 'A Reconsideration of Sexual Violence in German Colonial and Nazi Ideology and Its Representation in Holocaust Texts,' *The Palgrave Handbook of Holocaust Literature and Culture* (London: Palgrave Macmillan, 2020): pp. 651–668; Elisa von Joeden-Forgey, 'Women and the Herero Genocide,' in *Women and Genocide: Survivors, Victims, Perpetrators*, ed. by Elissa Bemoorad and Joyce W. Warren (Bloomington: Indiana University Press, 2018), pp. 36–57; Gesine Krüger, *Kriegsbewältigung und Geschichtsbewußtein: Realität, Deutung und Verarbietung des deutschen Kolonialkriegs in Namibia 1904 bis 1907* (Göttingen: Vandenhoeck & Ruprecht, 1999); 'Beasts and Victims: Women in the Colonial War,' *Genocide in German South-West Africa: The Colonial War (1904–1908) in Namibia and its Aftermath*, ed. by Jürgen Zimmerer and Joachim Zeller; trans. by Edward Neather. (Monmouth, Wales: Merlin Press, 2008), pp. 170–192.

erupted in late 1903 with widely reported Herero attacks on settlers, ultimately leading to war between German and Herero forces and the Nama peoples. We know at least 65,000 Herero people died, or 80 per cent of the population, and 10,000 Nama people, about 50 per cent of that population. Of the 19,000 German troops sent to southern Africa, 3000 engaged in active combat; 676 died fighting, and 689 died of disease. Most people now recognize the genocidal nature of these events.[7]

Our chapter's first case study uses this genocidal context to illustrate how sexual encounters between German soldiers and Indigenous Namibian women were often brutally violent and non-consensual. Even a fragmentary (written) archival record makes it clear that civilians faced violent and dangerous situations where the colonial power imbalance, coupled with anti-Black racism, and a military hierarchy of largely armed combatants and unarmed non-combatants, made them vulnerable to sexual violence. In short, male soldiers followed the precedent set by white settlers in seeing the sexual exploitation of Black female bodies as their prerogative, as evidenced by the widespread rape and sexual coercion perpetrated by German soldiers on women and girls in the battle zones, occupied territories, and concentration camps of the German-Herero and German-Nama Wars of 1904–1908.[8] At the same time, they used narratives of sexual danger against German men and women to defend the repression of local populations.

The period leading to 1904 saw increased conversations around the legal, moral, religious, and medical consequences of sexual relations between Indigenous peoples and European settlers.[9] While bureaucrats in

[7] David Olusoga and Casper W. Erichsen, *The Kaiser's Holocaust: Germany's Forgotten Genocide and the Colonial Roots of Nazism* (London: Faber and Faber, 2010); *Genocide in German South-West Africa.*

[8] Pamela Scully, 'Rape, Race, and Colonial Culture: The Sexual Politics of Identity in the Nineteenth-Century Cape Colony, South Africa,' *The American Historical Review* 100, no. 2 (April 1995): pp. 335–359.

[9] Ulrike Lindner, 'Contested Concepts of 'White'/'Native' and Mixed Marriages in German South-West Africa and the Cape Colony, 1900–1914: A Histoire croisée', *Journal of Namibian Studies* 6 (2009): pp. 57–79; Helmut Walser Smith, 'The Talk of Genocide, the Rhetoric of Miscegenation,' in *The Imperialist Imagination: German Colonialism and its Legacy*, ed. by Sara Friedrichsmeyer, Sara Lennox, and Susanne Zantop (Ann Arbor: University of Michigan Press, 1998), pp. 107–124; Lisa M. Todd, 'Studying Sexual and Racial 'Mixture': Eugen Fischer and the Rehoboth Basters of German Southwest Africa, 1908–1913,' in *After the Imperialist Imagination: Two Decades of Research on Global Germany and its Legacies*, ed. by Sara Pugach, David Pizzo, and Adam Blackler (Oxford: Peter Lang Publishing, 2020), pp. 79–92; Daniel J. Walther, *Sex and Control: Venereal Disease, Colonial Physicians, and Indigenous Agency in German Colonialism, 1884–1914* (New York: Berghahn Books, 2015).

the colony and metropole sought to determine the legal standing of so-called mixed marriages and the 'racial' implications of 'miscegenation', tensions were rising between the Germans and the Herero, especially around the new capital city at Windhoek. There were rumors of sexual violence against women on both sides.[10] The *New York Times* reported in March 1904 that the Herero 'rebelled partly on account of the oppressions of the traders, partly owing to the cruelties inflicted on them, and partly on account of the disregard of the Germans for the sacredness of the marital relations of the natives.'[11] Meanwhile, German media focused on the dangers Black men posed to white women in the colony. In language reminiscent of the so-called Black Peril myth propagated in other European colonies and the southern United States, eyewitnesses reported seeing mutilated female bodies, colonial fiction reveled in descriptions of helpless white women, and newspaper reporters decried such crimes as the work of 'black beasts'.[12] These stories fuelled existing worries that the colony was unstable and dangerous for white people. Increasingly, many settlers worried that Samuel Maherero was planning an uprising against the Germans. This rumor came true in late December 1903, when stories of murdered settlers and soldiers in the rural areas began to reach Windhoek. On 14 January 1904, a German official sent a telegram to the Foreign Office in Berlin that read: 'All farms in the vicinity of Windhoek plundered by Herero. Whites living on isolated farms murdered. Situation very grave.' Nationalists in the metropole saw in this 'native rebellion' a chance to exert more power over the territory. On January 18, Chancellor von Bulow demanded the *Reichstag* approve the military request to send reinforcements to the colony. Jingoistic voices were countered; however, by

[10] Wolfram Hartmann, 'Urges in the Colony. Men and women in colonial Windhoek, 1890–1905,' *Journal of Namibian Studies* 1 (2007): pp. 39–71; 'Making South West Africa German? Attempting Imperial, Juridical, Colonial, Conjugal and Moral Order,' *Journal of Namibian Studies* 2 (2007): pp. 51–84.

[11] 'Missionary Defends Natives: Says German Oppression and Cruelty Drove Hereroes to Revolt,' *New York Times*, 23 March 1904, p. 2. See also Krüger, *Kriegsbewältigung und Geschichtsbewußtein*, p. 171.

[12] Krüger, 'Beasts and Victims,' (p. 175); Maureen O. Gallagher, 'Fragile Whiteness: Women and Girls in German Colonial Fiction, 1900–1913,' *Women in German Yearbook: Feminist Studies in German Literature & Culture* 32 (2016): pp. 111–137 (p. 111); Medardus Brehl, 'Rassenmischung als Indiskretion. Textliche Re-Präsentationen des 'Mischlings' in der Deutschen Kolonialliterature über den Hererokrieg', *Rassenmischehen-Mischlinge-Rassentrennung: Zur Politik der Rasse im deutschen Kolonialreich*, ed. by Frank Becker (Stuttgart: Franz Steiner Verlag, 2004): pp. 254–267.

politicians such as August Bebel (Social Democrat), who remained opposed to military operations in the German colony.

While politicians in Berlin debated the merits of military action, fighting broke out in Southwest Africa. From what we can tell from existing historical documentation, and Herero oral histories, many colonial soldiers already on the ground operated *without* the greatest possible humanity and outside the boundaries of international law. Interpreter Manuel Timbu reported, 'the soldiers shot all natives we came across. It did not matter who they were. Some were peaceful people who had not gone into rebellion; others, such as old men and old women, had never left their homes; yet these were all shot.'[13] Centuries of racism had no doubt conditioned many of these men to believe in the racialized hierarchies of people being formulated by race scientists throughout Europe and North America.

After a series of military encounters, Herero military leadership changed tactics. It began to withdraw their soldiers from combat believing they had achieved their goal of standing up to colonial authorities. Gesine Krüger has illustrated how Herero women played a crucial part in military conflicts.[14] Community leaders began to move their people away from the German settlements to await the formal negotiations they believed to be forthcoming. By spring 1904, thousands of people gathered at the Waterberg, a series of mountain ranges that contained some of the last watering holes before the Omaheke or Kalahari Desert. At the same time, the Germans were assembling a colonial army of men and weapons led by General Lothar von Trotha. Von Trotha had much colonial warfare experience, and in taking up the position stated, 'I know enough tribes in Africa. They are all alike insofar as they only yield to violence. My policy was, and is, to exercise this violence with blatant terrorism and even cruelty. I finish off the rebellious tribes with rivers of blood and rivers of money. Only from these seeds will something new and permanent be able to grow.'[15] His

[13] *Report on the Natives of South-West Africa and their Treatment by Germany. Prepared in the Administrator's Office, Windhuk, South-West Africa, January 1918* (London: Published by his Majesty's Stationary Office, 1918), republished as *Words Cannot be Found: German Colonial Rule in Namibia*: An *Annotated Reprint of the 1918 Blue Book*, ed. by Jeremy Silvester and Jan-Bart Gewald (London: Brill Publishing, 2003), p. 114.

[14] Krüger, 'Beasts and Victims,' pp. 170–192.

[15] General Lothar von Trotha to Governor Theodor Leutwein, 5 November 1904, as quoted in George Steinmetz, *The Devil's Handwriting: Precoloniality and the German Colonial State in Qingdao, Samoa, and Southwest Africa* (Chicago: University of Chicago Press, 2007), p. 180.

military strategy was to encircle the peoples gathered at the Waterberg and attack. As part of the encircling action, von Trotha intentionally left one flank weaker, most military historians say, so that as the Herero soldiers fought, the frightened civilians, mainly women and children, would flee into the Omaheke desert. Strategically, von Trotha's plan was successful, and over a series of weeks, the fighting forced the people into the desert. The German soldiers then sealed up or guarded the watering holes so that the Herero could not gain access to water or food. Thousands died of starvation, thirst, and the brutal conditions of the desert. Survivors of the Waterberg later described seeing dead bodies across the area, including mothers lying beside their children.

We gain evidence of sexualized violence throughout the war from eyewitness accounts collected by a British commission for the 1918 *Report on the Natives of South-West Africa and their Treatment by Germany*. Originally published by the British as a testament to the unfit nature of Germans as colonial rulers, the so-called Blue Book was long dismissed as mere propaganda. However, when Jeremy Silvester and Jan-Bart Gewald republished an annotated edition in 2003, they stressed the importance of the forty-seven eyewitness testimonies as 'rare documentation of African voices describing the encounter of African communities with a colonial power.'[16] The Herero and Nama have strong oral traditions, but their narratives tend to focus on heroic events and victories; stories of murder, imprisonment, rape tend 'no longer to be found,'[17] though the trauma of survivors is still preserved in the social rituals, legends, and songs of Herero living in Botswana.[18] In the Blue Book, for instance, we read that Johannes Kruger stated under oath: 'I went with the German troops right through the Herero rebellion. [...] We refused to kill Herero women and children, but the Germans spared none. They killed thousands and thousands. I saw this bloody work for days and days, and every day. Often and especially at Waterberg, the young Herero women and girls were violated by the German soldiers before being killed.'[19] Another guide stated he saw a German soldier rape a 'young native girl about 16 to 18 years of age. I

[16] *Report on the Natives of South-West Africa*, p. xiii.

[17] Ibid.

[18] Kristen Alnaes, 'Living with the Past: Songs of the Herero in Botswana', *Africa* 59, no. 3 (1989): pp. 267–299 (p. 268).

[19] *Report on the Natives of South-West Africa*, p. 117.

came upon him in the veld. After he had raped her, he stuck his bayonet through her stomach and shot her.'[20]

The recently published *Mama Pence: Transcending the Genocide*, by Uazuvara Ewald Kapombo Katjivena, provides an invaluable source on wartime sexual violence. It relays the experiences of the young girl Jahohora (later Mama Pence), who was eleven years old when she witnessed German soldiers murder her parents. As she began to run toward their bodies, she was motioned away by a young German soldier. She was then on the run. Jahohora became an expert at avoiding German soldiers. 'She heard that in some places Ovaherero were being forced to do hard labor, and she heard that women were being raped by German soldiers. She was determined never to be captured alive let alone to be raped.'[21] Jahohora recounted how she made an oath that 'no German man would touch my body.' She rubbed painful nettles on her body so the resulting swelling and welts would make her diseased and unattractive. 'Not too surprisingly,' she said, 'the German men stayed away from me.'[22]

The German military celebrated their victory. But their work was not yet done. On 1 October 1904, von Trotha issued the infamous *Vernichtungsbefehl*—an 'annihilation order'—stating that all Herero must leave the land. He justified the decision to shoot all combatant and non-combatant Herero because they had allegedly 'murdered, stolen, cut off the ears and noses and other body parts from wounded soldiers.'[23] Three weeks later, Governor Leutwein sent a telegram describing von Trotha's actions to Berlin. There was shock and protest throughout Germany, and the mass murder in Southwest Africa triggered a national political scandal. Two months later, on 9 December, Kaiser Wilhelm II demanded that von Trotha retract the order and accept the formal surrender of the Herero people. By that time, tens of thousands of women, men, and children were dead.

[20] *Report on the Natives of South-West Africa*, p. 172.

[21] Uazuvara Ewald Kapombo Katjivena, *Mama Pence: Transcending the Genocide* (Windhoek: University of Namibia Press, 2020), p. 22.

[22] Katjivena, *Mama Pence*, pp. 34–35.

[23] The text reads in part: 'I, the great general of the German soldiers, send this letter to the Herero people. Herero are no longer German subjects. They have murdered, stolen, cut off the ears and noses and other body parts from wounded soldiers, and now out of cowardice, they refuse to fight. [...] All Herero must leave this land [...] Any Herero found within the German borders with or without a gun, with or without cattle, will be shot. I shall no longer receive any women or children; I will drive them back to their people or have them fired upon. These are my words to the Herero people.'

Von Trotha's references to the mutilation of soldiers' bodies echoed rumors circulating on the battlefield that German troops were threatened by attacks by 'barbaric' Black women. One soldier recorded in his diary: 'Woe betide the unfortunate enemy who falls into [the women's] hands. The most gruesome mutilations await him, such as gouging out his eyes, having his sex organs cut off, slowly and gradually having his head bashed in with a *kirri*. They are beasts and even worse than their male peers in the tribe.'[24] Another said men were 'provoked' to retaliate by these attacks.[25] These rumors played on stories of civilian *francs-tireurs* attacking soldiers during the Franco-Prussian War (and prefaced similar stories during the First World War) and were amplified by racialized stereotypes of 'wild' and 'uncivilized' Africans who did not understand 'modern' rules of war. Indeed, the Chief of the Army General Staff maintained that soldiers were 'provoked' to an 'understandable fury' by what they saw women do to their comrades. Indeed, he argued that Black women not only fought in battles but were the 'chief instigators of the cruel and vicious torture' attacks against wounded men. He continued that the sight of these victims put on display, with 'bestial intent,' led to women being shot.[26]

Black women were also viewed as dangerous sites of disease.[27] Throughout periods of active combat and into the camp period, the German military remained concerned not for the criminal actions of their men but that Indigenous women posed a threat to the health of German soldiers by infecting them with venereal diseases such as syphilis and gonorrhea; authorities put in place measures to ensure the availability of 'safe sex' for active troops. Indeed, the rapid rise in venereal disease rates after the arrival of German troops leads to the conclusion that rape crimes by soldiers must have run rampant. Indeed, one military official wrote that 'free and captive women' were dangerous sources of V.D. because of the 'great ease with which sexual relations could be undertaken with the female native population, even when they were not prostitutes.' He also wrote in April 1906 of 'the remarkable rise in V.D. in the last five months in the northern

[24] Krüger, 'Beasts and Victims,' p. 178. For a wider rhetoric of women combatants in colonial and irregular warfare as particularly violent and unnatural, see Greig, 'Counterinsurgency for Readers at Home: Chilling Encounters with Guerrillas in Post-Napoleonic Novels' and Menger, 'Racial Difference and the Limits of Fellow Feeling: British, German, and Dutch Soldiers Encounter the Enemy in Colonial Wars, c. 1900', this collection.

[25] Krüger, 'Beasts and Victims,' p. 173.

[26] Ibid.

[27] See Walther, *Sex and Control.*

lines. It obviously stands in a certain relation with the rise in prisoners.'[28] We know from medical statistics that sexually transmitted infections, spiked among German troops, indicating widespread sexual contact.[29]

Following the formal cessation of the German-Herero and German-Nama Wars, the German army began to patrol the area for survivors, often telling the emaciated people that they were free to return home. Instead, they sent people—perhaps as many as 13,000 men, women, and children—to concentration camps around the country. The first camps were built in the most significant areas of German settlement—Windhoek, Lüderitz, and Swakopmund. The conditions in the camps were notoriously brutal—food and water rations were scarce, and guards, for the slightest infractions, beat prisoners. The worst of the camps was at Shark Island, a barren stretch of rock in Lüderitz where the winds howled, and the cold dampness would have come as a profound shock to peoples used to arid conditions. Indeed, prisoners were treated so poorly at Shark Island, and the death rates were so high that many historians now refer to it as a death camp, the same term they give to the Auschwitz-Birkenau complex.[30]

The camps were also sites of rape, sexual exploitation, and violence.[31] From photographs, we can see the prisoner camp at the *Alte Feste* in Windhoek: rows of tents in front of a German fortress accompanied by few amenities. Two larger structures on the periphery may have been brothels.[32] In Gustav Frenssen's 1906 novel, *Peter Moor's Journey to Southwest Africa: A Narrative of the German Campaign*, which he based on interviews with soldiers, we read the following description of the main character's visit to the camp: 'when I took my leave after supper and went up to the fort, I saw some soldiers laughing with Hottentot women, and indeed one fellow said to me that these women were at our disposal at any time.'[33]

[28] Isabel Hull, *Absolute Destruction. Military Culture and the Practices of War in Imperial Germany* (Ithaca: Cornell University Press, 2005), pp. 150–151.

[29] Hull, pp. 150–151.

[30] For conditions in the camps, see Casper Wulff Erichsen, 'Forced Labour in the Concentration Camp on Shark Island', and Joachim Zeller, 'Ombepera I koza – the cold is killing me' Notes towards a history of the concentration camp at Swakopmund (1904–1908),' both in *Genocide in German South-West Africa*, pp. 84–99; pp. 64–83.

[31] von Joeden-Forgey, pp. 36–57.

[32] Baer, 'A Reconsideration', p. 657. See also Elizabeth R. Baer, *The Genocidal Gaze: From German Southwest Africa to the Third Reich* (Detroit: Wayne State University Press, 2017).

[33] Gustav Frenssen, *Peter Moor's Journey to Southwest Africa: A Narrative of the German Campaign*, p. 130, as quoted in Baer, 'A Reconsideration,' p. 656. See also Jeff Bowersox, 'Peter Moor Goes to Southwest Africa (1906)' at *Black Central Europe*, https://blackcentraleurope.com/sources/1850-1914/peter-moor-goes-to-southwest-africa-1906/

In her scholarly treatment of German colonial military violence, Susanne Kuss writes, 'The soldiers responsible for interning women in the prison camps set up during the Herero and Nama War—where they were registered, isolated, and in all probability raped—were able to choose whether to pursue intercourse with women and whether to take contraceptive precautions. The rape of the Herero and Nama women transformed these locations from prisoner of war camps ... to locations of violence directed specifically against nominally free women.'[34] A letter about the prison camp in Omaruru reads, 'One evening several white men from the railway traveled on the trolley to the Herero kraal and, with full knowledge of the guards, captured three Herero women and took them off as wives. Next evening the same hunt was repeated after dark. This time, the women who were sought out and lusted after, fled, and hid themselves'.[35] Historian Casper Wulff Erichsen writes of several rape-related charges in late 1906, after the arrival of Nama prisoners on Shark Island.[36]

In his diary, Rhenish missionary Johannes Carl Wilhelm Spiecker described the sex trade in the camps, the enforced gynecological examinations, and the yellow badges military doctors gave infected women.[37] Spiecker noted in his diary that the military 'doesn't even think any more about preventing the indecent contact of the soldiers with native women, but only tries to protect them from infections.'[38] Spiecker attended a November 1906 conference in Keetmanshoop where Brother Fenchel, who 'presided over the Nama mission,' stated in his Annual Report that imprisoned women were 'forced into a sinful life,' to which Spiecker responded by saying, 'If this is the case, then we have to speak up against these crimes,' but later in this same passage, accuses the women of too often consenting to 'sin' and then 'lying about it.'[39]

Sexual violence also accompanied the slave labor programs organized by soldiers and civilians; indeed, prisoners built large sections of the

[34] Susanne Kuss, *German Colonial Wars and the Context of Military* Violence, trans. by Andrew Smith (Cambridge: Harvard University Press, 2017), p. 201.

[35] Quoted in Krüger, 'Beasts and Victims,' p. 183.

[36] Erichsen, 'Forced Labour,' pp. 89–90.

[37] Johannes Carl Wilhelm Spiecker's 2000-page diary is held at the Archives of the Rhenish Missionary Society in Wuppertal, Germany. A condensed version was published in 2012. All quotations here are taken from Martin Siefkes, 'Discursive traces of genocide in Johannes Spiecker's Travel Diary (1905–1907),' *Journal of Namibian Studies* 16 (2014): pp. 83–114.

[38] Siefkes, 'Discursive traces,' p. 96.

[39] Siefkes, 'Discursive traces,' p. 97.

railway, several colonial-era buildings that tourists now take pictures of, and the parliament buildings in Windhoek. At the Swakopmund camps, prisoners unloaded ships as they came into port. Traugott Tjienda, Herero headman at Tsumeb, who was imprisoned as a forced laborer in the construction of a new railroad in the south, stated, 'When our women were prisoners on the railway work, they were compelled to cohabit with soldiers and white railway labourers. The fact that a woman was married was no protection. Young girls were raped and very badly used. They were taken out of the compounds into the bush and there assaulted.'[40]

Of central importance to this history are the Namibian women and girls who endured long-term impacts on their physical and mental health and were forced to navigate the effects of intergenerational trauma. In 1959, Mama Pence (Jahohora) met her friend Mukaatjauha, another Herero woman who had survived the genocide as a young girl. Mukaatjauha had been attacked by at least one soldier and became pregnant from the encounter. She bore a son. While gathering wood on that day in 1959, they had the following conversation. 'Did you let him rape you to save your life, or could you just not escape him?' She responded, 'At that time it was not so easy to say if it was consensual or by force. Force was always there, as was my need to survive. I didn't know if I refused him, whether he would kill me or not. Killing was accepted by the administration at that time. I just let him do what he wanted with me. Yet the pain was in my heart. But today, I am at peace. I love my son, Tommy. Yet in all he does, he resembles his father.' Mama Pence said in a 'tired soft voice.' 'Yes, there was much pain in my heart too. I couldn't imagine a German touching my body after what those men did to my family.' She continued by describing the women who had 'lost their minds' because of the sexual violence and forced pregnancies they endured during the war years. Some Herero and Nama women were prevented from working on settler farms because 'the German wives couldn't stomach what their husbands had done to the black women working there.' Mama Pence also cautioned readers to think about the white women who were 'helpless to control their husbands' bad behaviour.'[41] These very personal reckonings with a complicated and violent colonial past remind us of the ramifications of wartime sexual violence on girls and women in conflict zones and provide additional evidence that the exigencies of colonial warfare, when coupled with the anti-Black rac-

[40] *Report on the Natives of South-West Africa*, p. 178.

[41] Katjivena, *Mama Pence*, pp. 90–91.

ism of settler colonialism, resulted in a military structure that believed their enemies to be outside the laws of war.[42] As the following case study illustrates, a similar racialized power imbalance influenced intimate activities during the First World War in the Middle East.

Souvenirs from Wartime Egypt: Lieutenant Percy Deane's Love Letters

In August 1914, though not constitutionally bound, the Australian Imperial Force (A.I.F.) quickly organized and recruited men to supply troops to the British Empire in its war against Germany and the Central Powers, including the Ottoman Empire. On 1 November, approximately 30,000 troops left Australia and New Zealand for England via Egypt and spent six weeks at sea, confined in close quarters on troopships. The monotony of maritime travel played a critical role in facilitating curiosity, imagination, and preconceived notions about the unfamiliar landscapes and peoples they would soon encounter. These stereotypes, channeled through an Orientalist lens, met with reality when Australian and New Zealand troops were ordered to remain in Egypt, where they spent months training in the desert for their eventual participation in the Gallipoli campaign and guard of the Suez Canal.[43] Their proximity to major Egyptian cities like Cairo, Alexandria, and Port Said allowed troops on leave to become soldier-tourists and enjoy modern European entertainment, such

[42] For more information on the intergenerational effects of wartime violence, see Reinhart Koessler, *Namibia and Germany: Negotiating the Past* (Windhoek: University of Namibia Press, 2015).

[43] Anzac troops were initially ordered to encamp at Salisbury Plain in England but were redirected to Egypt partly because Salisbury could not accommodate forces from across the empire. Their presence in Egypt proved valuable because they supported British and Indian troops guarding the Suez Canal from enemy troops following the declaration of war between Britain and the Ottoman Empire on 5 November 1914. Not only was the Suez Canal a gateway to India, but it was strategically important for the movement of troops from Australia and New Zealand to the Western Front. John C. Mitcham, *Race and Imperial Defence in the British World, 1870–1914* (Cambridge: Cambridge University Press, 2016), pp. 214–215; C. E. W. Bean, *The Story of Anzac: The First Phase*, Volume I, The Official History of Australia in the War of 1914–1918, 11th ed. (Sydney: Angus & Robertson Ltd., 1941), pp. 89, 98, 110–116, 124–125, 142, 145; Daniel-Joseph MacArthur-Seal, *Britain's Levantine Empire, 1914–1923* (Oxford: Oxford University Press, 2021), pp. 40, 43; Justin Fantauzzo, *The Other Wars: The Experience and Memory of the First World War and Macedonia* (Cambridge: Cambridge University Press, 2020), pp. 52–53.

as theaters, opera houses, and ancient cultural and historical sites, like the Sphinx and the pyramids.[44] But this leave also granted the soldier-tourist access to Egypt's growing sex industry. This quickly became problematic when servicemen in Cairo broke camp at night and ventured into the city's entertainment district of Ezbekkiya to drink and meet women. In a letter, one Australian serviceman claimed 144 of 1045 men were declared missing after they were granted leave on Christmas Day 1914 and did not return.[45] On 27 December, Major-General Birdwood, Commander of the Australian and New Zealand Army Corps (Anzacs), wrote to Major-General Bridges, Commander of the 1st Australian Division, expressing concern for servicemen's behavior in Cairo, specifically regarding their drunkenness and increasing rates of V.D.: 'But there is no possibility whatever of our doing ourselves full justice unless we are every one of us absolutely physically fit, and this no man can possibly be if he allows his body to become sodden with drink or rotten from women … in the interests of our children and children's children it is necessary to keep a "clean Australia" as a "White Australia."'[46]

References to a 'White Australia' remind us that imperial and military occupations have a long history of imposing hierarchies of power based on gender, racism, sexuality, health, and other intersecting identities. In the British Empire, women of color were targeted by colonial policies that racialized and sexualized their bodies, and white women were targeted if they did not limit their sexual behavior only to reproduction. At the same time, non-white women were also thought to spread more virulent forms of venereal disease to white populations. These beliefs were based on

[44] The soldier-tourist is a concept historians use to explain servicemen's tourism from a military perspective. Historian Richard White notes that the Australian population had a natural tourist-ethos that servicemen applied during their wartime experiences. Servicemen positioned themselves as observers of the local environment, cultures, and peoples by engaging in soldier tourism while maintaining their military occupant identity. On peripheral fronts of the First World War, such as Egypt, soldier tourism emphasized the differences servicemen recognized between themselves and 'others,' often understood through an Orientalist gaze influenced by pre-war travel writings. Richard White, 'The Soldier as Tourist: The Australian Experience of the Great War,' *War & Society* 5, no. 1 (May 1987): pp. 63–64; Richard White, 'Sun, Sand and Syphilis: Australian Soldiers and the Orient, Egypt 1914,' *Australian Cultural History* 9 (June 1990): pp. 49–50.

[45] Letter from William McKenzie to Commissioner James Hay, 27 December 1914, PR85/185, AWM2018.785.85, Australian War Memorial.

[46] Letter from Major-General W. R. Birdwood Major-General W. T. Bridges, 28 December 1914, RCDIG0000113, 3DRL/3376 11/17, Australian War Memorial.

contemporary pseudoscience that placed 'othered' individuals who threatened or jeopardized 'whiteness' and did not embody an imperial ideal at the bottom of an imagined hierarchy. Military officials in First World War 'Oriental' Egypt feared that troops' sexual encounters with racialized and sexualized women threatened servicemen's health and 'whiteness' yet chose to continue a system of regulated prostitution traditionally practiced under British imperial and military occupation.[47] Not only did this system function to boost servicemen's morale, but it permitted men to act on their 'natural' sexual urges through the exploitation and medical control of civilian women and their bodies. Policymakers in colonial Australia adopted these beliefs to determine which women were appropriate or inappropriate sexual partners for white settler men. Upon federation in 1901, Australia was eager to maintain a racial link to empire via restrictive immigration and citizenship policies against non-white individuals, including Indigenous peoples. The Immigration Restrictions Act and the Pacific Island Labourers Act were the backbone of the 'white Australia policy' meant to racially exclude unwanted populations of color and those 'of suspect sexuality' to protect 'whiteness,' 'manhood,' and 'civilisation.' Rooted in the broader imperial and international (Western) eugenic discourses of public health and hygiene, and within the concepts of 'fit' or 'unfit' and 'pure' or 'impure,' the 'white Australia policy' framed Australian white masculinity.[48]

Despite warnings of the dangers V.D. posed to military efficiency and the future of the nation and empire,[49] many British, Australian, and New

[47] Systems of regulated prostitution throughout the British Empire in the late nineteenth and early twentieth centuries were influenced by the Contagious Diseases (CD) Acts which targeted suspected prostitutes in port and garrison towns in England and Ireland from 1864 to 1886 and forced them to undergo compulsory medical examinations and treatment if infected with V.D.

[48] As quoted in Marilyn Lake, 'Translating needs into rights: the discursive imperative of the Australian white man, 1901–30,' in *Masculinities in Politics and War: Gendering Modern History*, ed. by Stefan Dudink, Karen Hagermann, and John Tosh (Manchester; New York: Manchester University Press, 2004), pp. 199–201; Alison Bashford, *Imperial Hygiene: A Critical History of Colonialism, Nationalism and Public Health* (London: Palgrave Macmillan, 2004), pp. 2, 137–138; David C. Atkinson, 'The White Australia Policy, the British Empire, and the World', *Britain and the World* 8, no. 2 (2015): pp. 204, 207–209; Raelene Francis, 'Sex Workers or Citizens? Prostitution and the Shaping of "Settler" Society in Australia', *International Review of Social History* 44 (1999): pp. 101, 107, 109, 114–115.

[49] 'Warning to Soldiers Respecting Venereal Disease,' RCDI0000113, 3DRL/3376 11/17, Australian War Memorial.

Zealand servicemen engaged in sexual encounters with women in Egypt, often in the red-light district of cities like Cairo, Alexandria, and Port Said. These cities profited from Egypt's growing tourist industry in the late nineteenth and early twentieth centuries, with Cairo's downtown quarter of Ezbekkiya being the most notable for its nightlife where wealthy male visitors smoked, gambled, drank, and sought 'Oriental' pleasures in music halls, cabarets, and brothels. Cairo's brothel quarter, known as the Wass'ah (or Wazza), was especially popular among troops in Egypt. Here, Egyptian, Nubian, Sudanese, and European women solicited potential clients only a stone's throw from the Shepherd's Hotel, which served as Anzac headquarters. Prostitutes working in and around the Wazza could be seen soliciting troops from tenement balcony windows or in the doorway of their one-room 'shops.'

Sex work in wartime Egypt took several forms, including clandestine, amateur, and licensed prostitution. The regulation system, which long predated 1914, meant that licensed prostitutes were often segregated in the designated brothel district and subjected to surveillance through identification cards that were updated each time a woman underwent a medical examination.[50] Women worked in the sex industry for a variety of reasons; indeed, it can be at times difficult for the historian to discern the lines between agency, consent, and coercion. We do know that like in German-occupied Namibia, military and colonial authorities in First World War Egypt facilitated coercive sexual encounters between servicemen and civilian women and many women and girls were forced into prostitution through sex trafficking or due to wartime socio-economic factors. However, we must also recognize that some women—not exclusive to artistes and other entertainers—engaged in commercial sex by choice or, at the very least, with the ability to express a level of agency within this 'male-dominated repressive gender system.'[51]

To further curb servicemen's misbehavior, the military placed a guard outside of camp on the road leading into Cairo, and picquets were established throughout the city. However, these efforts proved ineffective as troops were still arrested for breaking camp and general misbehavior,

[50] Cairo Purification Committee, 18 April 1916, FO141/466/2, The National Archives of the UK.

[51] Francesca Biancani, '"Let Down the Curtain Around Us": Sex Work in Colonial Cairo, 1882–1952,' PhD diss., (London School of Economics and Political Science, 2012), pp. 133–134, 81, 127, 134, 157–158.

inevitably leading to 'a great deal of [venereal] disease.'[52] Yet, despite regular medical inspections, British, Australian, and New Zealand troops contracted V.D. at a significant rate. According to Major Dr. Bernhard Zwar of the Australian Army Medical Corps (A.A.M.C.), from the opening of the 2nd Australian Station Hospital (2nd A.S.H.) in late January 1915 and the Anzac's departure for Gallipoli in April 1915, 'between 500 and 600 patients were constantly in hosp[ital]' and upwards of 800 patients were either treated elsewhere, or convalescing on light duties following treatment.[53] A close reading of love letters sent to Australian serviceman Percy Deane by two women in wartime Egypt illustrates on a more personal level how anxieties around disease, when coupled with fears of interracial relationships, resulted in complicated questionings of power hierarchies.

Born Percival Edgar (Percy) Deane in Port Melbourne, Australia, in 1890, Deane won a scholarship to a private high school in his youth. He worked various jobs in early adulthood, including Methodist lay preacher, peddler, and clerk at the University of Melbourne, where he met Dr. James Barrett, an ophthalmologist and lecturer. He was also a journalist, publisher, editor, and skilled shorthand writer.[54] Deane enlisted as a private with the A.I.F. in October 1914 and was assigned to the 1st Australian General Hospital (1st A.G.H.). By the time he left Melbourne for Egypt in December, Deane had quickly moved up the ranks to the Lance Sergeant position as a non-combatant. By May 1915, he was granted honorary lieutenant status and served as quartermaster of the 1st A.G.H., located in Heliopolis outside Cairo. There, he worked alongside Barrett, the hospital's registrar, consulting oculist for the A.I.F., and an advocate for the use of prophylaxis to prevent V.D.[55] Dean's undated manuscript from the National Library of Australia reveals that he was in fact aware of the prob-

[52] Diary entry for 9–30 January 1915, Diary, January–March 1915, Official History, 1914–18 War: Records of C E W Bean, Official Historian, RCDIG1066573, AWM38 3DRL 606/2/1, Australian War Memorial.

[53] Paper written by Maj. Zwar on prevention of venereal disease, AWM2021.8.322, AWM27 376/175, Australian War Memorial; Ian Howie-Willis, *VD: The Australian Army's experience of sexually transmitted diseases during the twentieth century* (Newport, N.S.W.: Big Sky Publishing Pty Ltd, 2020), pp. 55–56.

[54] S. Murray-Smith, 'Deane, Percival Edgar (Percy) (1890–1946),' in *Australian Dictionary of Biography* Volume 8, 1981 <https://adb.anu.edu.au/biography/deane-percival-edgar-percy-5933> [accessed June 27, 2022]; Curriculum Vitae, Papers of Percy Deane, 1915–1946 [manuscript], ms10096, National Library of Australia.

[55] Attestation Papers of Percival Edgar Deane, N.A.A.: B2455, DEANE P. E., National Archives of Australia.

lem of V.D. within the A.I.F. and its potential negative impact on 'white Australia.'

In January 1915, the long journey from Australia to Egypt gave Deane a heightened awareness of his 'whiteness' amidst Egypt's ethnically and culturally diverse population. Deane categorized the local Egyptians in Suez according to racial stereotypes that linked skin color and criminal behavior and described Port Said as 'dirty' because of its poorly made streets, the public sale of 'smutty postcards,' and the French and 'native' women performing 'kan kan' dances to lure visitors into brothels, behavior that he noted left many servicemen with the 'Egyptian medal' (V.D.). It was also in Port Said that Deane saw veiled Arab women for the first time, their niqab emphasizing their 'beautiful,' 'huge ox-like eyes and olive skins'—an image that played into the trope of exotic, racialized sexuality that preoccupied the imagination of many white men. The town, according to Deane, was full of '[m]agnificent looking French and native women.' After dining with Barrett and Colonel Ramsey Smith, commander of the 1st A.G.H., Deane noted in his diary, '[m]ost of us expressed the opinion that Port Said contained the most beautiful women we had ever seen.'[56]

For Deane, Alexandria was a compromise between East and West. The best of the city was its rich architecture and predominantly European population who dressed in fine attire, and the French, Italian, and Greek shops. In contrast, the worst included the locals who carried their cattle and water about town and squatted in front of local confectionaries that sold 'greasy pastry,' and the prostitutes who solicited troops from brothel doorways. Deane noted that Australian servicemen were only too eager to engage with prostitutes in Egypt, primarily upon disembarkation and their long voyage from home. Troops were warned to avoid local brothels where tainted drink was served and women were 'loaded with disease.' This warning often went unheeded, as servicemen watched '[n]ative girls of 14 to 18 clad only in chemise … [dance] the 'kan kan' a species of voluptuous wriggle, in which naked women exhibit those alleged charms which have held man captive for years.' Deane blamed the prevalence and virulence of disease on both Egypt's hot environment and the 'immoral and deceitful' cultural practices of the local population. He was

[56] Undated, untitled manuscript, Papers of Percy Deane, 1915–1946 [manuscript], ms10096, National Library of Australia; Ronald Elmslie and Susan Nance, 'Smith, William Ramsay (1859–1937)', in *Australian Dictionary of Biography* Volume 11, 1988 <https://adb.anu.edu.au/biography/smith-william-ramsay-8493> [accessed July 17, 2022].

responsible for distributing calomel ointment and argyrol jelly to 1st A.G.H. to treat gonorrhea and syphilis, and credited Barrett's judgment in making these medicines accessible to troops. However, seemingly few servicemen were immune to the intrigue of Egypt's temptations, not even Percy Deane, who had not one but two girlfriends in wartime Egypt.[57]

A small file from the University of Melbourne Archives includes several photos and twenty letters written to Percy Deane in English by two women likely involved in Egypt's sex trade: Hamida, an Egyptian woman in Cairo, and Helene, a French woman in Port Said. It is unclear why Deane kept these letters—did he save them for sentimental purposes, or as a 'trophy' of his wartime experiences? We also don't know why these letters are in an archival collection. Though it is common for veterans' families to preserve wartime souvenirs, and eventually donate them to an archive, letters like the ones in this file are less common, especially from servicemen who fought outside the Western Front. The context in which Deane met Hamida and Helene is also unclear. Still, according to the dates on the letters, envelopes, photos, and postcards, Deane likely met Hamida shortly after he first arrived in Cairo in January 1915. Hamida's letters not only express affection and desire for Deane, but one note identifies her address as 'Hart el Hossani,' likely Haret el Hosseini Street, and on the back of a postcard from February 1915, Hamida identifies her address as No. 29 Gobran Bey House on Wagh el Birka Street. Haret el Hosseini was an off-street of Wagh el Birka, the main street in Cairo's red-light district. The file also includes a letter from an individual named Moussa (likely a procurer or pimp) who writes: 'Will you not go to Hameeda room tonight till after two days till I go with her to the Dr to see if she kleen or not.' The note's reference to Hamida being clean implies she had to be checked for V.D. This, with her address in Cairo's red-light district, makes it reasonable to suspect she was a licensed prostitute in Egypt's commercial sex trade.[58]

A single picture postcard (see Fig. 1) accompanies these letters and depicts Hamida wearing a European working-class dress and Middle

[57] Undated, untitled manuscript, Papers of Percy Deane, 1915–1946 [manuscript], ms10096, National Library of Australia.

[58] Letter from Hamida (Hameada) to Deane, 1915, 26 Letters from Cairo, 1914–1922, Percival E. Deane Collection, University of Melbourne Archives; Letter from Hamida (Hameeda) to Deane, 1915, 26 Letters from Cairo, 1914–1922, Percival E. Deane Collection, University of Melbourne Archives; Letter from Hamida to Deane, 11 March 1915, 26 Letters from Cairo, 1914–1922, Percival E. Deane Collection, University of Melbourne Archives; Letter from Moussa to Deane, n.d., 26 Letters from Cairo, 1914–1922, Percival E. Deane Collection, University of Melbourne Archives.

Fig. 1 A picture postcard of Amida (Hamida), Cairo, Egypt, February 1915. 26 Letters from Cairo, Percival E. Deane Collection, The University of Melbourne Archives

Eastern bangles on her wrists. This image demonstrates Hamida's attempt to embody an acceptable sexual partner for white men like Deane by defying 'Oriental' expectations in a colonial environment. Although Hamida's Western-style clothing broke down cultural barriers and helped make her identity more familiar, her Egyptian 'race' and 'illicit' sexuality as a prostitute meant that she still embodied the essence of 'Other.'[59] Letters from Hamida collectively suggest a short-lived, strictly sexual relationship between her and Deane.

The earliest love letter from Helene dates from 17 August 1915. Like Hamida, Helene expresses her affection and desire for Deane, writing that she sleeps with his photograph, that he makes her happy, and that she is faithful to him.[60] The nature of Deane's relationship with Helene is unclear, but photos in the file show that they courted in daytime outings to the park and the beach. This file's photo postcard (see Fig. 2) depicts Helene as a middle-class European woman; however, evidence suggests that she was also involved in Egypt's commercial sex trade. Helene lived and worked as a dancer in Port Said, a port town influenced by the French and with a large French population that had strong ties to the international sex trade via the French Mediterranean port town of Marseille. France's system of regulated prostitution made it relatively easy for women to be trafficked off the continent through the Mediterranean and into Egyptian cities like Port Said, Cairo, and Alexandria—sometimes before they were sent to the Ottoman Empire, India, Asia, or elsewhere.[61] Further, a second picture postcard (see Fig. 3) in the file shows Helene wearing stereotypical 'Oriental' dress likely to reflect her profession as a 'belly dancer.'[62] In the nineteenth and twentieth centuries, 'belly dancing' was a culturally appropriated form of entertainment typically for a white male audience in Egypt, Europe, and elsewhere. And in late nineteenth and early twentieth-century Egypt, this form of entertainment overlapped

[59] Analysis of Hamida's early twentieth-century clothing is based on Jane Ashelford, *The Art of Dress: Clothes and Society 1500–1914* (London: National Trust Enterprises Limited; New York: Harry N. Abrams, Incorporated, 1996), pp. 212–271 and Verity Wilson, 'Western modes and Asian clothing: Reflections on borrowing other people's dress', in *The Fashion History Reader: Global Perspectives*, ed. by Giogrio Riello and Peter McNeil (London; New York: Routledge, 2010), pp. 418–436.

[60] Letter from Helene to Deane, n.d., 26 Letters from Cairo, 1914–1922, Percival E. Deane Collection, University of Melbourne Archives.

[61] Letter from H[elene] to Deane, 17 August 1915, 26 Letters from Cairo, 1914–1922, Percival E. Deane Collection, University of Melbourne Archives; Undated, untitled manuscript, Papers of Percy Deane, 1915–1946 [manuscript], ms10096, National Library of Australia; See map of migration routes for prostitution (Fig. 1) in Liat Kozma, 'Women's Migration for Prostitution in the interwar Middle East and North Africa', *Journal of Women's History* 28, no. 3 (Fall 2016): p. 95.

[62] Ashelford, *The Art of Clothes*; Wilson, 'Western modes and Asian clothing'; Steele, 'Fashion and Eroticism'.

Fig. 2 A picture postcard photo of Helene, taken at Hiesinger's Photo Studio, 44 Sharia Magabegh, Cairo, n.d. 26 Letters from Cairo, Percival E. Deane Collection, The University of Melbourne Archives

with the local sex industry.[63] White women, like Helene, who performed 'belly dance,' perpetuated a colonial fantasy that exhibited 'exotic' and 'erotic' stereotypes of racially sexualized Middle Eastern women that

[63] What is referred to today in English as 'belly dancing' is a Middle Eastern and North African movement rooted in traditional forms of female dance. Ainsley Hawthorn, 'Middle Eastern Dance and What We Call It', *Dance Research* 37, no. 1 (May 2019): p. 2; Raphael Cormack, *Midnight in Cairo: The Divas of Egypt's Roaring '20s* (New York: W.W. Norton & Company, 2021), pp. 7, 15–16, 46.

Fig. 3 A picture postcard of Helene in 'Oriental' dress, n.d. 26 Letters from Cairo, Percival E. Deane Collection, The University of Melbourne Archives

appealed to men's Oriental imaginings. Helene's profession as a performer, her display of hypersexuality and racialized stereotypes, and her possible ties to the sex trade would likely have raised alarm bells for Deane, especially with his knowledge of the dangers of V.D. However, as a European woman, Helene was a more racially suitable sexual partner for Deane than non-white women in Egypt, such as Hamida.

Hamida's last dated love letter is from May 1916, when Deane returned to Australia. Hamida's letter hints the end of their romance as she conveys disappointment in the breakdown of their relationship. Deane was officially discharged that July and, in November, was appointed private secretary to Australian Prime Minister William (Billy) Morris Hughes, who advocated that Australia maintain its strong relation to the empire and its white, homogenous society.[64] As the Prime Minister's private secretary, Deane increased his social mobility and position of power through his close working relationship with Hughes. Deane also maintained a strong relationship with Barrett. Together, they published *The Australian Army Medical Corps in Egypt* in 1918, which dedicates a whole chapter to the problem of V.D. and how it was handled by military and medical personnel during the war. Of prostitution in Cairo, they write, 'the quarters are dirtier ... evil-smelling ... with narrow streets and objectionable places of entertainment ... for those who went to this quarter there is no excuse, since they acted deliberately,' thus placing responsibility on servicemen for their actions.[65]

The prevalence of V.D. and servicemen's access to prostitution was a significant problem for armies throughout the First World War. This case exemplifies that despite warnings to avoid vice, servicemen in wartime Egypt acted on their position of power by engaging in sexual encounters with civilian women. Like in the following account of sexual encounters between Canadian women and German P.O.W.s in the Second World War, Deane's wartime sexual encounters blurred understandings of 'enemies' and 'allies.' For Australian servicemen, these 'enemies' (V.D. and prostitution) not only jeopardized military efficiency but risked the future of 'white Australia.' Percy Deane, a white Australian occupying Egypt on behalf of the British Empire, advocated for the prevention of disease and was responsible for safeguarding his white colonial identity yet risked his own health and the future 'race' by engaging in sexual encounters with Hamida and Helene, two women whose 'race' and sexuality transgressed Orientalist expectations.

[64] 26 Letters from Cairo, 1914–1922, Percival E. Deane Collection, University of Melbourne Archives; Attestation Papers of Percival Edgar Deane, N.A.A.: B2455, DEANE P E, National Archives of Australia; Atkinson, 216, 220.

[65] James Barrett and Percy Deane, *The Australian Army Medical Corps in Egypt* (London: H. K. Lewis & Co. Ltd, 1918), pp. 184–185.

'Like the Boys from Town but More Polite': Encounters Between Canadian Women and German P.O.W.s, 1940–1945

Britain was responsible for the internment of 35,000 German Prisoners of War in Canada during the Second World War. When the war began on 3 September 1939, some 70,000 unnaturalized Germans and Austrians were living in Britain.[66] Many of these individuals were German Jews who fled Nazi persecution after Hitler became Chancellor in 1933.[67] Mounting levels of racism and war hysteria led the British to 'intern the lot' as enemy aliens. After the capture of German P.O.W.s during the British Expeditionary Force's retreat from France and in air and naval operations over the English Channel, North Sea, the Atlantic, or England itself, internment sites began to overflow.[68] In addition to this excess of prisoners, Britain feared a potential Nazi invasion. It would compromise the country's national security if the Nazis invaded Britain and freed their comrades from behind British lines.[69] Transatlantic cooperation and the dispersal of Britain's prisoners across the Empire was a straightforward solution to these security threats. After much deliberation, Canada, a Dominion known for its vast geography and ample resources, formally accepted the first 535 P.O.W.s from British internment camps on 19 June 1940.[70] Over the next five years, the British government transferred thousands of combatant prisoners, Enemy Merchant Seamen (E.M.S.), and civilian internees to Canada for safekeeping.

After arrival, prisoners were categorized based on the level of risk they represented and dispatched to Alberta, Ontario, Quebec, and New Brunswick. The province of Ontario had eleven P.O.W. camps constructed from former military training grounds, experimental farms, schools for

[66] Many of these individuals were German Jews who fled Nazi persecution after Adolf Hitler became Chancellor in 1933; Elizabeth A. Atkins, 'You must all be interned': Identity among internees in Great Britain during World War II,' *The Gettysburg Historical Journal* 4, Article 5. Available at: https://cupola.gettysburg.edu/ghj/vol4/iss1/5, 61.

[67] Ibid.

[68] David Carter, *P.O.W Behind Canadian Barbed Wire: Alien, Refugee, and Prisoner of War Camps in Canada, 1914–1946* (Elkwater, AB: Eagle Butte Press, 1998), p. 45.

[69] Ibid.

[70] The first shipment contained 2, 112 Civilian Internees and 535 German P.O.W.s; RG24, C-5368, HQS 7236, High Commissioner for Canada in Great Britain to The Secretary of State for External Affairs, Policy, Treatment of Enemy Aliens, 7 June 1940, Libraries and Archives Canada.

juvenile delinquents, abandoned sawmills, tuberculosis sanitoria, and prisons. For Ontario women, the German enemy landed on their doorsteps. This section explores intimate encounters between Canadian women and German P.O.W.s during the Second World War. It examines women not as victims of wartime sexual violence but as free agents who pursued sexual relationships with the enemy for their satisfaction. While far from 'normal' connections on the Canadian home front, broader society accepted these intimate encounters due to racial similarities between enemies. In contrast to German-occupied Namibia and wartime Egypt, these relationships present the historian with examples of largely consensual, mainly unpunished, intimate affairs between soldiers and civilians in modern warfare.

The Canadian government implemented the Defense of Canada Regulations (D.O.C.R.) immediately after the outbreak of the Second World War in September 1939 to keep physical contact between civilians and enemy soldiers to a minimum. Via a series of anti-fraternization regulations, the D.O.C.R. warned that any civilian caught entering P.O.W. grounds without cause, sending supplies to the prisoners, or accepting anything on behalf of a captive would be penalized.[71] Romantic or sexual relationships of any kind with Axis forces were also strictly forbidden. Canadian women were instructed to uphold standards of conduct through personal example and threatened with disciplinary action in infringement cases. When German P.O.W.s arrived in Canada in 1940, they were also given the rules of war and warned that intimate relationships with resident women were forbidden. Intercourse, dating, sexual advances, friendly gestures of goodwill, or even a simple conversation could all result in punitive action.[72] Most people seemed to abide by these rules at the beginning but broke them more often as the war dragged on.

Before Ontario women met the 'Nazi' enemy from overseas, most relied on the tabloids for their impressions of Germans in uniform. Portrayed as a horde of Huns who had ransacked Europe and slayed thousands of women and children, the P.O.W.s were allegedly unshaven, dirty men who wore oversized, rumpled uniforms.[73] 'They look like anything but good soldiers. Few of them are even good physical specimens ... they

[71] 'Location of P.O.W. Camp Kept Secret,' *Parry Sound North Star*, 8 March 1990.

[72] Michael O' Hagan, 'Beyond the Barbed Wire: P.O.W. Labour Projects in Canada during the Second World War', PhD diss. (University of Western Ontario, 2020). *Electronic Thesis and Dissertation Repository*. 6849. https://ir.lib.uwo.ca/etd/6849, p. 162.

[73] 'Security an Essential in Locating Prison Camp', *Toronto Telegram*, 3 October 1940.

look pale and tired and some of them even sickly', the *Globe and Mail* wrote in 1940.[74] Nevertheless, while this propaganda inflamed anxiety and depicted Germans as different from Canadians, the inundation of Ontario towns and cities with German P.O.W.s challenged its messaging.[75] The Germans' Anglo-European features—and the whiteness they presumed—reflected attributes then considered virtues in Canadian society and quickly depoliticized the enemy. This sentiment afforded the Germans individual identities, and women saw little reason not to admire or consort with them.

Evidence points to Ontario teenagers as one of the first groups to dispense with non-fraternization policies. As early as 1941, adolescent girls flocked to the perimeter of Internment Camp 20 after its establishment in Muskoka, Ontario, drawn by the prisoners' 'otherness' and exoticism. After seeing the Germans in the flesh, the girls searched for similarities between themselves and the enemy.[76] In Cecil Porter's *The Gilded Cage* (2003), a local summer camp attendee recalls rowing across Lake Muskoka after dark to the enclosure of Camp 20, stopping twenty feet from the shore to flirt with the German prisoners. She states, 'Many of them were young boys, maybe 20 or so, and we were all about 16 or 17, and things went along okay because some of them talked English and quite a few girls spoke German.' Nothing ever came of the practice, she says, because 'it was only in the summer, in the camp season about six weeks, and [they] were still prisoners, although they were just like boys from the towns, but more polite.'[77] In a similar situation, members of the Veterans Guard caught a gaggle of Espanola girls concealing messages into snowballs and throwing them over the Camp 21 fence to prisoners.[78] Another pre-teen hid love letters under a rock for an Espanola P.O.W. to find and answer as he did his camp chores. When a guard caught on to the arrangement and questioned both parties, authorities labeled this needless meddling into an entirely harmless affair.[79] Like Hamida and Helene's letters to Percy Deane, these love letters served to maintain a personal relationship with

[74] Ibid.

[75] Jordyn Bailey, 'Arrival of the Fittest: German P.O.W.s in Ontario during the Second World War' M.A. thesis (University of Western Ontario, 2019), *Electronic Thesis and Dissertation Repository*. 6279. https://ir.lib.uwo.ca/etd/6279

[76] Barry Broadfoot, *Six War Years, 1939–1945: Memories of Canadians at Home and Abroad* (Toronto: Paper Jacks Ltd., 1976), p. 61.

[77] Broadfoot, Six War Years, p. 127.

[78] George Morrison, *Espanola on the Spanish* (Meaford, ON: Oliver Graphics, 1989), p. 129.

[79] Morrison, Espanola on the Spanish, p. 9.

the enemy. However, while the communication of the former group brought the future of 'white Australia' into question, correspondence between Canadian youth and German P.O.W.s served a morale-boosting function in society. Though these letters were not legally permitted under D.O.C.R., their mix of ordinary chat with gentle flirtation was accepted as emotional relief rather than treasonous activity.

With striking regularity, the Canadian public forgave teenage girls for fraternizing with the German enemy in wartime Ontario. When members of the Veterans Guard caught them communicating with P.O.W.s, regional and national newspapers published their stories. These articles affirm how acceptable it seemed for Ontario women to find German lovers at the time. For instance, when five young girls went to court in Sudbury, ON, over the exchange of love letters with German P.O.W.s inside Camp 21 on 19 March 1941, an article published in the *Ottawa Journal* titled 'These Five Silly Girls Are not the Real Problem,' called the charges trivial and thoughtless. The young teens should not be punished, it argued, because they had 'looked upon the whole thing as a romantic adventure' and did not have malicious intentions.[80] Likewise, the *Globe and Mail* called the incident 'a silly love affair thing' and reduced it to a simple case of natural attraction. Instead of young women in peril, they were 'silly and stupid young girls who likely [collaborated with the enemy] in the spirit of romance and may be forgiven for knowing no better.'[81] The real culprit, by widespread agreement, was the male guard who allowed such events to transpire under his nose.[82]

While some teenagers defied fraternization bans with impunity in 1941, most private citizens' adherence to non-fraternization policies began to break down in 1943. This increase in defiance occurred for two reasons. First, when members of the Afrika Korps, a force remembered for its devotion to National Socialism, were captured in Tunisia and sent to Ontario in 1943, they triggered, in the words of Matthias Reiss, an 'erotic shock' amongst civilians on the Canadian home front.[83] The allure of the Afrika Korps' came from their tanned skin and sun-bleached uniforms, and citi-

[80] *Ottawa Journal*, 21 March 1942.

[81] The judge gave them suspended sentences for two years or the duration of the D.O.C.R.—'Sent Prisoners Camera, Love Notes, Girls Admit,' 1942, RG24, Reel C-5365, File 9139-4-33 Conditions at Internment Camps, Espanola and Gravenhurst, Ont: Sudbury Detachment Investigation, March 22, 1942, p. 2, Libraries and Archives Canada.

[82] 'Sent Prisoners Camera, Love Notes, Girls Admit,' *Globe and Mail*, 1942.

[83] Reiss, *Controlling Sex*, pp. 17–18.

zens believed that the Germans' strength, veteran status, and physiques exceeded Canadian ideals of manhood. When a Northern Ontario man ran into forty Afrika Korps men out of uniform, he decreed: 'every bloody one was a giant, 6'2, 6'4. All fair-haired, blue-eyed, thighs like hams and tanned like Charles Atlas ... we have to fight men like that?'[84] It is particularly telling that this man chose to liken the prisoners' physical appearances to bodybuilder Charles Atlas. He was named the 'World's Most Perfectly Developed Man' in 1922 and revered as an icon of virile strength and masculinity.[85] The Afrika Korps had built their recent careers around Nazi advancement, but their ability to attract admiration impelled many free women to rank personal over political judgements and pursue intimate relations with the enemy.

The year 1943 also saw the authorization of captive labor, which allowed P.O.W.s to interact with resident women on a more intimate level. The fact that the German soldiers were European had made them instantly appealing to Ontario farmers, who had begged the government for captive labor since 1940. When the Mackenzie King Administration passed Order-in-Council P.C. 2326 on 10 May 1943, it finally approved the employment of able-bodied, rank-and-file P.O.W.s on non-war-related labor projects located outside of their camps. Northern Ontario's forestry industry employed 8000 P.O.W.s, with another several thousand assuming labor positions on private farms.[86] P.O.W.s also worked in urban environments, such as leather tanneries or conservatories, for they were tasked with construction, road work, peat-cutting, and general repair and maintenance tasks.[87] Prisoners' increased mobility outside barbed wire confines afforded local populations unprecedented access to the enemy. Women used the opportunity to speak to work parties on their way to and from projects. Women also visited P.O.W.s at their job sites, sometimes for sexual intercourse. In 1944, for instance, a mother-daughter duo routinely approached the boundary of Chatham's Camp 10 to solicit sex from

[84] Broadfoot, Six War Years, p. 61.

[85] Ibid.

[86] The government did not allow German officers to work as a sign of respect for their status, but rank-and-file soldiers were encouraged. Stefania H. Cepuch, 'The Public and the P.O.W.s: Reaction to the Release of German Prisoners of War for Agricultural Labour' in *Canadian Papers in Rural History* IX, ed. by Donald H. Akenson. (Gananoque, ON: Langdale Press, 1994), p. 327.

[87] O'Hagan, 'Beyond the Barbed Wire,' p. 4.

P.O.W.s.[88] When members of the Veterans Guard of Canada threatened to turn the pair over to the Royal Canadian Mounted Police (R.C.M.P.), the mother became visibly agitated. She reprimanded the military personnel, arguing she 'didn't care about any Mounties' and 'would do as she damned well pleased.'[89] A police investigation confirmed the women had 'been following the Prisoners from work project to work project and have been seen talking to and have left the fields with the Prisoners for sex.'[90] Even after Camp 10 was shut down and the P.O.W.s were relocated to northern Ontario, their relationship continued through a different channel: the exchange of love letters.[91] Similarly, a twenty-one-year-old married woman made daily attempts to visit P.O.W.s working in and around Blenheim, ON, in the fall of 1945.[92] When police searched her vehicle, they found two hurriedly penned letters written by P.O.W.s. One said, 'I know that we belong together for all times. ... My dearest darling, we'll be the happiest couple in the world as long as we live.' She confessed to meeting the P.O.W.s in a field a year ago and promised to end her infidelity.[93]

The government's war labor policy also allowed some Canadian women to mingle with German P.O.W.s at their workplaces. In 1944, for instance, relations developed between female civilian employees of New Toronto's Donnell & Mudge leather tannery, located fifteen kilometers from downtown Toronto, and the thirty-two enemy prisoners who worked there.[94] Though the Department of National Defence told guards to prevent improper associations in the tannery, this was not an easy task to enforce

[88] Ibid.

[89] Q.M.S. Macfarlane, 'Intelligence Summary for October', November 1, 1944, H.Q.S. 7236-94-6-10, T.E.A. - Intelligence Reports - Chatham, C-5416, RG24, Libraries and Archives Canada.

[90] 8A/Cpl. S.W. Green, 'Mrs. Mary VanNeck et al., Blenheim, Ont., Defence of Canada Regulations', October 30, 1944, H.Q.S. 7236-94-5 - Correspondence with R.C.M.P. re Canadian civilians, C-5415, RG24, Libraries and Archives Canada.

[91] H.Q.S. 7236-94-5 - Correspondence with R.C.M.P. re Canadian civilians, C-5415, RG24, Libraries and Archives Canada.

[92] Cst. D.B. McCannell, 'Unknown Person Attempting to Contact German Prisoners of War,' October 26, 1945, H.Q.S. 7236-34-3-55 - Treatment of Enemy Aliens - Dept. of Labour Project - Farming Operations - Dresden, ONT., C-5385, RG24, Libraries and Archives Canada.

[93] Cpl. S.W. Green, 'Unknown Person Attempting to Contact German PW', November 12, 1945, H.Q.S. 7236-34-3-55 - Treatment of Enemy Aliens - Dept. of Labour Project - Farming Operations - Dresden, ONT., C-5385, RG24, Libraries and Archives Canada.

[94] O'Hagan, 'Beyond the Barbed Wire,' p. 280.

on the ground, and fraternization occurred, nonetheless. Canadian postal censors flagged and seized a letter sent from the tannery by P.O.W. Guenter Höppner to the main internment camp, Camp 22, which was located nearby. In it, Höppner told a fellow P.O.W. about a woman named 'Marjory' 'whom he met every afternoon in a hidden corner of their workplace.'[95] Historian Michael O'Hagan states, '[W]hether or not Höppner was exaggerating, POWs were working in close proximity to civilian employees, including women like Marjory, and authorities ... were considerably concerned with the dangers of fraternization.'[96] Several months later in December 1944, officials confiscated fifty-five photographs during an unannounced, facility-wide surprise sweep of the tannery. Thirty-six of the pictures featured young women between the ages of fifteen and nineteen, and ten featured past or present female employees of the tannery.[97] Military authorities had recently fired one woman in the latter group of photos for getting too close with the enemy. Unlike the picture postcards sent by Helene and Hamida, these pictures were taken by amateur photographers, suggesting an additional degree of closeness and familiarity between enemies.

The R.C.M.P. reiterated the potentially negative consequences of enemy encounters to the girls, and three P.O.W.s were stripped of their working privileges. Having prohibited fraternization, internment officials could not tolerate civilian-soldier relationships and did their best to keep German P.O.W.s and local women far apart from one another. In September 1945, the R.C.M.P. ended a romance between Doris Cameron and Heinz Von Haefen by reassigning him to a different work project. Doris was warned not to mingle with any other German P.O.W.s. One month later, Canadian authorities ended another relationship between Ethel Hoffmann and a German P.O.W. in the eastern irrigation district. Camp personnel reassigned the P.O.W. to a Rainier, ON, farm, and warned Hoffmann.[98]

In some instances, these relationships had reproductive consequences. For example, there were several reported pregnancies across Canada. Alberta's Strathmore Legion, for instance, took issue with the fact that

[95] O' Hagan, 'Beyond the Barbed Wire,' p. 309.

[96] Ibid.

[97] O'Hagan, 'Beyond the Barbed Wire,' p. 311.

[98] RG24, Reel C-5415, Letter, H.N. Streight to Commissioner RCMP, 13 December 1945, Libraries and Archives Canada.

two local girls were now pregnant with the children of German P.O.W.s.[99] Likewise, in Medicine Hat, Alberta, a concerned mother sent a letter to the Department of National Defence detailing that 'several girls in or around Medicine Hat have become pregnant through associating with German Prisoners of War' as they labored in surrounding areas.[100] While Canadian military authorities confiscated condoms from P.O.W.s like A. Schiffman when they returned from work in Northern Ontario labor camps, the fact that pregnancies occurred throughout the country suggests the use of contraception was inconsistent among P.O.W.s.[101] However, unlike previous examples in this chapter, we have not found compelling evidence that German prisoners, or their captors, worried about the spread of venereal disease in these relationships. This surprising absence of fear around disease points to how authorities treated sexual interactions between 'racially equal' peoples.

In wartime Ontario, the day-to-day experience of civilians was tightly intertwined with those of German soldiers, sailors, and aviators. When the Second World War ended in Canada on 7 May 1945, Ontario girls and women who had had forbidden contact with the enemy, including those who became pregnant, were not punished. The debasement, gender violence, and public humiliation that plagued women in postwar Europe had no place in Ontario communities. To our knowledge, no woman was publicly marked as a collaborator. We attribute this cultural difference to three things. First, there were no fears of racial defilement or miscegenation where German P.O.W.s were concerned in Canada. Since German P.O.W.s conformed to the widely publicized wartime ideal of Anglo-Saxon masculinity, they were viewed as racially suitable sexual partners instead of threats to Canadian whiteness. This racial similarity allowed women who engaged in sexual activity with German prisoners to be branded not as prostitutes or sexual deviants but seen as young women seeking adventure and casual sex. Second, Ontarians experienced the German military as a defeated force, not an occupying one. Without the brutality and violence of German occupation, citizens felt the Germans in their hometowns were no

[99] R.C.M.P. Report, 'Labour Project - Strathmore, Alberta,' 5 March 1945, H.Q.S. 7236-34-3-58 - Dept. of Labour - Work Project, Strathmore, Alta., C-5386, RG24, Libraries and Archives Canada; Lieberwirth, Alter Mann und Corned Beef, 216.

[100] RG24, Reel C-5396, Letter, Camp 132 to D.N.D., Army, 7 October 1944, Libraries and Archives Canada.

[101] LAC RG24 Reel C-8250, Intelligence Report No. 155, 16 October 1945, Libraries and Archives Canada; Bailey, p. 77.

different from their husbands and brothers. Just like men in the Allied Forces, the Germans had volunteered when their country entered the conflict and fell into enemy hands in the field of duty.[102] Lastly, male leadership understood the appeal of the German soldier in uniform. Canadian men were so impressed by the prisoners' fit, masculine, disciplined demeanor that a handful even pretended to be German P.O.W.s between 1941 and 1945.[103] That despite the knowledge of ongoing Nazi crimes in Europe, Canadians could view the friendly relationships with German prisoners as 'natural' signals the complicated potential of wartime sexual encounters.

Conclusion

This chapter has examined the distinct interplay between war, heterosexual encounters, and racialized thinking in three global conflicts. Sexual encounters have always been part of war, brought about by the movement of people and the relaxing of gender expectations. Encounters between peoples from warring nations occurred in brothels, bars, cafés, homes, and prisoner-of-war camps and intimate encounters took on many forms: some were incredibly violent and constituted rape, and some were deadly. Others were fuelled by passion, loneliness, or by curiosity. Many included a financial transaction. Most wartime encounters were brief and short-lived; others resulted in lasting relationships. Whatever the nature of the contact, sexual relationships between combatants and non-combatants provide a window into a facet of the history of warfare that has, until recently, remained underexamined. They also offer valuable insight for historians of sexuality. As Clare Makepeace maintained, because peoples at war tend to generate more historical sources (letters, diaries, government reports), military events 'can offer the historian a unique insight into behavior and self-conceptions that we simply cannot 'get at' during peacetime.'[104]

[102] Bailey, 'Arrival of the Fittest,' p. 30.

[103] 'Poses as Fleeing Nazi Prisoner, Youth Can't Even Talk German,' *Globe and Mail*, 24 April 1941; 'Reformatory Term for Posing as Nazi,' *Ottawa Journal*, 5 April 1945.

[104] Clare Makepeace, 'Punters and Their Prostitutes: British Soldiers, Masculinity and *Maisons Tolérées* in the First World War,' in *What is Masculinity? Historical Dynamics from Antiquity to the Contemporary World*, ed. by John Arnold and Sean Brady (New York: Palgrave Macmillan, 2011), pp. 413–414.

The three case studies in this chapter expand our understanding of interactions between military forces and local populations and remind us that intersectional power categorizations can operate differently across time and space. In German-occupied Namibia, colonial soldiers used their status as soldiers and white Europeans to perpetrate sexual violence against the Black women they considered their 'racial' inferiors. With these acts, soldiers sought to reinforce military and political dominance over the 'enemy' within a burgeoning settler society. In First World War Egypt, Australian soldier Percy Deane used his position as military and imperial occupier to engage in forbidden sexual relationships with an Egyptian and a French sex worker. In both cases, Deane ignored his own military's warnings about the personal dangers of venereal infections and the national problems they caused, especially in building a 'white Australia.' Lastly, when Canadian women's illegal encounters with German prisoners of war went largely unpunished, this reiterated that the identity of enemy combatants could be overlooked when their 'whiteness' made them acceptable partners. To a Canadian audience, the men's enemy status (even as Nazis) could make them appealing and exciting; in Egypt, soldiers often viewed sex trade workers as 'exotic' yet dangerous; and in Namibia, pre-existing prejudices encouraged German men to consider enemy women as less than human. In short, 'racial' identities often determined how wartime sexual contacts between combatants and non-combatants were received, managed, interpreted, rewarded, or punished across the twentieth century.

'Yoors truly German': A Micro-History of Poetic Enemy Encounters of the 1917 Retreat

Julia Ribeiro S. C. Thomaz

A Different Kind of Encounter—Mediation Through Objects

In Sam Mendes's film *1917* (2019), one of the most memorable scenes sees the character of Tom (played by Dean-Charles Chapman) telling Will (George MacKay) the story of Wilco, their unfortunate friend who had slathered a sweet-smelling hair product on his head only to find his ear bitten off by a trench rat. While this is a very engaging tale, which

This article deals with a poetic source containing racist language. While the use of this wording accurately depicts the colonialist background in which the enemy encounters of the First World War took place, some readers may find the language distressing. It is important to acknowledge the work that still needs to be done in documenting how colonial troops and racialised combatants and civilians experienced the Great War. It is also important to continue to diversify the voices in the field, including more BIPOC scholars and researchers from the Global South in conferences and in publishing opportunities—those speaking about the First World War should do justice to its global and totalising character.

J. R. S. C. Thomaz (✉)
School of Advanced Study, London, United Kingdom

H. Furneaux, M. Greig (eds.), *Enemy Encounters in Modern Warfare*, https://doi.org/10.1007/978-3-031-56748-3_7

familiarises the viewer with the more scatological aspects of the First World War, there is something equally interesting happening in the background: the protagonists are encountering the German enemies for the first time. This encounter does not take place in combat or even through a meeting in person. Rather, Tom and Will arrive in a space previously inhabited by the German troops and encounter the Germans through the objects left behind during the retreat to the Hindenburg line in the spring of 1917. This kind of enemy encounter, mediated by objects left (deliberately or not) to be found, forms the object of the present study. More specifically, this chapter examines a cupboard door with a poem written on it and signed simply 'German'. This artefact is stored in the permanent collection of the *Historial de la Grande Guerre*, a museum and international research centre founded by the cultural historians of the Great War in the town of Péronne, in the Somme region. This choice to focus on a poem held in a museum, valued therefore for its importance as material culture, is part of a larger defence of the study of poetry as a wartime practice that

However, the author would like to highlight some of the most recent works on BIPOC experiences of the war, the majority of which were published after the effervescence of the Centenary commemorations began to dissipate, and which have been added to the First World War Studies Bibliography since 2019: Ella St. George Carrey, 'Red, White, and Black: The Debate over the Active Service of Black Nurses in the United States during the First World War', *Nursing History Review*, 30.1 (2022), 119–132. Anna Branach-Kallas, 'Tirailleurs Sénégalais, Savagery, and War Trauma in *At Night All Blood Is Black* by David Diop', *Journal of War & Culture Studies*, Online First (2021). Bakary Diallo and Senghor Lamine, *White War, Black Soldiers: Two African Accounts of World War I* (Indianapolis: Hackett Publishing, 2021). Manel Herat, *Epistolary Constructions of Post-World War I Identity The Invisibility of Minority Groups* (Springer Nature, 2021). Michael Joseph, 'Black Women, Separation Allowances and Citizenship in the French Caribbean during the First World War', *French History*, (2020). Dan Black, *Harry Livingstone's Forgotten Men: Canadians and the Chinese Labour Corps in the First World War* (James Lorimer & Company, 2019). Kinshasha Conwill, Lonnie G. Bunch, Lisa M. Budreau, and Philippe Étienne, eds., *We Return Fighting: World War I and the Shaping of Modern Black Identity* (Washington: Smithsonian Books, 2019). Santanu Das, Anna Maguire, and Daniel Steinbach, eds., *Colonial Encounters in a Time of Global Conflict, 1914–1918* (London: Routledge, 2021). Tim Grady and Hannah Ewence, eds., *Minorities and the First World War: From War to Peace* (Palgrave Macmillan, 2017).

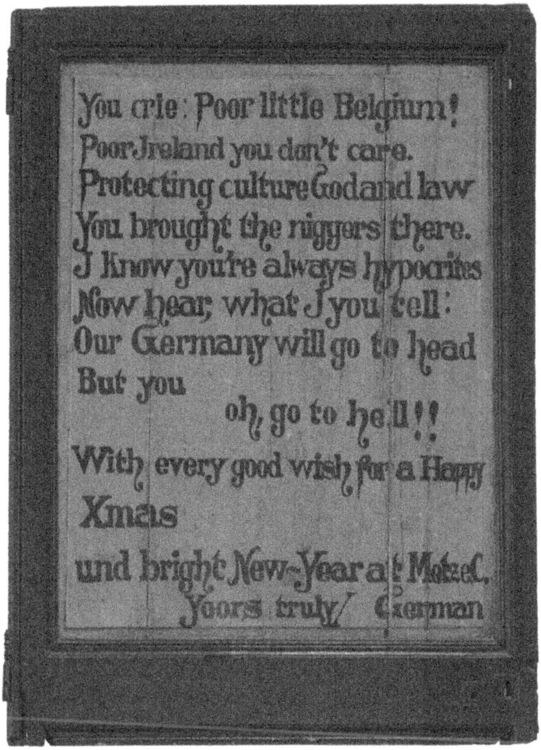

Fig. 1 Wooden cupboard door with a carved poem, Somme c. 1917. Historial de la Grande Guerre—Péronne (Somme) © Yazid Medmoun

accomplished things that reach beyond its aesthetics and literary value.[1] Analysing an anonymous poem 'published' on a cupboard door, in a language and form that were clearly foreign to those who wrote it, is taking a stance against canonicity (Fig. 1). This shows that poetry can mediate enemy encounters not when it functions as a literary genre, but rather

[1] For a larger discussion on canonicity and this approach of poetry as practice, see Julia Ribeiro S C Thomaz, '"Knowing You Will Understand": The Usage of Poetry as a Historical Source about the Experience of the First World War', *Alicante Journal of English Studies / Revista Alicantina de Estudios Ingleses*, 31 (2018), 109–124 and Julia Ribeiro S C Thomaz, 'Poésie Grande Guerre : les humanités numériques et la construction d'un corpus', *Materiaux pour l'histoire de notre temps*, N° 133–134 (2019), 80–85.

when it functions as a wartime practice and as enunciation. Poetry will therefore be examined from an anthropological perspective, instead of primarily from a literary one. This sheds light on what one specific enemy encounter, mediated through poetry, during the First World War can say about enemy encounters in general between 1914 and 1918.

The curatorial team of the *Historial de la Grande Guerre* had access to little information regarding this object, whose sole description was that it had been found during a German retreat. Carved on wood in English using German calligraphy, the poem, which includes a racist slur, reads:

> You cry: Poor little Belgium !
> Poor Ireland you don't care.
> Protecting culture God and law
> You brought the [n-word] there
> I know you are always hypocrites
> Now hear what I you tell:
> Our Germany will go to head
> But you
> oh, go to hell!
> With every good wish for a Happy
> Xmas
> und bright New Year at Metz.e.C.
> yoors truly German[2]

[2] The volume's editors and I thoughtfully debated whether to reproduce this poem in full given its inclusion of an extreme racial slur. As a scholar from the Global South and a direct descendant of the ethnocide perpetrated against the indigenous peoples of Brazil, I am particularly sensitive to the use of racist language and I can understand the argument for not perpetuating racist ideologies in the name of historical accuracy. However, while the last thing I wish for is for my work to be a platform for language that should no longer be in use, I felt it was essential for a full understanding of the effects of this poem that it be seen in full. Seeing such racist language in a poem makes my stomach turn and it should make my stomach turn, because that is what the German poet intended: it is there to signify that his own stomach turned at the sight of colonial troopers. This is a racist text, not only because of the language used or because colonial soldiers are used in a game of accusations between German and British soldiers. It is racist because the Germans equate the use of colonial troopers to the atrocities they are accused of, not as a way of denouncing imperialism but because the poet believed colonial troopers were inferior to white combatants. The affective reaction to the n-word and to all of these layers of racism it carries is an important part of what the poem is meant to do, but it is also an important part of encountering this poem in the twenty-first century, and an integral part of my own research.

Upon encountering the door, one must ask 'why was it written?' and 'what does it do?' This is central to determining the meaning of the poem for, as Peter Stockwell argues, '[w]hen I ask what does the poem mean, I am really asking what the poem does, which is another way of asking what it is being used for. Meaning, then, is what literature does. Meaning is usage'.[3] There are two possible (albeit over-simplified) answers: it was either written to be read or written to be written. In other words, it was either an act of enunciation and a message addressed to the enemy—an enemy encounter—or a cathartic writing practice that the Germans did for fun or for some kind of quasi-therapeutic release. Part of the answer lies in the poem's signature, which indicates where it was written: 'Metz. E. C'. In the French database of urban areas (*communes*), two parishes have names roughly corresponding to these initials: Metz-le-Comte, located in the present-day Bourgogne-Franche-Comté region, and Metz-en-Couture, in the present Pas-de-Calais.[4] Not only does the latter correspond more exactly to the signature of 'Metz E. C' but also there were no First World War combats in Bourgogne, so it is safe to assume that the door was carved in Metz-en-Couture. The town had been under German occupation from the first months of the war until the 10th and 11th King's Royal Rifle Corps captured it in April 1917. It was later evacuated by allied troops in March 1918, during the German offensive known as Operation Michael, and then captured again in September.[5] This poem was therefore likely left behind during the 1917 retreat to the Hindenburg Line, also known as Operation Alberich, and is therefore analogous to the *1917* scene that opens this section. Due to the well wishes for the holidays and the reference to the Easter Rising, it is also safe to assume that this poem

[3] Peter Stockwell, *Cognitive Poetics - An Introduction* (London: Routledge, 2002), p. 4.

[4] Mohamed Badaoui, 'Communes de France - Base Des Codes Postaux - Data.Gouv.Fr' <https://www.data.gouv.fr/en/datasets/communes-de-france-base-des-codes-postaux/> [accessed 17 August 2022].

[5] This information is available through the *Commonwealth War Graves Commission*'s information about the Metz-en-Couture communal cemetery: CWGC, 'Metz-En-Couture Communal Cemetery British Extension | Cemetery Details', *CWGC* <https://www.cwgc.org/visit-us/find-cemeteries-memorials/cemetery-details/32700/metz-en-couture-communal-cemetery-british-extension/> [accessed 17 August 2022]. For contextual elements that place this history within the larger context of the First World War combats, cf. Michael S. Neiberg, '1917: Global War', in *The Cambridge History of the First World War*, ed. by Jay M. Winter, 3 vols (Cambridge - New York: Cambridge University Press, 2014), 1. Global War, 110–132 and Robin Prior, 'Le Front de l'Ouest', in *La Première Guerre Mondiale. Volume I. Combats*, by Jay Winter (Paris: Fayard, 2013), 223–256.

was written towards the end of 1916. The Hindenburg line to which the Germans retreated started being built during the winter of that year, after the Battle of the Somme was halted due to the weather. The German soldiers who carved this poem onto a cupboard door, likely taken from a home under occupation, probably knew that the poem was going to be found by the British troops engaged in the Battle of the Somme. In solving the mystery of the artefact's origin, it is possible to define why it was written: in all likelihood, this poem was written to be read, it is meant to mediate an enemy encounter.

The understanding of this mediation relies on the relationship between the anthropological character of cultural history and micro-historical approaches. This articulation between cultural history and a micro-historical approach of a source perceived as subjective is not unheard of. In the 2002 celebration of the *Historial de la Grande Guerre*'s 10th anniversary, one of its founders, Stéphane Audoin-Rouzeau, argued that shifts in scale are an important heuristic tool (echoing Giovanni Levi, one of the Italian founders of *microstoria*). He claims that the micro-historical approach has been hiding in plain sight in the works of cultural historians of the Great War—constituting what Audoin-Rouzeau calls an '*impensé*'.[6] By prioritising a broadening of the field of First World War studies, cultural history may initially appear to have little to gain from a reduced scale of analysis. The construction of a macro-analytical and transnational category such as the '*culture de guerre*' (war culture) owes a lot, however, to an acute attention to detail, according to Audoin-Rouzeau, especially pertaining to the very singular act of writing. In many ways, one could look at the cultural history of the First World War as an anthropological collection of many different studies that adopt a micro-analytical and ethnographic posture in dealing with the symbolic repertoire of the war. With regards to enemy encounters specifically, this approach is best exemplified in Richard Van Emden's book *Meeting the Enemy*, where, by means of a collection of cases followed through multiple sources (letters, reports, diaries, etc.), the author illustrates how the social, cultural and military ties between Britain and Germany deteriorated but never ceased to exist during the war, being reconfigured by the conflict. Audoin-Rouzeau concludes that the micro-historical approach could deepen or even rectify

[6] Stéphane Audoin-Rouzeau, 'Micro-Histoire et Histoire Culturelle de La Grande Guerre : Apports et Limites d'une Approche', in *Histoire Culturelle de La Grande Guerre*, ed. by Jean-Jacques Becker (Paris: Armand Colin, 2005), 231–238.

some of the concepts forming the cultural history paradigm, while providing the tools for a finer analysis of the totalising character of 1914–1918, testing the validity of explicatory models and determining the levels of variability within the war culture paradigm, which can only benefit from being re-examined from points of view that include long-forgotten voices, such as colonial soldiers and women. The present chapter is inspired by both Audoin-Rouzeau's paradigmatic thought and the case-study approach deployed in works such as Richard Van Emden's, looking at the *culture de guerre* not as an all-encompassing term that ignores marginal experiences of the war but rather as a repertoire that found multiple expressions, including poetic. This 'degeneralisation' can deepen our understanding of the war, but it is even more advantageous in the study of poetic sources, since these are situated in the intersection between the very private realms of inspiration and prosodic expression and the collective poetic culture and literary field.[7]

In the same spirit, the present study of the German poem and the material support on which it was written is used to validate much broader, anthropological hypotheses pertaining to a corpus of approximately one thousand French poets present on the fighting fronts between 1914 and 1918. By examining this corpus, which is anything but micro, one concludes that poetry did things for those writing and reading it during the war, just as it can do things for the historians reading it over a century later. One of the things the poems in the French corpus did was to mediate encounters between those on the fighting front and those at the home front, between combatants and civilians, and even imagined encounters between the living and the dead.[8] From this point of departure of poetry as mediation, it is easy to imagine poetry written in the enemy's language could be used to mediate enemy encounters. By examining an anonymous poem written in English by a German soldier (or soldiers) on a cupboard door, the analysis departs as far as it can from the corpus of published

[7] For a discussion on the relationship between the singularity of the poetic voice and the cultural repertoire or historical events, as pertaining to French prosody, see Laurent Jenny, *La Parole Singulière* (Paris: Belin, 1990) and Jerôme Thélot, *Le Travail Vivant de La Poésie* (Paris: Éditions Les Belles Lettres, 2013).

[8] Jay Winter actually claimed that war poetry was a set of meditations about the dead and their passing, but he also claims that this poetry helped 'keep the fallen alive, by speaking for them, to them, about them'. From this reflection, it is easy to explore the wordplay between meditation and mediation. *Cf.* Jay Winter, *Sites of Memory, Sites of Mourning. The Great War in European Cultural History* (Cambridge: Cambridge University Press, 1995) p. 204.

French poets without abandoning poetry altogether. It, therefore, provides an excellent laboratory to isolate poetic enunciation and validate the poet's role as a mediator, in the sense that poetic form provides a common ground with the reader and an element of historical anchoring, while figurative language allows for a thick description of a kind of warfare that had never been experienced before. This micro-historical approach therefore reveals a lot about what poetry did in, during and to the conduct of war, but it also reveals that not all enemy encounters need to take place 'in the flesh'. It shows that an enemy encounter deferred by writing (which is therefore unilaterally constructed, being determined exclusively by the German poet) draws upon cultural representations of the enemy and thus renders the mechanisms of the construction of this enemy more explicit. Finally, it demonstrates that, within the context of a totalising experience of industrial warfare, where both sides are subject to the same shock when confronting new forms of violence, enemy encounters can create identity as much as they create alterity, corroborating the existence of a shared culture of war, but also demonstrating the cultural underpinnings of enemy encounters in general.

A Deliberate Mediation of an Enemy Encounter

This desire to promote an enemy encounter is corroborated by both the poem's form and content since it has a double intent. On the one hand, there is clearly a mocking tone, but there is also a more profound *jeu de mirroirs* where the German combatants point out that everything they are accused of, especially the atrocities committed during the invasion of Belgium in 1914, has also been committed by enemy troops.[9] It is interesting, therefore, to note that this particular enemy encounter ends up creating identification rather than only reinforcing alterity. Furthermore, in line with an approach stemming from cognitive poetics, the tone used and the formal choices reveal a lot about the author(s) and the expected reader, and this enemy encounter indicates who the German soldiers perceived as being the enemy.[10]

[9] For a deeper discussion on the mythology surrounding the German atrocities as a means to forge the enemy's identity, see the ever-relevant John Horne and Alan Kramer, *German Atrocities, 1914* (New Haven/London: Yale University Press, 2001).

[10] Arnaud Schmitt, 'De la poétique cognitive et de ses (possibles) usages', *Poétique*, 170 (2013), 143–162.

A closer look at the mocking tone employed by the German soldier-writer reveals the elements of the British culture that were accessible to the enemy troops, and which would have formed the cultural background of enemy interactions during the First World War. The language and tone of the poem show that the person or people who wrote it were not familiar with the English language or metre: 'Now hear what I you tell', for example, is a literal translation of German. The tonic syllables, moreover, seem to be placed at random (the first verse, for example, appears to be composed of two iambs and one amphibrach: 'You **cry** : Poor **lit**tle **Bel**gium'). This mixture of types of feet is not common in the English language, and neither is the placement of a tonic syllable outside of a foot, occasionally used by canonical English poets in an attempt to break with monotony.[11] The first verse's metre is, therefore, unexpected. It is, however, very important to note that verse and metre are relational categories, so a verse and its rhythm can only be defined in relation to the other verses of the poem. With this in mind, the second verse is an iambic trimeter ('Poor **Ire**land **you** don't **care**'). It would therefore be possible to interpret the final syllable of the first verse ('gium') as the first syllable of an incomplete iamb (or one can also consider that the German poet chose to consider the word as having three syllables: Bel-gi-um), with the initial two verses being therefore an iambic tetrameter and an iambic trimeter. A stanza with four verses, alternating rimes in two couplets composed of a tetrameter and a trimeter is the basic structure of nursery rhymes and of the English ballad, a type of fixed-form poetry that knew great success during the Great War, especially in trench newspapers and amongst soldiers who had little poetic experience.[12] The same newspapers from which the German soldiers extracted the phrase 'Poor Little Belgium' may have also familiarised them with the ballad structure, which they have attempted to reproduce in the poem. The choice of this musical and quasi-childish metre shows the kind of poetry in English that had been accessible to the German combatants at the start of the war and testifies to the renewal of oral tradition after the start of the conflict.[13] Furthermore, it also plays into the culture of war and

[11] Richard Moore, 'Poetic Meter in English: Roots and Possibilities', *Kansas Quarterly*, 24/25.4/1 (1992), 233–246.

[12] Santanu Das, 'Reframing First World War Poetry: An Introduction', in *The Cambridge Companion to the Poetry of the First World War*, ed. by Santanu Das (New York: Cambridge University Press, 2013), 3–34.

[13] Marc Bloch, 'Réflexions d'un Historien Sur Les Fausses Nouvelles de La Guerre', *Revue de Synthèse Historique*, 7 (1921), 13–35.

the idea that the Germans were fighting for *kultur* because it implicitly criticises British poetry as lacking creativity and artistic maturity, especially when comparing nursery rhymes and ballads to the German exponents of romanticism and expressionism.

This reading of the poem as an approximate ballad offers new interpretive possibilities. According to Benoît de Cornulier, rhymes not only mark the end of each verse but also indicate meaning by arranging a poem into semantic blocks that may or may not correspond to stanzas.[14] Although there are no clear typographically separate stanzas marked on the material support of the door, the rhymes of 'care' and 'there' in the first four verses and then 'tell' and 'hell' in the subsequent ones allow for a division of the poem into two quatrains. This sheds light on the argumentative organisation of the text: the first four verses denounce the British hypocrisy and the contradictory nature of the British involvement in the war, while the next four anticipate a German victory precisely because of this hypocrisy. The German soldiers point out two contradictions: on the one hand, Britain claims to have joined the war to defend Belgium from the German atrocities; all the while, the repression in Ireland is particularly violent. Secondly, Britain claims to be fighting for values such as God, Culture, and Law, but at the same time employs colonial troopers, perceived as being lawless, Godless, and cultureless.

These different racist discourses (as illustrated by the use of the racist slur in the poem) founded on racial hierarchies were used, as Nicolas Beaupré has argued, to mask one's own violence.[15] This corroborates the idea that depicting the enemy in poetry was a way to come to term with one's own contradictory feelings and guilt in the First World War.[16] Despite the mocking tone (or even because of it), this particular enemy encounter, mediated by an object from an occupied region in the North

[14] Benoît de Cornulier, 'Rime « riche » et fonction de la rime : le développement de la rime riche chez les romantiques', *Littérature*, 59.3 (1985), 115–125 <https://doi.org/10.3406/litt.1985.2256>.

[15] Nicolas Beaupré, *Écrits de Guerre* (Paris: CNRS Éditions, 2006), 203. 'Ultimately, these different types of racist discourses, founded on the implicit and widespread idea of a hierarchy between races, can be understood as masking one's own violence. This shift is also a displacement of violence from one's enemy to ones allies, those fighting alongside you. It can therefore, dialectically, be interpreted as an interrogation about one's own violence, not only as a means of attributing the responsibility for this violence to the enemy' [my translation].

[16] David Poynor, 'Meeting the Enemy in World War I Poetry: Cognitive Dissonance as a Vehicle for Theme', *Humanities*, 8.1 (2019), *s.p.*

of France and by poetry, ends up creating a sense of identification between enemies, as the German atrocities seem to find a reflection in the British repression of the Easter Rising and in colonial practices. This is not, however, a fraternising text. Not only is the metre used to mock the enemy and the enemy's inconsistencies placed at the forefront of the poem, but once the Germans establish that the British are not as 'civilised' as they pretend to be, the real argument of the text arises: their hypocrisy will be the cause of the British downfall. The phrase 'oh, go to hell' occupies a central position in the text, both semantically and typographically. In terms of Peter Stockwell's cognitive poetics, the curse 'go to hell' is the real figure of the text, while the rest is simply grounding this dominant message, the focal point from which stems the structural cohesion of the written work.[17] This interpretation of the poem as having two stanzas clarifies both what it says (that Germany deserves to win because England is hypocritical) and what it does (mediate an enemy encounter).

Ordinary Enemy, Extraordinary Encounter

This text inscribes itself in a German tradition of Anglophobia dating back to, at least, the 1880s nationalist movements. As Matthew Stibbe's study of German Anglophobia demonstrates, this intensified in the South African War (or Boer War) of 1899–1902.[18] Stibbe demonstrates that, from the early days of the First World War, England went from being perceived as an enemy amongst many to being blamed as solely responsible for the conflict and an enemy whose defeat was the war's main objective. Indeed,

> [i]n contrast to this bellicose attitude towards England, Germany's traditional enemies, Russia and France, were extended a kind of "benevolent sympathy" for the duration of the war, as the left-wing pacifist Hellmut von Gerlach later put it. The French in particular were often admired as patriots fighting to defend themselves against inevitable decline as a great power. More often than not they were depicted as the mercenaries of England and only very rarely as the instigators of the war.[19]

[17] Peter Stockwell, *Cognitive Poetics - An Introduction* (London: Routledge, 2002), p. 15.
[18] Matthew Stibbe, *German Anglophobia and the Great War, 1914–1918*, Studies in the Social and Cultural History of Modern Warfare, 11 (Cambridge; New York: Cambridge University Press, 2001).
[19] Idem, pp. 16–17.

The violation of Belgian neutrality, which led to the British ultimatum in August 1914 and which is referenced in the poem's opening verse (using the propagandistic phrase 'Poor Little Belgium', which made the headlines of British newspapers immediately before the declaration of war) was presented in the German press as an excuse to set in motion the British project of a continental war. One of the main arguments used to attack the English was the slogan 'Perfidious Albion': the accusation that Britain is historically an unreliable ally, which led to the sacrifice of French and Russian lives and territory for a cause that was not theirs. The notion that the defining trait of English foreign policy is hypocrisy, illustrated in the poem by the contrast between the defence of Belgium and the repression in Ireland, represents a view widely held in Germany. Furthermore, while the collective construction of Anglophobia pre-dated the war and saw a recrudescence in the Autumn of 1914, the writing of this poem coincided not only with the retreat to the Hindenburg line but also with a new Anglophobic surge, marked by an increase in popular demand for unrestricted naval war against the United Kingdom in the winter of 1916–1917. This text is, therefore, the product of a long tradition of anti-Englishness in Germany, while also very much in line with the most recent shifts in the public opinion after two years of fighting.

The holiday wishes at the end of the poem, when read as an expression of German sarcasm, reinforce these shifts. Whereas both armies had engaged in fraternisation—which entered posterity in the English-speaking world as a defining moment of enemy encounters despite being less acknowledged in Germany and France, during the Christmas celebrations of 1914 and during the lesser known, unauthorised but nonetheless present Christmas truces of 1915—this kind of encounter can no longer be envisaged in 1916.[20] As Malcolm Brown puts it,

> [i]f we look at the Western Front canvas overall, it is impossible not to be aware that a conflict that began in the relatively innocent days of 1914 ended in the grimmer, harsher days of the all-out assaults of 1918. The concept of enemy in close contact with enemy, which might mean a shaking of hands in 1914, was by this time far more likely to revolve around the mutual

[20] For a more detailed account of the Christmas truces, see Anthony Richards, and Eva Burke, *The True Story of the Christmas Truce: British and German Eyewitness Accounts from World War I* (Greenhill Books, 2021). For further discussions about how this truce relates to a larger context of enemy encounters during the First World War, see Richard van Emden, *Meeting the Enemy: The Human Face of the Great War* (A&C Black, 2013), pp. 78–83.

> thrusting of bayonets. No poem of that later period catches the harsh irony of this sea-change better than the famous *Strange Meeting* by Wilfred Owen in which two enemies recently caught in the destructiveness of mortal combat meet afterwards in the macabre fraternization of death.[21]

Written in the first person, Owen's poem adopts the point of view of a soldier who believes he has managed to escape the battle by going down a tunnel. In the second stanza, however, the soldier realises the truth: he is in hell, along with other fallen combatants. The protagonist tries to console one of the other dead he meets in hell, a 'strange friend' who then replies (in direct reported speech) that the war, in all its pity, has destroyed the unlived lives of both men. This speech creates an identification between the soldiers, who are mourned not only for the past to which they look with melancholy but also for a future that will never take place. This identification, however, takes an unexpected turn in the final stanza of the poem, where it is revealed that the 'strange friend' the soldier meets in hell is in fact the enemy he had killed the day before.[22] While this is indeed representative of the escalation of the hatred of the enemy and of the pitiful nature of the conflict in final years of the war, the poem examined in this chapter is a testament of the transitional moments of 1916–1917. The paradigm-altering battles of 1916 (the Somme and Verdun) had already taken place, as had the gas attacks of 1915, but while fraternising as these armies had known in 1914 and 1915 was no longer possible, a certain level of irony and 'banter' between the German and the British could still take place. In this sense, poetry can offer a framework in which enemy encounters are contextualised in all their complexity.

The racial dimension of the poem is also a part of larger context of propaganda efforts present in Germany even before the official declaration of war. One of the underlying elements of German Anglophobia, according to Matthew Stibbe, was racism. Before the war, this was manifested in an amalgamation of anti-Englishness and anti-Semitism, with journalists and intellectuals denouncing an alleged conspiracy led by the rich Jews of the City of London to destroy continental Europe, thereby halting German economic progress. After the arrival of colonial troopers on

[21] Marc Ferro, Malcolm Brown, Rémy Cazals, and Olaf Mueller, *Meetings in No Man's Land: Christmas 1914 and Fraternisation in the Great War* (Hachette UK, 2013).

[22] Michael Sarnowski, 'Enemy Encounters in the War Poetry of Wilfred Owen, Keith Douglas, and Randall Jarrell', *Humanities*, 7.3 (2018), 89, *s.p.*

European soil during the First World War, the racial hatred's focus became the idea of England as a traitor to the white race. Furthermore, by allying itself with Russia and supporting Japan's declaration of war against Germany, England became the target of accusations of lacking a sort of 'racial scruple' in its quest to dominate Europe economically, as illustrated by numerous racist cartoons in the German press. One of these cartoons, published in the satirical journal *Kladderadatsch* in June 1915, shows the figure of John Bull (frequently used to represent Britain in the German press) as a dark-skinned, terrifying man, wearing both adornments traditionally associated with Africa (such as a nose cuff) and an earring shaped as a coin bearing the image of George V. This image highlights Britain's alleged treason of the white race in deploying colonial troops on European soil, which seemed to violate a taboo, but also denounces the greed that has led Britain into war. The cartoon's caption reads 'In view of the colour of the English who face the Germans in Flanders, it would appear to be time to correct the figure of the nation personified'.[23] In the opening chapters of his book *The World's War*, David Olusoga demonstrates that colonial troopers were indeed present in the Ypres region of Flanders in 1915 when chemical weapons were first used. To bring colonial troopers to fight in Europe was perceived as a treason to the white race in the sense that, in a context influenced by social Darwinism, it put white superiority at risk. A German wartime essay written by Bernhard Dernburg, former Secretary of State for Colonial Affairs, states:

> Just as the belligerents in Europe are divided by nationality, so people are divided by race in the colonies; and, just as closer ties bind nationalities and nations, so there is a community of races… in the colonial domain, every member of the white race is answerable to every other for the maintenance of his purity, culture and prestige of this greater community… But as, in the colonies, it is a question of dealing with great masses of undeveloped beings, far superior to the whites in number and not united among themselves, this task of the colonizer can be accomplished only if he succeeds in maintaining the prestige of the white race morally and culturally. If the white man is looked upon as mentally superior, on a higher plane economically, superior in weapons and power, the natives will decide that to render obedience to him is not only necessary, but wise. That is what is called the prestige of the white race. It is based on the native's belief that the will of the white man is good, unshakable, unconquerable.[24]

[23] Cited in Matthew Stibbe, *German Anglophobia and the Great War, 1914–1918*, Illustrated edition (Cambridge ; New York: Cambridge University Press, 2001), pp. 41–42.

[24] David Olusoga, *The World's War: Forgotten Soldiers of Empire* (Head of Zeus, 2014), p. 119.

The title of the essay could not be clearer: *England: Traitor to the White Race*. Bringing these natives to Europe and allowing them to share in the Europeans' misery imposed by trench warfare risked removing the mystique with which Dernburg believed the indigenous peoples of the colonies should regard the colonisers. A similar argument is made by Woldemar Schütze in his essay *Englands Blutschuld gegen die weiße Rasse* (England's blood debt towards the white race), published in Berlin in 1914: 'In this way [involving colonial troopers in the conflict] the English have burdened themselves with a terrible guilt, not only towards Germany and Austria-Hungary, but also towards the entire white race, including their own allies...[They] have transformed the struggle between nations of one and the same race into a struggle between races, and the European war into a world war'.[25] The representation of the enemy in the poem as a greedy traitor is, therefore, not novel. Indeed, it is perfectly in line with a broader, Anglophobic trend that was part of the German culture of war. The Englishman the German soldier or soldiers who wrote this poem imagined they were encountering was the ordinary, stereotypical archenemy of the German Reich.

Poetry had traditionally been used in anti-English propaganda in Germany. The warning and apocalyptic tone of the second half of the poem has been present in popular German poetry since, at least, November 1914, when Friedrich Weidig published the following stanzas:

> Beware England, beware, for waves of fire are closing in on you,
> The sea is surging at your shores,
> The sea which you basely abuse
> To commit crimes and robberies in all corners of the earth.
>
> Woe to you England, a day of judgement awaits you,
> It will crush you, destroy your bloody empire,
> Which you so treacherously created –
> From the life and property, the possessions or murdered nations.[26]

The final verse makes for an interesting object of analysis, as Germany seems to take an anti-colonial stance when it maintained several colonies in Africa: Togo, Cameroon, German Southwest Africa, and German East

[25] Waldemar Schütze, *Englands Blutschuld gegen die weiße Rasse* (Berlin, 1914), cited in Matthew Stibbe, *German Anglophobia and the Great War*, *op. cit.*, pp. 39–40.

[26] Cited in Stibbe, Matthew, *German Anglophobia and the Great War*, *op. cit.*, p. 37.

Africa. The final combat of the Askari (German colonial troopers) actually took place in Zambia on 12 November 1918, a day after the armistice in Europe.[27] Both the poem analysed in the present chapter and Weidig's are works embedded in colonialist rhetoric, not only because of the clearly racist tone of the 1916 poem but also because the situation of colonial troopers is instrumentalised in the definition of two European nemeses. Soldiers would have had access to the poetry collection *Gewittersegen*, a 1915 reprint of Otto Ernst's 1914 collection *Deutschland an England*, subtitled 'war book'. In this collection, aimed specifically at those on the fighting fronts, Germans would have read poems dehumanising the British and denouncing their pride and lack of loyalty to their allies. The poem left behind by the German soldiers, though unique as a mediation between enemies and because of its ironic and playful tone, stems therefore from a cultural and propagandistic context that portrayed the English as Germany's main enemy and accused them of turning a European war into a world war and therefore betraying the white race.

This does not mean, however, that this artefact is a mere expression of German Anglophobia and that the micro-historical approach simply corroborates what cultural historians already knew. Contrary to the political discourses and to what was being said in the German press, by intellectuals and in the arts (Stibbe describes a poem that was recited in schools from as early as the first months of the war and which states 'Fly Zeppelin! Fly to England! England shall be destroyed by fire!), this poem is not straightforwardly a work of propaganda, as it is not meant to influence public opinion. Fiona Houston has identified the First World War as a pivotal moment, when the notion of propaganda lost its theological connotations and started to integrate the political lexicon.[28] She claims that 'it is important to acknowledge that what many authors wrote during the First World War was propaganda, it was produced with the intention of influencing public opinion and supporting official aims'.[29] Her literary examination of the works of authors who fought in 1914–1918 is also a conceptual one, as she offers a definition of propaganda as it was understood during the

[27] Brian Digre, 'Colonial Warfare and Occupation (Africa) | International Encyclopedia of the First World War (WW1)' <https://encyclopedia.1914-1918-online.net/article/colonial_warfare_and_occupation_africa> [accessed 22 October 2022].

[28] Fiona Houston, '"Seducers of the People": Mapping the Linguistic Shift', *Alicante Journal of English Studies/Revista Alicantina de Estudios Ingleses*, 31 (2018), 33–52 <https://doi.org/10.14198/raei.2018.31.03>

[29] Idem, 50.

Great War: work produced to either influence the public opinion or support the war effort. The enemy encounter studied above is therefore unique in the sense that it is unilaterally determined by propaganda (and does not react to the enemy himself, who could have nuanced cultural stereotypes if encountered in person). Furthermore, it uses the same medium of many of the propaganda efforts disseminating Anglophobia in Germany in the early years of the twentieth century (writing and, more specifically, poetry, as illustrated above). It was not, however, meant to influence Germans in taking an anti-English stance. On the contrary, it was addressed to the English themselves. While it is easy to imagine that the catharsis provided by the composition of this poem did support official aims in the sense that it had an effect on combatant morale, it was an enunciation directed towards the enemy, a deferred enemy encounter. It attests not only to the effectiveness of propaganda in determining the vision of Britain that shapes this enemy encounter (a hypocritical racial traitor) but also in determining the support through which the encounter is mediated. On the one hand, it shows that enemy encounters struggle to depart from the content and the form of a propagandistic hatred of the enemy, which has been identified as one of the bases of the culture of war. On the other hand, this poem also demonstrates that, while enemy encounters may be influenced by propaganda, they also tend to recuperate codes that are familiar to the enemy addressed, in this case prosody. This is made evident by the fact that this particular encounter takes place through a written medium, but it may be true of enemy encounters in general. The ballad structure of the poem reinforces the mocking tone of the text as a whole, making it understandable to the English reader, but also demonstrates that, despite the prevailing Anglophobia, Germany was not hermetically sealed to British cultural production during the war. Indeed, the poem's form seems to depart from the Romantic poetry (Wordsworth, but also Burns, Byron, Moore, Scott, Coleridge, and Felicia Hemans) which circulated in Germany before the conflict, while reproducing the structure of ballads, often published by amateur poets in trench newspapers.[30] This hypothesis, coupled with the content of the poem,

[30] For a case study of Wordsworth as exemplifying the German reception of British poetry in general (and the Romantic poets cited in particular) in the nineteenth century and immediately before the First World War, see John Williams, *Wordsworth Translated: A Case Study in the Reception of British Romantic Poetry in Germany 1804–1914*, 1st edition (Continuum, 2011).

where the German soldiers merely point out that everything they are accused of also applies to the British, renders this an extraordinary encounter, albeit with an ordinary enemy. It sheds light on the fact that even the most propagandistic, hate-filled, and unilateral enemy encounter rests on some kind of symbolic common ground from which messages can be transmitted. In other words, enemy encounters establish as much identity as they do alterity, and a case study of an encounter mediated by poetic prosody renders this particularly visible.

* * *

What do these thirteen lines tell us about enemy encounters? First of all, that they are always mediated. In this particular case, this mediation is more evident, as it happens both through an object where one would not expect to see a poem and that was deliberately left behind to be found by the enemy, and because it involves prosodic language. At the intersection of materiality and textuality, the elements of typography and meter have proven central to interpreting this particular enemy encounter while also contextualising it within larger usages of poetic models during the First World War, which open the field for a transnational study of the circulation of poetry across no-man's land. However, this particular case study sheds light on more general elements of enemy encounters: even when the models being used are less explicit than the mocking of the enemy through a nursery-rhyme poetic form, every enemy encounter is semantically built through a framework of meaning that those encountering an enemy 'other' use to configure identity and alterity. This framework is both contextual and pertaining to the ongoing war and stems from previous traditions.

If all enemy encounters are therefore mediated by language and through models that are socially and culturally shared through language, the written character of this particular enemy encounter ensures that it is not constrained to a specific 'here and now'. Therefore, the instant in which the Germans encounter the British through writing is not the same as that in which the British encounter the Germans through the objects and poems left behind. The German soldier or soldiers who wrote the poem were thereby in full control of this intentionally deferred enemy encounter, rendering this object an interesting source to examine how enemies build the significance of these encounters. Furthermore, poetry has the particularity of rendering the structures that create meaning visible, so one can clearly see how these structures are used to create identity and alterity.[31] The poem actually suspends the otherness, supporting George Mosse's thesis that brutalisation was shared across both sides of no-man's land.[32] The explicit way in which poetry in general and this poem in particular attribute meaning to the enemy sheds light on how these encounters reinforce differences between enemies but also establish identification to a certain degree.

The 13 lines of this poem have the power of telling the story of how Germans perceived British poetry, of how they perceived the atrocities they were accused of, and of this very particular, unilaterally determined, enemy encounter which took place during the 1917 retreat. More importantly, they show that enemy encounters are plural in form and do not depend on adversaries occupying the same place at the same time. They are dependant, however, on two different culturally determined elements: the vision one has of the enemy as an 'other' and all that which is shared with this 'other' and that allows the encounter (understood as communication) to take place. Poems, when read outside of the framework of canonicity and literary value, are powerful sources that render visible the cultural scaffolding of enemy encounters through prosodic uses of language. This structure is, however, present any time enemies encounter each other.

Acknowledgements The author would like to acknowledge the *Historial de la Grande Guerre* for rendering this study possible and for their continuous support of her work, particularly through the Gerda Henkel Scholarship. She would also like to thank Nicolas Beaupré for the discussions surrounding the door and for introducing her to Stibbe's work. Finally, she would like to acknowledge Stéphane Audoin-Rouzeau, who was the first reader of this piece, when he was the author's MA supervisor at the *Ecole des Hautes Études en Sciences Sociales*.

31 Jerôme Thélot, *Le Travail Vivant de La Poésie* (Paris: Éditions Les Belles Lettres, 2013).

32 George L. Mosse, *De La Grande Guerre Au Totalitarisme. La Brutalisation Des Sociétés Européennes* (Paris: Hachette Littératures, 1999).

Dying with the Enemy: Prisoners of War Deaths in First World War Britain

Tim Grady

In 1906, William Le Queux penned his sensational story of the German military overrunning an ill-prepared and vulnerable Britain. After landing in the Yorkshire towns of Goole and Hull, the German troops advanced rapidly, storming through villages and towns across the country. They raised the 'German ensign' over Sheffield's Victorian town hall, before sweeping west and then south to seize Manchester and Birmingham in quick succession.[1] The genre of invasion literature, of which Le Queux was the master, certainly helped to build an image of Germany as a potential enemy.[2] At the same time, diplomatic and political rivalries with Germany, whether through colonial expansion, international alliances or

[1] William Le Queux, *The Invasion of 1910* (London: Macmillan, 1906), pp. 154, 276, 287.

[2] Ailise Bulfin, *Gothic Invasions: Imperialism, War and Fin-de-Siecle Popular Fiction* (Cardiff: University of Wales Press, 2018); Joseph Meisel, '"The Germans are Coming!" British Fiction of a German Invasion, 1871–1913', *War, Literature and the Arts*, 2 (1990), 41–79.

T. Grady (✉)
University of Chester, Chester, UK
e-mail: t.grady@chester.ac.uk

H. Furneaux, M. Greig (eds.), *Enemy Encounters in Modern Warfare*, https://doi.org/10.1007/978-3-031-56748-3_8

the naval race, all cemented the growing 'Anglo-German antagonism'.[3] However, public fears of an enemy invasion had very little to do with actual Germans. Instead, their concerns were always about an imagined enemy, one that had been drawn from sensationalist novels, newspaper reports and political machinations. In everyday life, where 'zones of contact' existed between the British and German populations, relations could even be constructive, even friendly.[4]

All of this changed as soon as the first salvos of the First World War were fired. The German military never followed Le Queux's premonition to the letter and failed to land their troops in Goole or Hull, which meant that the enemy still remained a largely imagined one. Yet, while the enemy may never have set foot on British soil in the way that Le Queux and others predicted, the start of hostilities did nonetheless provide the public with an opportunity to hunt out actual Germans at home. Spy fever, which was particularly rife in the first months of the war, saw innocent people accosted for supposedly acting in a suspect manner. 'Anything German, and anyone thought to have the least sympathy with the Germans', as Catriona Pennell rightly observes, 'was accused of espionage'.[5] Such widespread fears of *The Enemy in our Midst* also had a dramatic effect on the daily lives of civilians of German heritage.[6] By autumn 1915, some 32,000 men, many of whom had lived in Britain for decades, had been arrested and placed behind wire.[7] As the conflict dragged on, military prisoners taken on the front-

[3] Paul Kennedy, *The Rise of the Anglo-German Antagonism, 1860–1914* (London: Allen & Unwin, 1980); James Joll, *The Origins of the First World War* (London: Longman, 1984).

[4] David Blackbourn, '"As dependent on each other as man and wife": Cultural Contacts and Transfers', in Dominik Geppert and Robert Gerwarth (eds), *Wilhelmine Germany and Edwardian Britain: Essays on Cultural Affinity* (Oxford: Oxford University Press, 2008), pp. 15–37.

[5] Catriona Pennell, *A Kingdom United: Popular Responses to the Outbreak of the First World War in Britain and Ireland* (Oxford: Oxford University Press, 2012), p.102. See also: David French, 'Spy Fever in Britain, 1900–1915', *The Historical Journal*, 21 (1978), 355–370.

[6] Panikos Panayi, *The Enemy in our Midst: Germans in Britain during the First World War* (Oxford: Berg, 1991).

[7] Panikos Panayi, *Prisoners of Britain: German Civilian and Combatant Internees during the First World War* (Manchester: Manchester University Press, 2012), p. 51. On civilian internment, more generally, see: Matthew Stibbe, *Civilian Internment during the First World War: A European and Global History, 1914–1920* (Basingstoke: Palgrave Macmillan, 2019); Stefan Manz, *Migranten und Internierte: Deutsche in Glasgow, 1864–1914* (Stuttgart: Franz Steiner, 2003).

lines joined these so-called enemy aliens in the entangled webs of Britain's wartime internment system.[8]

The image of the enemy, as the existing literature stresses, was largely an imagined one; externally, Germans were ready to invade Britain, while at home a dangerous group of spies were all set to sap Britain's strength from within.[9] This chapter takes these imagined encounters a stage further and explores what happened when the enemy became fleetingly real, but then expired. In short, it considers the treatment of the German dead in First World War Britain. Perhaps because of their relatively small number, with only around 2700 deaths recorded, historians have largely overlooked this aspect of the war.[10] The one exception comes from Panikos Panayi's seminal work on internment, which discusses prisoner deaths, but largely in statistical terms.[11] This chapter, in contrast, argues that the presence of the dead played a crucial role in altering British perceptions of the enemy, helping people to rehumanise their previously imagined foe. As such the German dead, to follow the words of Thomas Laqueur, remained 'active agents in this history' of British–German relations.[12] The chapter begins with the moment of death itself, moving then through the mortuary stages of burials and funerals, naming and inscription, before finally considering the care of the enemy graves.

Death

In an edition from October 1915, *The Illustrated London News* published 'remarkable photographs' of recent French offensives on the Western Front. To emphasise the strength of these 'victories', the newspaper included several images of ravaged German trenches. The most striking of

[8] Brian Feltman, *The Stigma of Surrender: German Prisoners, British Captors and Manhood in the Great War and Beyond* (Chapel Hill: University of North Carolina Press, 2015); Heather Jones, *Violence against Prisoners of War in the First World War: Britain, France and Germany, 1914–1920* (Cambridge: Cambridge University Press, 2013).

[9] Pennell, *Kingdom*, p. 230.

[10] The exact number of deaths differs between British and German records; see: 'German Graves of 1914/18 in the United Kingdom', 27 August 1953, The National Archives, Kew [Hereafter, TNA], FO369/5162; 'Zahlenmässige Zusammenstellung der deutschen Kriegsgräber in England', 22 February 1956, Volksbund Deutsche Kriegsgräber archive, Kassel [Hereafter, VDK], A100/975.

[11] Panayi, *Prisoners of Britain*, pp. 129–130.

[12] Thomas Laqueur, *The Work of the Dead: A Cultural History of Mortal Remains* (Princeton: Princeton University Press, 2015), p. 18.

these shots depicted the body of a German soldier, apparently killed 'while aiming at a French officer'. His body lay slumped forwards, firmly wedged between the sides of the narrow-fortified position, with his rifle, no longer a threat, strewn on the floor in the foreground.[13] This visual illustration of a dead soldier, bearing a striking resemblance to the images of a slain 'sharpshooter' in the American Civil War, helped to place wartime death at a safe distance.[14] In the British photographs of the dead German soldiers, this occurred in two ways. First, it was the enemy and not the British or their Allies that had suffered losses. Second, the dead were spatially confined to the battlefields. What *The Illustrated London News* and other publications conveniently ignored was the death of the enemy at home. Yet, throughout the conflict, British civilians encountered the dying enemy not on some distant battlefield, but in their midst, meeting not a vicious fighting force, but fellow humans gasping for their last breath.

Death on British soil took multiple forms. Some German combatants were killed in the air or at sea around Britain, but far more common was the death of prisoners of war (POWs) who arrived in Britain already sporting combat wounds and never recovered. War hospitals located around the country dealt with both allied and enemy soldiers, split, where possible, into separate wards for treatment. In the largest of these facilities, Dartford War Hospital in Kent and Nell Lane Military Hospital in Manchester, there was a steady stream of German deaths throughout the war. Gerhard Pohl, like many of those to die, succumbed quickly to severe wounds. In Pohl's case, he had shrapnel in the lungs and gunshot wounds to his jaw, neck, and back.[15] On a few rare occasions, POWs even died from British bullets, shot at the fences surrounding internment camps while trying to escape their captivity or during moments of unrest.[16] One of the earliest incidents occurred in the Leigh internment camp, west of Manchester. One evening, Wilhelm Karl Schmidt, originally from Spandau,

[13] 'Remarkable Photographs of the French Offensive', *The Illustrated London News*, 16 October 1915, pp. 492–493.

[14] Anna Jardine (ed.), *Photography and the American Civil War* (New York: Metropolitan Museum of Art, 2013), pp. 96–97.

[15] Prisoners of War Information Bureau, 24 September 1918, Bundesarchiv Berlin [Hereafter: BArch Berlin], R67/852.

[16] The most serious disturbance incident resulting in the death of five internees occurred on the Isle of Man in 1914: 'Disturbance at the Aliens Detention Camp at Douglas on Thursday, November 19, 1914. Official Report of the Proceedings' (Douglas, 1914), Manx National Heritage, Douglas, M06104/71.

had climbed onto a corrugated steel roof when a guard spotted him. After getting no response to his calls, the guard picked up his rifle and shot Schmidt through the side. The British called it an escape attempt; the Germans, in contrast, claimed Schmidt had merely been trying to clamber back to his side of the camp.[17]

Shootings, however, were rare. It was far more common for military combatants, as well as German civilian internees, to die from sickness and disease that spread through the confines of the camps. The influenza pandemic of 1918–1919 in particular quickly spread from prisoner to prisoner, leaving a trail of death in its wake.[18] The Donington Hall internment camp for German officers and the neighbouring Sutton Bonington Prisoner of War camp were fairly typical in this regard. After passing through the barbed wire gates in late October, the virus had taken its first victim by the month's end. Erwin Kazenmaier, a Leutnant der Reserve, died within days of being first infected.[19] Over the following two weeks, a further 33 men followed Kazenmaier to the grave. Finally, accidents at work, while labouring in fields or in mines, accounted for another tranche of enemy deaths in wartime Britain. Typical in this regard was the case of Albert Krämer, a 29-year-old German prisoner, who died in a Northumberland quarry in 1917. Krämer had been captured the previous year with gunshot wounds to his legs and chest. After being nursed back to health, he was then crushed to death by a falling pile of rocks in the quarry. With little sympathy, the coroner declared the accident was 'entirely' Krämer's 'own fault'.[20]

Regardless of the actual cause of death, what united all of these cases was their very visibility. When the enemy died at home, they did so with British civilians often looking on. The most publicised German losses occurred with the downing of four giant Zeppelin airships over Britain in 1916 and 1917. Gliding silently over the North Sea to their mainland

[17] 'German Prisoner Shot Dead', *Wigan Observer*, 1 June 1915, p. 3. Interview with Otto Schlagk, 12 December 1917, Bundesarchiv-Militärarchiv, Freiburg, PH2/588.

[18] In Dorchester POW camp, 26 men died during the main waves of the influenza pandemic: Brian Bates, *Living with the Enemy: Dorchester Great War Prison Camp* (Frampton: Roving Press, 2016), pp. 71–72.

[19] 'List No. 230 of German Prisoners of War', A34171, 'International Committee of the Red Cross': https://grandeguerre.icrc.org/en/List/5557524/898/34171/

[20] 'German Prisoner Killed in Stone Quarry', *Shields Daily News*, 25 June 1917, p. 3. 'List No. 98 of German Prisoners of War', A2125, 'International Committee of the Red Cross': https://grandeguerre.icrc.org/en/List/4005325/898/2125/

targets, the Zeppelins had been a menace to urban areas mainly in the south and east of the country since the second year of the war. By the war's end, Zeppelin raids over Britain had cost the lives of some 557 people and injured many more. However, it was not just fear that accompanied the Zeppelin raiders. Whenever people spotted the airships overhead, there was also fascination. During a raid in September 1916, Londoners jostled for the best spot, reportedly wanting 'to see as much of the show as possible'.[21] If nothing else these dangerous raids offered civilians a rare opportunity to view the war and the enemy itself up close.

Public interest in the Zeppelins reached ever greater heights whenever one of the giant machines was seen floundering. In Cuffley, Hertfordshire, the father of five-year-old Doris Cobham, even woke her in the middle of the night just so she could see a German airship being attacked in the sky above.[22] Further east in Suffolk, 'hundreds of people [also] left their beds', remembered one local resident, 'to watch and cheer the fateful descent' of another German airship.[23] There was understandable relief at the Zeppelin's demise, but its destruction also marked the moment of death for the airship's crew. The first person to reach the field outside of Theberton, where the Zeppelin hit the ground, was confronted with a burning wreck of entangled metal and a lone body in the field. 'He was an awful sight', the local man later recalled, the body 'literally fell to pieces'. With the help of a policeman, he managed to drag three German airmen from the flames, burnt with broken limbs, but just about alive. When other locals reached the field, they also came face to face with the brutal injury and death of the enemy, particularly as the charred bodies of the Zeppelin's 15 remaining crew members were freed from the wrecked airship.[24]

The three survivors of the Theberton crash were taken to a military hospital in Ipswich for treatment.[25] Carl Mieth, one of the Zeppelin's officers and one of the airship's three survivors, later recalled coming round on the ward to discover his legs were wrapped in 'thick bandages' and

[21] 'The Public and the Zeppelin Raids', *Lancashire Evening Post*, 26 September 1916, p. 2.

[22] 'Cuffley Hertfordshire: First German Airship Shot Down on British Soil', BBC Radio 4, 'World War One at Home', (2014): https://www.bbc.co.uk/sounds/play/p02b8qgg

[23] Notes on Destruction of L48, Suffolk Archives, Ipswich, HD1295/2/33.

[24] Untitled Newspaper, June 1961, Suffolk Archives, HD1295/2/33.

[25] Mark Mower, *Zeppelin over Suffolk: the Final Raid of the L48* (Barnsley: Pen & Sword, 2008), pp. 89–90.

burns covered his head and limbs.[26] From Ipswich, Mieth was taken to an officers ward in the Dartford War Hospital to recuperate. It was in Kent that the Swiss Legation, who inspected the POW camps, later discovered Mieth apparently 'doing very well' and with 'no complaints to make about anything'.[27] The British, mainly female, nurses charged with caring for Mieth and the other enemy soldiers generally embarked on their task with some trepidation. It was one thing to treat the British wounded, but quite another to nurse the enemy. After all, repairing the broken bodies of German soldiers seemed to undermine their patriotic wartime duty. However, as Alison Fell observes, the confined spaces of war hospitals acted as 'contact zones' in which nurses gradually viewed the 'German patients as suffering bodies'.[28]

The nurses' role, however, often extended far beyond basic medical care, requiring them also to provide comfort for the dying German soldiers too. Such was the severity of wounds that nurses could do little more than to prepare the patient, even if he was the enemy, for death. This process can be discerned from the POW medical notes that accompanied every patient. In the Oswestry Prisoner of War Hospital in Shropshire, for example, the notes for one young soldier explained that he had 'no chance of recovery'. Confronted with the hopelessness of his situation, the nurses reported that they did everything to 'relieve his pain and make him comfortable'.[29] In a different hospital, one nurse even recalled her final hours with a German soldier whose condition deteriorated rapidly after an operation. 'The hand that still held mine grew colder', she wrote, 'the breathing stopped, and I knew my prisoner had been released'.[30]

It was not just the war wounds that needed treating; British medical personnel also encountered the enemy because of sickness and disease. Most POW camps had their own dedicated sick bays for treating more minor ailments, which local doctors often visited as part of their rounds. In the Brooker Hall POW camp in Hove, for example, a Dr Smith passed

[26] Carl Mieth, 'L48 stirbt', in Werner von Langsdorff (ed.), *Flieger am Feind. Einundsiebzig deutsche Luftfahrer erzählen* (Gütersloh: Bertelsmann, 1934), pp. 33–34.

[27] Swiss Legation, 'Dartford', 30 July 1917, BArch Berlin, R69/909.

[28] Alison Fell, 'Nursing the Enemy in the First World War', *European Journal for Nursing History and Ethics*, (2022), 1–23, (pp. 14–18). For the origins of this influential term, see: Mary Louise Pratt, 'Arts of the Contact Zone', *Profession*, (1991), 33–40, and for its legacies see the introduction to this volume.

[29] Prisoners of War Information Bureau, BArch Berlin, R67/909.

[30] 'Nursing the Enemy', *Hull Daily Mail*, 19 December 1914, p. 3.

by three times each week to check on the German patients.[31] During the 1918 influenza pandemic, when medical services were particularly stretched, these local doctors also ended up tending to the dying enemy. The close confines of the camp environment meant that the virus spread rapidly between the prisoners, leaving a trail of death in its wake. One of those to die in this way was the third survivor of the Theberton Zeppelin crash, an engineer named Wilhelm Uecker. After several months of recuperation, Uecker had started to recover, but then in November 1918, he was struck down by the pandemic, dying in the Brocton War Hospital, Staffordshire, some 18 months after the initial crash.[32]

The influenza deaths occurred under the watchful eyes of medical personnel either in camp sick bays or military hospitals. More rarely, British civilians also found themselves on the scene at the moment of the enemy's passing. Fatal work accidents, for example, could happen anywhere from the field to the factory floor, particularly where safety measures were lax or entirely absent. At the National Shipyard Beachley, a mast being lowered onto a newbuild ship fell striking a German prisoner and British worker below, killing the pair instantly, while in Frimley Surrey, home to a large POW camp, another German died in a public setting.[33] The prisoner had been in the back of a lorry that was taking timber to the town's railway station, when it swerved to avoid an oncoming milk cart. The sudden jolt was enough to throw the German prisoner off balance, as he fell his head then struck a lamppost on the side of the road. A telegraph boy and local farmhand who were first on the scene tried to help, but the man died later in hospital.[34] Throughout the conflict, British civilians witnessed such suffering in camps, hospitals and in the open. Death wherever it came transformed their once imagined enemy into a vulnerable body, whose life had also been ripped apart in a global conflict. In these circumstances, at least, the difference between 'grievable and ungrievable lives' that apparently separated friend and foe in war time was briefly bridged.[35]

[31] Swiss Legation, 'Brooker Hall', 15 March 1918, BArch Berlin, R901/83129.

[32] 'List No. 221 of German Prisoners of War', A28982, 'International Committee of the Red Cross': https://grandeguerre.icrc.org/en/List/83863/898/28982/

[33] 'Fatality at Beachley Shipyard', *Gloucestershire Echo*, 1 April 1919, p. 4.

[34] 'Frimley. German Prisoner Killed', *Surrey Advertiser*, 24 October 1914, p. 7.

[35] Judith Butler, *Frames of War: When is Life Grievable?* (London: Verso, 2010), p. xxii.

Funerals

In January 1915, the German admiralty sent a small flotilla of cruisers to intercept British ships at Dogger Bank. The mission, though, proved to be very short-lived, as the numerically superior British battlecruisers outraced and outgunned their German counterparts, sending the weakest of the German cruisers, Blücher, to the seabed. The Blücher's captain, Alexander Erdmann, and some 250 of a crew of at least 1000 men were fished from the freezing water and taken into captivity. Erdmann, cold, wet, and suffering from exposure, struggled on for a few weeks but eventually died from hypothermia while interned in Edinburgh Castle. His family immediately made plans to have the body returned, requesting the remains 'be placed in a zinc lined coffin' for repatriation. The British were less enamoured at this proposal, rejecting it as 'an awful nuisance' and 'a practical impossibility'.[36] Their main concern, though, was that it would be rather awkward to repatriate the enemy dead, while at the same time refusing to return the bodies of British officers killed on the battlefield overseas.[37] With Erdmann and the other German dead, therefore, destined to remain on British soil, the public often ended up encountering the enemy as he was laid to rest in their midst.

In burying the enemy dead at home, the British had to adhere to the early iterations of international law. The 1907 Hague Convention stipulated that POWs were to be buried in the same way as other soldiers with 'due regard [...] to their grade and rank'.[38] For practical reasons, the British generally fulfilled these obligations by making use of the closest burial ground, whether a parish churchyard or civic cemetery, to the place of death. Erdmann himself, who had died in the hospital in Edinburgh Castle, was buried in Newington Cemetery, some three kilometres further south. Similar arrangements occurred elsewhere too. The 45 Germans

[36] American Embassy London, 9 March 1915; War Office, 15 March 1915, TNA, FO 383/55.

[37] In April 1915, the War Office outlawed the return of the British dead back home. On this policy, see: Dominiek Dendooven, '"Bringing the Dead Home": Repatriation, Illegal Repatriation and Expatriation of British Bodies during and after the First World War', in Paul Cornish and Nicholas Saunders (eds) *Bodies in Conflict: Corporeality, Materiality and Transformation* (New York: Routledge, 2014), pp. 66–79.

[38] 'Convention (IV) respecting the Laws and Customs of War on Land and its Annex: Regulations concerning the Laws and Customs of War on Land', 18 October 1907. See also: Anna Petrig, 'The War Dead and their Gravesites', *International Review of the Red Cross*, 874 (91) (2009), 341–369, (p. 366).

who died in internment in Dorchester were laid to rest in the churchyard of St George's in Fordington, just south of the centre, while Manchester Southern Cemetery received the bodies of the German patients who succumbed to wounds or disease in the city's Nell Lane hospital, situated across the ground from the burial ground. Many of the enemy Catholic dead to die in the city, meanwhile, were taken to a separate site laid out in St Joseph's Cemetery in Moston.

Generally, local people had little option but to accept these new mortuary arrivals. Nonetheless, there were occasional complaints that forced the authorities to hunt out alternative burial sites. The commandant of the Stobs internment camp in southern Scotland experienced this difficulty following the burial of an internee in the nearby Kirkton parish churchyard. After this first interment, the landowners refused to accept any further enemy bodies. Left with no alternative, the commandant arranged for the construction of a new cemetery alongside the camp itself.[39] The situation was very similar further south in Shropshire, where local councillors banned the burial of enemy soldiers from the Park Hall camp in Oswestry's town cemetery, 'unless the death occurred [explicitly] in the town and parish of Oswestry'.[40] With nowhere to put the dead, the camp commandant was also forced to develop a new POW cemetery on open farmland, slightly removed from both the camp and the centre of the town.

However, the distance between the camps and the designated cemeteries proved something of a problem. It raised the awkward question of how to transport the body from the place of death to the final resting place. In the case of Stobs camp and Nell Lane hospital, where the cemeteries bordered the main buildings, this was a straightforward process. However, in most other areas, the enemy bodies had to be transported through urban centres. In Handforth in Cheshire, for example, which housed one of Britain's largest internment camps, the dead had to be moved to Wilmslow Cemetery, a little over two kilometres from the gates of the camp. This final journey placed British civilians once again in direct contact with their former enemy, turning the procession into a shared experience between wartime foes.

[39] Chief Maintenance Surveyor, 25 March 1921, Commonwealth War Graves Commission Archive, Maidenhead, CWGC/7/4/2/1137.

[40] Oswestry Town Council Minutes, 27 June 1918, Oswestry Town Council Archive, OTC 10/12/2/2/10.

The camp commandants generally allowed the POWs to arrange the procession and funeral for their comrades, so long as it was done ‘at their own expense’.[41] There was after all considerable propaganda value in providing a visual demonstration of apparent British tolerance. Most German funerals started at the gates of the camps, where small groups of POWs formed a cortege to accompany the coffin to the grave. At the Frith Hill internment camp, only 16 POWs processed, while in Handforth it was common for up to 200 German prisoners to march behind the coffin.[42] With uniformed POWs in attendance, there could have been no mistaking the identity of the deceased, particularly as the Imperial German flag was often ceremoniously draped over the coffin.

The funeral of the Blücher's commanding officer, Erdmann, captures not only the formality of the processions held for the enemy dead but also their impact on local residents (see Fig. 1). Although Erdmann was given a British rather than a POW guard of honour for the march through Edinburgh, the other characteristics of the procession mirrored those seen in other areas of the country. Local people thronged together across both sides of the street, eight to nine deep in places. These appear to be curious crowds who had taken to the street hoping to get a glimpse of their wartime foe. Later on in the conflict, as death at home and at the front became more familiar, the crowds accompanying enemy funerals diminished in size considerably.

For the most part, the body of the expired enemy elicited more sympathy than hate. This was the case when the funeral was held for two German airmen in Kent. The men had died in a farmer's field after their Gotha bomber had crash landed in 1918. The farmer's family, whose own son was in the British airforce, reportedly had particular ‘sympathy with the German mothers whose boys’ died on their land.[43] Such sentiments, however, were not shared by everyone. When the deceased had been directly involved in a deadly military attack, then funerals could also provide a space for personal revenge. This was the case during the funeral of a Zeppelin crew in Potters Bar in 1916. As the procession passed through the crowd-filled streets, a ‘stale egg’ flew out hitting one of the coffins, followed by a barrage of ‘insulting language’. After being arrested, the

[41] War Office, ‘Report on the Directorate of Prisoners of War’, 24 September 1920, TNA, FO369/1450.

[42] ‘Funeral of German Prisoner’, *Surrey Advertiser*, 12 September 1914, p. 7.

[43] Isobel Whitehead to Mrs Sthamer, 11 September 1923, VDK, R13.

Fig. 1 Funeral of Alexander Erdmann, Edinburgh, 18 February 1915. Illustrated War News, 24 February 1915, p. 23 (The Stapleton Collection/Bridgeman Images, STC312638)

suspect—a 40-year-old Londoner—defended her actions, explaining that she had 'lost relatives by bombs thrown during Zeppelin raids'.[44]

Once a cortege had made its way through the sea of onlookers, the actual interment took on a more familiar pattern. Erdmann's funeral in Edinburgh, which was more spectacular than most, ended with three volleys over the grave. A German pastor, who was himself a civilian internee, then gave a short sermon and recited the Lord's Prayer in German.[45] While a German liturgy was often given at other funerals, the honour of fired volleys was not always given. Their use in Edinburgh reflected both Erdmann's status as an officer and the timing of the funeral relatively early in the war. By the time of the Zeppelin crew's interment in Potters Bar 18 months later, the Metropolitan Police stated explicitly that there will be

[44] Metropolitan Police, 'Funeral of the crew of the Zeppelin destroyed by fire', 7 September 1916, TNA, MEPO 2/1652.

[45] 'Funeral of the Captain of the Blucher', *The Scotsman*, 19 February 1915, p. 7.

'no band, firing party or flowers'.[46] Regardless of the overall scale of these ceremonies, the funerals of the German dead brought British civilians into contact with the enemy. As the remains were laid to rest, onlookers were forced to confront wartime death as a shared experience, rather than as solely a British one.

Naming

Once in the ground, of course, the bodies of the dead became indistinguishable, as the remains slowly decomposed in the soil. Yet, while the body vanished below ground, the living retained a 'very long commitment' to the names of the dead. During the First World War, names took on an even greater significance as the Imperial War Graves Commission (IWGC) and its national equivalents made it their mission to give each military casualty a named grave. The aim of naming the dead, as Rudyard Kipling so poignantly put it, was to ensure 'their name liveth forevermore'.[47] The German POWs of course already had their own names, biographies and lives. Yet, with some 115,000 internees ensconced in British camps during the war, the individual POWs ended up being viewed as merely one homogenous enemy group. In death, however, the names of the Germans came to the fore. No longer were they an indistinguishable mass, but rather individuals with their own rounded identities.

Following the death of the enemy at home, the first official mention of their name and detailed discussion of their lives occurred during a coroner's inquest. The British generally held these for unusual deaths where the circumstances remained unclear. The Hertfordshire coroner, for example, commandeered the Plough Inn in Cuffley, close to where one of the Zeppelins had crashed, to conduct his investigation into the fate of the crew.[48] On the Isle of Man, where British guards had shot five German civilians during a 'disturbance' in the Douglas camp over food and conditions, the coroner held an inquest over two days in November 1914. The course of the investigation not only named the dead, but witnesses also revealed details about their lives. Richard Fohs had been a waiter at the Grand Hotel in Brighton, the public discovered; Richard Matthias,

[46] Metropolitan Police, 5 September 1916, TNA, MEPO 2/1652.

[47] Laqueur, *Work of the Dead*, pp. 365, 426.

[48] 'The Wrecked Zeppelin', *Aberdeen Press and Journal*, 6 September 1916, p. 5.

another of those killed, was a 22-year-old sailor from Hamburg, while Bernhard Waring lived and worked in London as a labourer in the docks.[49]

Once the identities of the dead had been firmly established, then information about their fate could be passed on to relatives at home. There were several methods for transferring news. The Prisoner of War Information Bureau, a branch of the War Office, forwarded death certificates, which were crucial for later pension claims, while the International Committee of the Red Cross collated its own records which it shared with relatives.[50] Meanwhile, the British and German governments reached an agreement in early 1917 to supply the names of prisoners who suffered a violent death, with regular reports crossing the divide via neutral intermediaries.[51] In these cases, knowledge of the names of the individual dead helped both the Germans and British to draw a distinction between their own care of the POWs and the other side's apparent neglect. When Claus Carstens, a 28-year-old soldier from Schleswig-Holstein died in the Brocton camp in 1917, for example, the German government made a formal request to the British Foreign Office, demanding it 'furnish details respecting [his] violent death'. The subsequent details revealed only that Carstens had taken his own life by hanging.[52]

As information about the deceased seeped out either through inquests or through the probing of foreign governments, people also started to discover more about the lives of their German enemy. After Carstens was found dead in Brocton, the local Staffordshire newspaper provided its readers with full details of his life and sudden death. Carstens had been captured in the fighting around Arras and then brought across to Britain. He had seemingly become very isolated behind wire, was 'always very quiet and rarely spoke to anyone'.[53] Similar biographical descriptions also surfaced elsewhere in the wake of enemy deaths. In Buckinghamshire, the press reported that Carl Siebenhühner, an unmarried 27-year-old infantryman, died while bathing in the river Colne, near Denham. Siebenhühner, who had never learnt to swim, got out his depth and lost sight of his friends in the flowing water.[54] When Carstens and Siebenhühner were

[49] 'Disturbance at the Aliens Detention Camp'.

[50] War Office, 15 September 1917, TNA, FO383/305.

[51] German Foreign Office, 15 January 1917, TNA, FO383/292.

[52] 'Note Verbale', 10 October 1917, TNA, FO 83/292.

[53] 'German Prisoner's Suicide', *Staffordshire Advertiser*, 28 July 1917, p. 5.

[54] 'German Prisoner Drowned at Denham', *Middlesex and Buckinghamshire Advertiser*, 15 August 1919, p. 5.

eventually buried, local residents in Staffordshire and Buckinghamshire already knew something of the enemy bodies being interred in their communities.

Knowledge of the individual war dead, however, was always partial. The regional press—sometimes even national newspapers—gave much greater coverage to the deaths of prominent German internees than to the ordinary German soldiers who died in Britain. The *Daily Mirror*, for example, saw fit to print three photographs and to dedicate almost an entire page to the funeral of Erdmann in Edinburgh.[55] In stark contrast, the frequency of influenza deaths towards the end of the conflict meant that the press simply ignored many of them. By the time that the seventh German prisoner had succumbed to the virus in Dorchester, the local newspaper simply noted that the 'body of another German prisoner of war was buried [...] on Saturday'.[56] This rather blasé response certainly reveals how the individual lives of the German dead were overlooked later in the war, but it also suggests a comfortable familiarity with the enemy, whose presence no longer required a lengthy explanation.

However in death, the enemy prisoners gradually became visible again. The process of marking the graves ensured the public visibility of the names of the enemy dead, even in cases where there had been minimal discussion of the deceased prior to the interment. In accordance with the stipulations of the Hague Convention, the British agreed to place a wooden cross over each German POW grave. As a further sign that the enemy was being given the 'same treatment' as the British dead, each grave was supposed to receive a metal plaque giving details of the name, regiment and date of death.[57] When a member of the German Embassy in London visited the POW graves in St John's parish churchyard in the village of Sutton Veny after the war, he discovered a sea of wooden crosses, by now somewhat weather beaten, but still standing that reminded visitors of the names of the 38 Germans who had died in rural Wiltshire.[58] Such silent markers not only gave the dead a name but also made it clear that the enemy had also died in towns and villages across Britain.

[55] 'The Funeral of the Bluecher's Captain', *Daily Mirror*, 20 February 1915, p. 4.

[56] 'Funerals of German Prisoners', *Western Gazette*, 8 November 1918, p. 10.

[57] Director of Graves Registration and Enquiries to Foreign Office, 1 June 1916, TNA, FO382/122.

[58] German Embassy London, 'Besuch der Kriegergräber Nr. 27', 19 November 1927, VDK, R13.

The German civilian dead, however, were not always quite as visible. International law prioritised the correct treatment of military combatants over the fate of civilians in wartime. One consequence of this framework was that relatives of civilian internees had greater freedom to erect their own headstones over their loved one's grave. In Hawarden, North Wales, which bordered the Queensferry internment camp, three German prisoners were laid to rest. The family of one of these men, Friedrich Segelken a sea captain from Bremen, managed to erect a white marble headstone over the grave. Yet, while Segelken received a stone headstone, the two civilian internees buried nearby were not so fortunate; the pair ended up sharing both a grave and a roughly painted white wooden cross.[59] The situation was even worse for the small number of civilians who died in asylums. Often, they were buried in little more than an unmarked pauper's grave, which meant that their names quickly vanished from their British surroundings.

Communities

When the Central Powers sued for peace in November 1918, fighting in the west came to a grinding halt. The armistice, though, did little to alter the routines of men interned in POW camps. Work duties, daily roll calls and rationing continued, only now with an even deeper longing to return home. 'It is in no way clear when the little hour of liberation will strike, at Easter, or Whitsun, or later', bemoaned the German POWs in the Stobs camp, 'only one thing is certain: it is the final winter'.[60] With POWs still trapped behind wire on British soil, accidents, sickness and deaths also continued to rise. In Dorchester, for example, Franz Radojewski, a 21-year-old soldier, was fatally shot in May 1919 after allegedly attempting to escape. According to eyewitnesses, the guards had spotted a man trying to crawl under the wire; they shouted 'halt', but after getting no response fired off two shots, striking Radojewski in the back.[61] A similar incident occurred two months later in the Park Hall camp, when Willy Oster, a young soldier from Stuttgart was shot through the head. On the day of his death, relations between the POWs and guards were already tense. The POWs, who were increasingly restless with the war over, started

[59] German Embassy London, 'Hawarden Cemetery', 9 May 1933, VDK, R13.
[60] 'Der letzte Winter', *Stobsiade*, December 1918, p. 1.
[61] 'Prisoner of War Fatally Shot', *Dorset County Chronicle*, 22 May 1919, p. 2.

shouting and throwing stones. Amidst the turmoil, Oster had clambered to the top of a toilet block to retrieve something that had been thrown up there, when the fatal shot rang out.[62]

It was only after the new German government had signed a formal peace treaty in June 1919 that the British started to arrange for the repatriation of the POWs.[63] As the internees packed their few possessions and embarked on the slow journey back to friends and family, the dead remained stuck in Britain, far from home. All that the POWs could do was to put in place measures that they hoped would ensure the long-term care of their former comrades' graves. The internees in Park Hall raised funds in 1919 for a large mound memorial to be placed in the German cemetery, where Oster and 101 other prisoners lay. A local firm completed the groundwork, which still allowed funds to be set aside for a future maintenance regime.[64] On the Isle of Man, home to the largest 'enemy alien' camps, internees made similar arrangements before their eventual departure. Prisoners from the huge Knockaloe camp raised £500 from their canteen that they left with the vicar and churchwardens of the neighbouring Patrick parish church for repairs and maintenance of the graves.[65]

The Treaty of Versailles emphasised the importance of the POWs' forward planning. Article 225, which dealt specifically with war graves, stated that the Allied and German governments were responsible for any graves 'in their respective territory'.[66] As a result, the British, and not the Germans, took control of the enemy war dead buried throughout the country. What this meant in practice was that the German dead were to remain forever on British soil, with little chance of reparation home. The family of Otto Günther, for example, had sought to bring his body back to Bremen at the war's end, but their plans fell foul of British bureaucracy.[67] The reason for the IWGC's reluctance to countenance the repatriation of the dead was that it ran counter to the policy it had imposed on British families. As the British dead were to remain overseas in new national

62 Foreign Office, 25 October 1919, BArch Berlin, R901/83132.

63 Reinhard Nachtigal, 'The Repatriation and Reception of Returning Prisoners of War, 1918–22', *Immigrants & Minorities*, 26, (1/2) (2008), 157–184, (p. 175).

64 German Embassy London to Zentral-Nachweise-Amt, 30 July 1920, VDK, R13.

65 IWGC, 'Patrick Parish Churchyard', 14 May 1923, Commonwealth War Graves Commission Archive, CWGC/7/4/2/3778-1.

66 Treaty of Versailles, 28 June 1919. Available, United States Library of Congress: https://www.loc.gov/law/help/us-treaties/bevans/m-ust000002-0043.pdf

67 Swiss Minister, 4 December 1919, BArch Berlin, R901/85256.

cemeteries, to have allowed the return of the enemy would have undermined the policies introduced at home.

Yet, while there were similarities in non-reparations, there was one area of clear policy difference. In post-war Germany, the IWGC gradually consolidated the British and Empire dead into four new war cemeteries in Hamburg, Berlin, Cologne and Kassel. It had no such plans for the German dead in Britain and was instead content to leave these bodies scattered around the country in 216 separate sites.[68] The IWGC did, however, agree contracts with local landowners for the care of the individual German war graves, although tellingly the civilian dead were excluded from these agreements. In Shropshire, the Oswestry Rural District Council agreed to take on the maintenance of the Park Hall German cemetery for an annual fee of £10, while in North Mymms, Hertfordshire, it was the local vicar who signed an agreement with the IWGC to maintain the graves of three enemy combatants buried in his churchyard.[69] Caring for the enemy dead, therefore, may have come under national jurisdiction through the auspices of the IWGC, but as a result of their delegation, it effectively became a local matter. The ties between Britain's former enemy and local communities became even closer as a result.

It did not take long for the German graves to blend fairly unobtrusively into their surroundings. Over time, the original grave markers started to fade and vegetation slowly reclaimed the burial plots. In 1926, a German army officer from Itzehoe in Schleswig-Holstein took it upon himself to make a tour of German military graves in England and Scotland and then to report back on their condition. He started his visit at the Zeppelin graves in Great Burstead, where he found the inscriptions on the crosses 'so faded that they were no longer legible'. From Essex, he travelled west to a cemetery in Richmond, Surrey. While the grave markers were in better condition here, he did note that some of the graves had settled a little. The report's author was careful to stress that none of the German graves had been wilfully neglected. As he pointed out, they were 'not worse cared for than the majority of the other English graves'.[70] The gradual

[68] 'Deutsche Kriegsgräber in England', *Kriegsgräberfürsorge*, August 1936, p. 114.

[69] Oswestry Rural District Council, Minutes, 7 November 1921, Shropshire Archives, Shrewsbury, DA24/100/8; IWGC, 'Agreement for Maintenance of War Graves in North Mymms Churchyard', 16 September 1922, Hertfordshire Archives, Hertford, DP69/3/20.

[70] Oberleutnant to Zentralnachweiseamt, 3 November 1926, VDK, R13.

weathering of the graves and wooden crosses was simply a result of time and the environment.

On receiving the officer's findings and other similar reports, the German Embassy in London decided it need to act. In consultation with the IWGC, it devised a seemingly simple plan for replacing all of the existing wooden crosses with concrete markers, known as ZAK tablets.[71] The face of these new stones, which contained the deceased's details, was shaped as an Iron Cross and placed low to the ground at a slight angle. Installation of the ZAK tablets, which were all delivered direct from Germany, proceeded at pace through the mid-1920s. However, despite these efforts, the results were disappointing. Thanks to their low profile, the new stones had a tendency to collect rainwater on their surface, which quickly eroded the inscriptions. What had started as a means to make the German dead more visible, ended up doing the reverse.[72] The German graves in Britain, with badly weathered, low-lying, markers, became ever more inconspicuous, as they slowly sunk into the ground.

In their current setting and in their unassuming state, the former enemy graves provided no threat; they simply took on the appearance of other weather-beaten burials in British churchyards and cemeteries. Although the German Embassy may have hoped the German war dead would be given greater prominence, their very ordinariness actually helped the dead to 'work' towards British–German reconciliation. Their main role in this regard was in bringing the former foes together. With the silent German dead laid to rest far from home, British residents living near the burial grounds often saw themselves as the custodians of the German graves. Local people would often tend the sites and sometimes even give the bereaved relatives in Germany updates on the state of their loved one's grave. In Bedford, boys from the town's Bunyan Sunday School started a campaign to write to 'parents or friends of the German soldiers' buried in the cemetery with a 'short report on the position and condition of each grave'.[73] Elsewhere, local people often provided relatives with photographs of the German graves in their midst as a means of connecting them to the distant body. In rural Gloucestershire, the wife of the churchwarden

[71] The abbreviation of the German body responsible for registering war graves: Das Zentralnachweiseamt für Kriegerverluste und Kriegergräber.

[72] Deutsche Botschaft London to Zentralnachweiseamt für Kriegerverluste und Kriegergräber, 7 April 1930, VDK, R13/25.

[73] Bunyan Sunday School to German Embassy, 12 October 1926; Mrs Knight to German Embassy, 25 September 1931, VDK, R13.

even took it upon herself to arrange photographs of the ZAK stones again as a means of comforting the bereaved.[74]

The presence of the war dead, therefore, never intimidated the British communities that surrounded the graves. Indeed, local people often drew the former enemy into wider commemorations for those killed in the conflict. In Dorchester, once home to a large POW camp, a local scout group tried to embrace all sides in its remembrance activity. Each year at Christmas, the troop laid one wreath on the town's central war memorial, followed by a second wreath on the German memorial in the Fordington cemetery, where 45 POWs had been buried. The inscription on the wreath bore the simple inscription: 'Peace on Earth, good will towards men'.[75] A similar narrative of reconciliation between former foes came from young people in Birmingham. From 1926 onwards, members of the city's Shenley Fields children's home laid wreaths on both the British and German graves in the Lodge Hill cemetery to mark Armistice Day. The superintendent of the children's home explained that as wartime 'bitterness' between the British and Germans was fading, it was important to come together. 'Death the leveller should soften all animosities', he added.[76]

Conclusion

Both before and during the First World War, sensationalist writers, such as Le Queux, built up a simple image of a dangerous, marauding enemy who was on the verge of invading Britain. The physical absence of the German forces from Britain made it all the easier for people to imagine the threat to come. Maintaining this imagined construct, though, proved much trickier when the enemy himself was weak and sapped of his strength. The British public started to see a different side of their wartime foe very early on in the conflict. As captured German prisoners started to be shipped across the English Channel to internment camps in Britain, much of the original threat had dissipated. However, it was only really when some of these captured POWs started to die, most commonly from battlefield wounds or disease, that sympathies towards the Germans started to change. Dealing with the dying body forced people to view the enemy not

[74] P. Knight to German Embassy, London, 25 September 1931, VDK, R13.

[75] 'Peace on Earth', *Daily Herald*, 23 December 1925, p. 1.

[76] Children's Homes, Shenley Fields to German Consul, 18 October 1926, VDK, R13.

as a dangerous foe, but rather as vulnerable humans with their own complex life stories, made up of friends, relatives and loved ones back at home. This encounter, therefore, was very different to Judith Butler's discussion of the dehumanisation of certain lives in America's war on terror that rendered them 'ungrievable'.[77]

Following the moment of death, funeral corteges and burials all brought local people face-to-face with the German dead. Other than in the Stobs camp in Scottish borderlands and the Park Hall camp near Oswestry, where separate POW cemeteries were constructed, the Germans were all buried in municipal burial grounds and parish churchyards throughout the country. In these sites, the enemy was no longer a faceless hostile soldier; he had instead become a named casualty of war. Visitors to these cemeteries were, therefore, confronted not just with the graves of the British and other allied dead but also with the graves of their former foe. On many occasions, this confrontation went a stage further with local people laying flowers, taking photographs and generally caring for the German graves. In death, the enemy had been rehumanised, transformed into a more benign figure and eventually into just another casualty of a global war.

[77] Judith Butler, *Precarious Life: The Powers of Mourning and Violence* (London: Verso, 2004), pp. 19–49.

Enemies, Subjects and Allies in Wartime Indian Northeast

Lily Tekseng

For the indigenous peoples from the borderland Indian Northeast, cases of 'mistaken identities' based on their race are common occurrences. At the start of the Covid pandemic in 2020, Northeastern Indians were frequently reported to have been discriminated against or attacked after being mistaken for being Chinese.[1] A year earlier in 2019, around 40 people from the region, mistaken for Tibetans, were detained in anticipation of protests against the imminent visit of the Chinese President Xi Jinping to the country.[2] Writing many years earlier of a different side to what is another kind of racial confusion, the Manipuri author and a native

[1] See Thongkholal Haokip, 'From "Chinky" to "Coronavirus": Racism against Northeast Indians during the Covid-19 Pandemic', *Asian Ethnicity*, vol. 22, no. 2, Apr. 2021, pp. 353–373.

[2] See Pallavi Pundir. 'Police Detained Northeast Indians Thinking They're Tibetans Who Might Protest Against Chinese President's Visit', *Vice*, 18 Oct. 2019. <https://www.vice.com/en/article/8xwemz/police-detained-northeast-indians-thinking-theyre-tibetans-who-might-protest-against-chinese-presidents-visit>.

L. Tekseng (✉)
University of Cambridge, Cambridge, UK
e-mail: lt501@cam.ac.uk

H. Furneaux, M. Greig (eds.), *Enemy Encounters in Modern Warfare*, https://doi.org/10.1007/978-3-031-56748-3_9

of its capital city Imphal, Akoijam, recalled a passing, whimsical 'compliment' given to her by a shopkeeper in Kathmandu, Nepal: 'You have wasted your looks by being born an Indian'.[3] At the time of writing the article, Akoijam divided her time between Nepal and India, flitting between 'looking Nepali' and not looking Indian, between fighting 'a bitter battle against exclusion from India' and entertaining the tempting offers of belonging and being embraced as one's own.[4] Highlighting the everyday disjunction between the notions of national attachment that her race elicits and the nationality that her passport matter-of-factly states, she wrote, 'Chinese, Nepali, Japanese [...] I was offered a long list of citizenships from the Mongoloid-looking world'.[5]

These currents of racial confusion are underpinned by national policies, the most direct of which are special legislations which grant the Indian military extraordinary powers in the areas designated as 'disturbed'. Rechristened and modified since their inheritance from the British colonial laws, many of these legislations, especially the Armed Forces Special Powers Act (AFSPA), have been used as indispensable instruments by the post-independence Indian state to manage its relationship with the racialised Northeastern borderland.[6] The national logic of battling insurgency in a region that has historically witnessed several secessionist movements has also reinforced the transformation of the internal enemy into a racialised enemy. This was recently highlighted by what has locally come to be known as the 'Oting Massacre'. In 2021, just weeks before Christmas, in the village of Oting in Nagaland, a truck carrying eight Naga civilians who worked in a coal mine nearby were shot at indiscriminately, without prior warning by the Indian army on the pretext of a counter-insurgency operation.[7] Six people died on the spot and two were severely injured. On

[3] Sunita Akoijam, 'Feeling Nepali', 1 August 2011: *Himal Southasian*, <https://www.himalmag.com/feeling-nepali/> [Accessed 1 January 2023].

[4] Akoijam, 'Feeling Nepali'.

[5] Akoijam, 'Feeling Nepali'.

[6] Laws such as Unlawful Activities Prevention Act and Sedition law have been used widely in the region to target separatist movements as well as curb dissent.

[7] See Niha Masih and Sadiq Naqvi, 'Indian Troops Kill 14 Civilians in Weekend Incidents, Spurring Demands for Repeal of Special Powers in Some Regions' (*Washington Post*), <https://www.washingtonpost.com/world/2021/12/06/india-nagaland-encounter/> [Accessed 26 Sept. 2022]; Staff, 'Nagaland Killings: Rioting as Indian Security Forces Shoot Dozen Civilians' (*The Guardian*, 5 Dec. 2021), <https://www.theguardian.com/world/2021/dec/05/nagaland-killings-india-security-forces-shot-civilians-mistaken-for-militants>.

hearing the gunshots, some residents of the village arrived at the site, who were then shot at too.[8] Seven more were shot dead and several more injured in the second round of firing. A protestor present at the scene, who was also a member of the ruling Hindu nationalist party BJP,[9] despairingly observed, 'How will the world run if the *Hindustani*[10] Army kills civilians like this?'[11] Although the identity of the soldiers present at the ambush is not known—various reports suggest that by the time the protestors arrived at the scene, the party who had killed the civilians had been replaced by a new army reinforcement—the choice of the word 'Hindustani' is telling. Two days later, the Home Minster of India called the incident a case of 'mistaken identity'—a pre-planned 'ambush' that was meant to have been one of the many 'operations against insurgents' in the region.[12] A few months later, two local civilians were shot and injured in Tirap district in the province of Arunachal Pradesh near the Indo-Myanmar border.[13] The two men, part of a group of four, were returning home after fishing when they were ambushed by the army. The official statement, yet again, was that the Army had mistaken them for insurgents.[14]

Several other cases of 'mistaken identity' shootings have been recorded in the region, but the state's explanations have not been

[8] See Rokibuz Zaman, 'A Nagaland Village Mourns Its Dead: "How Can the Army Kill My Innocent Sons?"' (*Scroll.In*, 7 Dec 2021) <https://scroll.in/article/1012235/a-nagaland-village-mourns-its-dead-how-can-the-army-kill-my-innocent-sons> [Accessed 27 Sept. 2022].

[9] The Bhartiya Janata Party.

[10] Emphasis added. The word 'Hindustan' means 'the land of the Hindus' in Persian, historically used to refer to the Indian subcontinent. 'Hindustani', by extension, denotes the peoples of the Indian subcontinent.

[11] Rokibuz Zaman, 'A Nagaland village Mourns Its Dead'.

[12] See 'Nagaland Killings: Full Statement Made by Home Minister Amit Shah in Parliament', *The Indian Express*, 6 Dec. 2021, <https://indianexpress.com/article/india/nagaland-killings-amit-shah-full-statement-parliament-7658950/>.

[13] See Tora Agarwala, 'As AFSPA Lifted in Parts of NE, 2 Injured in Army Firing in Arunachal', *The Indian Express*, 2 Apr. 2022, <https://indianexpress.com/article/north-east-india/arunachal-pradesh/arunachal-pradesh-army-civilians-injured-mistaken-identity-7849784/>; Bureau, The Hindu, 'Soldiers Shoot at Two Civilians in Arunachal' (*The Hindu*, 2 Apr. 2022), <https://www.thehindu.com/news/national/other-states/two-civilians-shot-at-in-arunachal-by-soldiers-mistaking-them-for-extremists/article65284466.ece>.

[14] Quote by the Deputy Commissioner of the Tirap District, Arunachal Pradesh. For more, see Agarwala.

revised.[15] This is an interesting, even unique admission on the part of the Indian state since the preferred terminology for extrajudicial killings is usually that of 'encounter killings'.[16] The continual failure of 'credible intelligence' at such a catastrophic scale owing to racial confusion is in fact rendered possible because of the entrenched colonial and postcolonial architecture of racialisation in place in the region. Race was and is arguably the most important organising factor for the borderland region, and at no point of time were the 'global processes'[17] that enabled its racialisation more magnified than during the Second World War when the region became embroiled as a strategically integral part of the China–Burma–India theatre. Wars, as a rule, demand a neat delineation of the category of the 'enemy', and conversely, 'friend'. However, the wartime transformed the borderland region into a potpourri of murky categories: empires both old and new, colonial subjects, witnesses, friends, allies and enemies. These categories were destabilised and blurred amid the liminal porosity of the borderworld, and race was a centrepiece of this story. By bringing together writings and testimonials from the region, archival sources and veterans writings by the Allies, this chapter will interrogate how the categories of 'enemy', 'friend' and 'ally' were imagined and re/produced. It will also explore the ways in which indigenous host communities engaged with and shaped these ideas.

The Fascinating Enemy and the Fascinating Ally

Although seemingly disparate, the Japanese writer and Nobel-laureate Kenzaburō Ōe's short story 'The Catch' (1958)[18] and the Sahitya Akademi-awarded Naga writer Temsüla Ao's short story 'The Boy Who

[15] See Bikash Singh, 'NHRC Issues Show Cause Notice to Defence Ministry over Relief Amount for Kin of Civilian Killed in Firing', *The Economic Times*, 27 Feb. 2021, <https://economictimes.indiatimes.com/news/politics-and-nation/nhrc-issues-show-cause-notice--to-defence-ministry-over-relief-amount-for-death-of-arunachal-resident-in-army-firing/articleshow/81235024.cms?from=mdr>; Staff, 'Army Kills Villager in "Mistaken Identity"', *Morung Express*, 15 June 2017, <https://morungexpress.com/army-kills-villager-mistaken-identity>, [Accessed 27 Sept. 2022].

[16] See Suchitra Kartikeyan, 'Explained | Police "Encounters" in India - Cases, Conviction & Court Orders', *The Hindu*, 3 June 2022, <https://www.thehindu.com/news/national/explained-police-encounters-in-india-cases-conviction-court-orders/article65463140.ece>.

[17] Mabel D. Gergan, and Sara H. Smith, 'Theorizing Racialization through India's "Mongolian Fringe"', *Ethnic and Racial Studies*, vol. 45, no. 2, Jan. 2022, p. 363.

[18] Kenzaburō Ōe, 'The Catch', in *The Catch and Other War Stories*. Ed Shōichi Saeki (Tokyo: Kodansha International, 1981).

Sold an Airfield' (2009)[19] have several parallels and overlaps. Reading the two stories together is especially productive as they reveal the expectations and surprises in the encounter between purported enemies and allies and the ideas about race which structure these encounters. 'The Catch', narrated by a young Japanese boy, is about an African American airman who survives a plane crash in a remote, unnamed village in Japan but is captured by the villagers, becoming their titular 'catch'. Ao's short story follows Pokenmong, the young Naga protagonist's stint working as an errand boy at a US military camp in the town of Jorhat in Assam towards the end of the Second World War. Both locate their stories in a setting where the tremors of war experienced are low, the insulation from active sites of war providing the necessary conditions for the destabilising surprise encoded in the stories. The remoteness of the two settings creates a deceptive sense of distance from and complacency about the violent realities of the war, consequently engendering a friendly fascination for the racial Other, only for the stories to both disrupt *and* reinstate the simplified categories of the enemy, ally and one's own. Because, after all, the stories point out, the Americans were still a dubious ally in the borderland Indian Northeast irrespective of their overwhelming, captivating material and racial presence, and the Allies—even represented by a fictional, friendly, fascinating Black airman in a racially fraught US army—for all intents and purposes, were still an enemy in Japan. Just as we settle into rethinking these categories, Ao and Ōe reveal that the fissures in fact do not lie beyond enemy lines at all but within the intimacies among your own.

When the Black American prisoner is discovered alive after the crash in 'The Catch', the villagers are taken aback. At one level, their surprise seems to be compounded by the remoteness of their village. The child-narrator describes the war as an event occurring 'in the town beyond the rolling hills',[20] as something that had 'gone on too long' and was 'vast and cumbersome' like a 'legend'.[21] And like legends, its immediacy to their lives was spatially and temporally distanced, it 'meant nothing more than the absence of the young men from the village and an occasional announcement of death in battle delivered by the postman'.[22] Before the arrival of

[19] Temsüla Ao, *Laburnum for My Head: Stories* (New Delhi, India: Penguin Books India, 2009).

[20] Kenzaburō Ōe, p. 19.

[21] Kenzaburō Ōe, p. 19.

[22] Kenzaburō Ōe, p. 19.

the airman, the sightings of 'enemy planes'[23] are described as 'very uncommon'[24] and as 'no more than a kind of rare bird'.[25] The extent of the insulation of this village is denoted further by its administrative distance from the intermediary mayoral office and the main prefectural office where decisions regarding the village are taken, including that of the fate of the indefinitely captive Black airman. So, when the 'enemy' plane crashes into 'the hills during the night',[26] offering the possibility of surviving 'enemy fliers' and 'enemy soldiers',[27] the war finally enters the village in the manner of an inescapably intimate and unexpected encounter.

Ao's story—set at the end of the war at a time of haphazard but relatively peaceful evacuation—also distances itself spatially and temporally from the battlefields of war. The teenage protagonist Pokenmong first encounters the 'foreign'[28] American soldiers, who had 'come to set up camp in the perimeter of the barely functional airfield to oversee the final evacuation of men and materials from the last allied command-post of the Indo-Burma campaign',[29] at the railway crossing managed by his then Oxomiya[30] employer. Utterly fascinated by the 'fifty trucks laden with men and materials [...] [who] were all white soldiers, some singing and some talking loudly in a strange language'[31] and 'unable to contain his curiosity',[32] he surreptitiously climbs onto one of the trucks and follows them to their camp. He ends up working and living at the American camp, becoming the 'commander's Man Friday',[33] and eventually inherits the material leftovers in the abandoned airfield after the departure of the Americans.

Of the eight stories in the collection that contains 'The Boy...', it is the only story about the Second World War in the region. Most of the rest of the collection deals with encounters between the Naga people and the Indian military in post-independence India. The singularity of 'The Boy'

[23] Kenzaburō Ōe, p. 19.
[24] Kenzaburō Ōe, p. 18.
[25] Kenzaburō Ōe, p. 19.
[26] Kenzaburō Ōe, p. 20.
[27] Kenzaburō Ōe, p. 20.
[28] Temsüla Ao, p. 44.
[29] Temsüla Ao, p. 44.
[30] Alternative for 'Assamese'.
[31] Temsüla Ao, p. 43.
[32] Temsüla Ao, pp. 43–44.
[33] Temsüla Ao, p. 47.

in the collection gestures towards a paratextual discursive context coded with contradictions. First, it teases the expectations of where a story about the Second World War might be located. Although the region was central to the China–Burma–India theatre in several ways, it has often been referred to as the 'forgotten'[34] war due to both national and international public amnesia regarding the theatre. The region was a witness to two strategically crucial battles which halted the Japanese advance into India, namely, the Battle of Imphal (7 March–17 July 1944) and the Battle of Kohima (4 April–22 June 1944); a host to the overnight exodus of refugees from Burma after its capture by Japan; and the locale for intense and frantic preparation by the Allies to create and maintain a supply route to China, the most infamous of which was the construction of the Stilwell road, a 1726 km road connecting India to China via Burma. Secondly, Ao signals towards a regional post-independence condition that she has explicitly called 'a war zone'[35] in her other works. In other words, the story highlights the gap between the thriving vernacular culture of Second World War memories in circulation and the historical silence about the China–Burma–India theatre; further, the rendering of the Second World War in the region as a relatively genial time characterised by curious encounters, in contrast to the everyday violence of post-independence India, echoes the collective nostalgia for the war in the region.[36]

There is, however, another facet on which the authors inscribe surprise. The village adults and children in Ōe's story are not just taken aback by their sudden exposure to war; it is, in effect, the surviving airman's

[34] See Anbarasan Ethirajan, 'Kohima: Britain's "forgotten" Battle That Changed the Course of WWII'. *BBC News*, 14 Feb. 2021 <https://www.bbc.com/news/world-asia-india-55625447>; Debendranath Acharya, trans. Amit R Biashya. *Jangam: A Forgotten Exodus in which Thousands Died* (New Delhi: Vitasta, 1983); 'China-Burma-India: WWII's Forgotten Theater | World War II (1941–1945) | Serving: Our Voices | Veterans History Project Collection | Digital Collections | Library of Congress'. *Library of Congress, Washington, D.C. 20540 USA*, <https://www.loc.gov/collections/veterans-history-project-collection/serving-our-voices/world-war-ii/china-burma-india/>. Accessed 28 Feb. 2023; and broadly: C.A. Bayly & Tim Harper, *Forgotten Armies: The Fall of British Asia, 1941–1945* (Cambridge, Mass.: Belknap Press of Harvard University Press, 2005), and C.A. Bayly & Tim Harper, *Forgotten Wars: The End of Britain's Asian Empire* (London: Allen Lane, 2007).

[35] Temsüla Ao, *These Hills Called Home: Stories From a War Zone* (New Delhi: Zubaan, 2013).

[36] Writers such as Easterine Kire (Nagaland), Mamang Dai (Arunachal Pradesh) and Janice Pariat (Meghalaya) have engaged with the subject of nostalgia about the war in the borderland region in their works.

Blackness that catches them completely off-guard. The child-narrator and his coterie of friends are swept away with excitement and fascination about the Black airman. When it is decided that the airman is to be kept in the cellar of the narrator's family home, he describes his proximity to the airman as 'splendid, adventurous [...] almost unbelievable'.[37] As the children await the prisoner's fate while progressively befriending him, they animatedly discuss his status as an enemy: one 'triumphantly' proclaims, 'Well, they are not killing him are they? [...] A [N-word]'s not an enemy, just as I said!'; another adds, 'He's too good to kill'.[38] Although he increasingly becomes 'a component of village life, in the same way as the hunting dogs, or the children or the trees',[39] his racial attributes remain a subject of fascination. His body parts are described as 'perfect, and beautiful',[40] yet also animalised, as in, 'We looked on him as on some rare, wonderful domestic animal, a genius of an animal',[41] often making for a discomforting read. Similarly, Pokenmong too is fascinated by 'the strange looking men'.[42] Overwhelmed by their whiteness, he compulsively sets out to 'find out [...] if they [white Americans] were at all like ordinary people or were a species apart from anything he knew'.[43] Both the stories highlight the instability of the purported categories of enemy and friend by pivoting the stories around the encounters of the self with the racial Other, highlighting the fascination and sympathies generated in lateral interactions in the otherwise hierarchically and vertically organised imperial ideas of race. Ōe's story calls attention to the anti-empire impulse to see the Black airman as a special enemy who really ought to be an ally, and Ao's story draws out the fascinating proximity to an ally who really is a power broker and has been imagined as the coloniser or a coloniser adjacent during this time. These encounters proliferate and become especially heightened in the context of the Indian Northeast after the Japanese incursion into the region as Japan leverages its status as a representative of Asian modernity and its ideological opposition to Western imperialism. Strikingly, racial affinities are invoked to cement these alliances.

[37] Kenzaburō Ōe, p. 25.
[38] Kenzaburō Ōe, p. 37
[39] Kenzaburō Ōe, p. 45.
[40] Kenzaburō Ōe, p. 47.
[41] Kenzaburō Ōe, p. 48.
[42] Temsüla Ao, p. 47.
[43] Temsüla Ao, p. 44.

The Indian Northeast, today a cluster of eight provinces and home to around 200 indigenous communities usually called the 'Scheduled Tribes' as per the categorisation in the Indian constitution, has historically been produced as a place apart. The British began expanding into the region during the mid-nineteenth and early twentieth centuries in what would become its final phase of expansion. Epistemically and administratively constructed as a 'frontier'[44] and 'the Mongolian Fringe',[45] the region was administered through a complex network of shifting alliances with the several 'hill tribes' and regional kingdoms. Unlike in the Indian mainland, the political category of 'tribe' did not only designate a status within the colonial logic of the civilisational hierarchy or the progressive stages of development in human social evolution; in the region, 'tribes' also represented a distinct racialised category that did not easily fit the classifications previously used to manage populations in the Indian mainland. The war enters the region in this context—where empire-building and empire-resisting activities are still active and ongoing, and nationalist movements are gaining unprecedented grounds in both Myanmar and India. In this complicated setting, the British and their American allies are tasked with defending India, keeping China supplied with resources and recapturing Burma. Within this mosaic of shifting loyalties and resentments, as Ao and Ōe reveal, the colonial logic of racial difference and affinities—components of imperial ideology previously deployed advantageously in the colonised zones—begin to malfunction, instead creating confusion and suspicion.

'You couldn't tell a friend from a foe'—Racial Camouflaging and Confusion

Among the peoples of the region, the war has come to mean several things, but most importantly, as has been succinctly summarised by a veteran witness from Manipur: 'It was not our war'.[46] Reiterating this sentiment, Devi writes in her compilatory essay on the collective memories of

[44] Thomas Simpson, *The Frontier in British India: Space, Science, and Power in the Nineteenth Century* (Cambridge: Cambridge University Press, 2021).

[45] Sanjib Baruah, 'The Mongolian Fringe', *Himal Southasian*, 14 Jan 2013, <https://www.himalmag.com/the-mongolian-fringe/>.

[46] Leisangthem Gitarani Devi, 'How Women Remember War: Unearthing Memories of the Second World War in Manipur', *Zubaan E-Essays from the Northeast*, 2018–2019, <https://zubaanprojects.org/projects/fragrance-of-peace/e-essays/>, p. 5.

Manipuri women witnesses of the war, 'It was a war not to have been fought [...] least of all in Manipur and Kohima'.[47] Caught between two warring sides, a Naga witness who was 'a driver and mechanic for the British'[48] and a part of the Naga delegation to the Allied victory Parade in Delhi in 1945 recalled having to work with both sides, declaring: 'What to do, we had to learn to be friends with both British and Japanese in order to survive!'[49] For others still, the war incites nostalgia. The Naga writer Easterine Kire points out that the war has several local referents: *Japan Rüwho* or *Japan kepar tei*[50] for the Nagas and *Japan Lan* for the Manipuris, meaning 'the Japanese War', or simply 'The War'.[51] Suggesting an irrevocable rupture between the pre- and the post-war periods, she describes the war as 'almost equivalent to the big bang, the beginning of all life'[52] and that 'people have very little memory of what they were doing before the war years'.[53] In unearthing this strange nostalgia, she discovers, 'No Naga family is without its personal wartime story to tell',[54] and 'even ninety-year-olds came to life [...] eyes glowing with excitement'[55] when recounting their wartime memories.

It was not only the host communities who had to contend with surprises. In Borpujari's 2015 documentary, 'Memories of a Forgotten War', Graham, an army veteran recalls his feeling of displacement, stating, 'Like thousands of other young British men I had joined the Army to defend our homeland against the Germans. But we never saw the Germans. We found ourselves in a different war'.[56] Several veteran writings[57] repeatedly

[47] Leisangthem Gitarani Devi, p. 5.

[48] Charles Chasie and Harry Fecitt, p. 61.

[49] Charles Chasie and Harry Fecitt, pp. 61–62.

[50] Charles Chasie and Harry Fecitt, p. 9.

[51] Easterine Kire, *Mari.* (HarperCollins India, 2010), p. viii.

[52] Easterine Kire, p. viii.

[53] Easterine Kire, p. viii.

[54] Charles Chasie and Harry Fecitt, p. 9.

[55] Easterine Kire, p. viii.

[56] *Memories of a Forgotten War.* Utpal Borpujari. 2015.

[57] This is an illustrative rather than an exhaustive list of veteran writings on the war in the region: John McCann, *Echoes of Kohima* (Chadderton: J. McCann, 1989), D. D. Rooney, *Burma Victory: Imphal and Kohima, March 1944 to May 1945* (London: Cassell, 2000, c1992), Arthur Swinson, *Kohima* (London: Head of Zeus, 2015), R. E. S. Tanner, *Burma 1942: Memories of a Retreat: The Diary of Ralph Tanner, 2nd Battalion the King's Own Yorkshire Light Infantry* (Stroud: History Press, 2009), and Geoffrey Tyson, *Forgotten Frontier* (Calcutta: W. H. Targett & Co Ltd, 1945).

refer to the war in the region as the 'forgotten war'. In part, it was due to the fact that as the China–Burma–India theatre veterans returned home after the war, and as many of the previously colonised countries became independent states, the impetus for the public recalling of the experiences of the war in the region had little to no currency both at home and in the ex-colonies. But even during the war, it was widely felt that the resources allocated to the theatre weren't adequate, the staff on the ground often having to contend with a dangerous scarcity of essentials such as clothing and food. Serving in a distant corner of the empire where the empire itself grappled with its geographic, administrative and epistemic limits, also highlighted the 'jarring disjunction between ideology and lived experience'.[58] Many of these otherwise military history-focused accounts reveal anxieties arising from the absence of a clear official justification for the war as was the case in Europe.

The Indo-Myanmar region, constituted even today as a borderland, was the erstwhile colonial 'frontier' where British control diminished and frayed. Military reconnaissance against the 'enemy' had to be carried out in a terrain that was still largely unmapped. Oatts, the Commander of the Chin Levies, a special wartime force made up of Kachin people, describes one such march:

> We were in the northern Naga Hills, that high, wild country, most of it unexplored and unmapped, which lies between the sources of the Upper Chindwin and the Patkoi Bum—the great mountain range which forms a natural barrier between Upper Burma and Assam. Although I did have a map of sorts, it had two large, blank spaces on it marked 'unsurveyed', and we were now in about the centre of the most northerly of these spaces, into which no white man had ever penetrated before.[59]

Oatts' prior work in the region, in fact, had been related to 'empire-building activities',[60] wherein he was tasked with penetrating 'the whole Naga territory in order to proclaim that it would henceforth be taken into the administration of the Burma Government'.[61] Yet overnight, the object

[58] Nancy Martin, '"And all because it is war!": First World War diaries, authenticity and combatant identity', in *Writing War, Writing Lives*, ed. by Kate McLoughlin, Lara Fiegel and Nancy Martin (London : Routledge, 2018), pp. 27–45, p. 39.

[59] Lewis Balfour Oatts, *The Jungle in Arms* (London: W. Kimber, 1962), p. 13.

[60] Lewis Balfour Oatts, p. 26.

[61] Lewis Balfour Oatts, p. 26.

of his adversarial surveillance shifts from the Nagas—described by him as 'a sore trial to the Burma Government [...] and to a lesser extent to the Indian Government'[62]—to the intruding Japanese army. His trepidation about the now murky category of the 'enemy' is revealed more the farther he leads his troops into the territories of the peoples who were previously enemies who had to be disciplined through various 'punitive expeditions' but who had now become the host communities whose cooperation and goodwill were essential to the success of their military campaign. He writes anxiously, 'The inhabitants of these unmapped territories had [...] a very bad reputation, even among the Nagas. No expedition had ever been there before, and I was not at all certain that they would obey the laws of chivalry'.[63] It is unclear which particular community he is referring to here but it is interesting to note that the notions about the internal enemy merges with the invading enemy, i.e. by this time, the Japanese army were infamous for their perceived brutality and for disregarding the conventions of military and international law. In contrast to the new enemy, he now describes the Nagas as 'most chivalrous people'[64] who would 'never attack an armed force moving openly against them without first talking things over—or at least offering the opportunity for doing so'.[65]

Previously the object of subduing and administrative management, the peoples of the region are transformed into harmless onlookers caught in the frontline, who in many ways injected a sense of normalcy to their extraordinary wartime routines. Writing about the Arakan campaign in *We Fought at Kohima* (2015), Street recalls that 'many strange things happen[ed] during the war'.[66] He remembers how 'young village lads about ten or twelve years old [...] herding [...] cattle close by'[67] would make 'several trips to our lines selling things'[68] such as wild bananas, jaggery and sheets of bamboo matting, indifferent to the fact that this was 'the front line, with both sides ready to fire at any time'.[69] Baffled, even amused, by the appearance of quotidian commerce in a war front, he

[62] Lewis Balfour Oatts, p. 26.
[63] Lewis Balfour Oatts, p. 14.
[64] Lewis Balfour Oatts, p. 14.
[65] Lewis Balfour Oatts, p. 14.
[66] Raymond Street, *We Fought at Kohima: A Veteran's Account* (Barnsley, South Yorkshire: Pen & Sword Military, 2015), p. 35.
[67] Raymond Street, p. 35.
[68] Raymond Street, p. 35.
[69] Raymond Street, p. 35.

notes, 'Some blokes bought them ["six-foot sheets of bamboo matting"] to sleep on during the two-hour rest periods in guard duty',[70] and adds, 'We admired the bravery of these young lads',[71] marking a turn in the perception of the natives.

In fact, in 1941, still a couple of years away from the confrontation between the British and the Japanese forces, Mackrell, a tea garden manager who volunteered to rescue the refugees from Burma in what came to be known as the Chaukhan rescue operation, notes in his diary, 'I close this diary with the statement that I have already made to everyone [...] too much credit has been given to me [...] not nearly enough to the Rifles, Porters, Mahouts and Boatmen [...] without the latter splendid fellows, little or nothing could have been done in time to save the bulk of this party'.[72] It becomes apparent early on to both the Allies and the Japanese that the cooperation of the host communities—their knowledge of their territory, their conscripted labour and their skills in manoeuvring non-human 'resources' such as elephants[73] as was illustrated in Mackrell's rescue operation—was going to be a determining factor in the trajectory of the war. They also realise that the region's liminality was going to be both a challenge and an opportunity for the intruding parties.

The wartime rendered the enemy into an indispensable friend. The peripheral and partial colonial subjects who were 'a sore trial'[74] to the British empire overnight became 'volunteers' and 'splendid fellows'[75] who displayed incomparable 'gallantry and endurance', and without whom 'both in the construction of the [Stilwell] road and in supplying the troops [...] the Japanese would have reached India'.[76] So vital were the new tribal friends that 'when the medals were handed out at the end of the war', it was retrospectively perceived as shocking that 'tea garden labourers [...] Nagas, Abors, Mishmis and others did not dangle MBEs from their

[70] Raymond Street, p. 35.

[71] Raymond Street, p. 35.

[72] Andrew Martin, *Flight by Elephant: The Untold Story of World War Two's Most Daring Jungle Rescue* (London: Fourth Estate, 2013), p. 305.

[73] For a detailed discussion on the entanglements between the empire and non-human animals, see: Jonathan Saha, *Colonizing Animals* (Cambridge: Cambridge University Press, 2021).

[74] Lewis Balfour Oatts, p. 26.

[75] Andrew Martin, p. 305.

[76] Alan Macfarlane, *Green Gold: The Empire of Tea* (London: Ebury, 2003), p. 222.

chests'.[77] Captain Ralengnao Khathing, who had led the local intelligence gathering unit, V Force, was in fact awarded an MBE[78] and the Military Cross 'for highly distinguished service and gallantry in action'.[79] He is credited to have

> organised and controlled a most efficient system of obtaining information from villagers [...] imbued them with the spirit to fight against the Japanese and to give troops every assistance [...] organised the collection of supplies for the column. Without him it would NOT have been possible for the column to live on the country and so carry out its task.[80]

Khathing is described to have carried out this 'successful' operation and, as a result, caused 'great embarrassment to the enemy's 15th Division'[81] by cutting off the 'Japanese sources of supply [...] [and] Line of Communication' while 'information of the greatest value was obtained'.[82] Additionally, about 150 Japanese soldiers were estimated to have been killed and '43 Jiff (INA)[83] Prisoners of War [...] captured'.

However, for every debatably 'friendly' gesture towards the Allies, the host communities also mounted several obstructions. The so-called volunteers, who were usually conscripted workers, were continually 'absconding'. The diary of the Chaukhan rescue operations between June and October 1942,[84] for instance, is replete with mentions of 'intractable' local conscripts. An entry from 8 June reads, 'the Mahouts were on the verge of mutiny and small blame to them'.[85] Elephants and their mahouts or elephant drivers were indispensable to the rescue operations. Without them, it would have been impossible to transport supplies and frail refugees through the dense, uncharted tracks of the Patkai mountain range. In

[77] Alan Macfarlane, p. 222.

[78] Member of the Order of the British Empire (MBE) is awarded by the British government for services to the local community as per their official website.

[79] Charles Chasie and Harry Fecitt, p. 234.

[80] Charles Chasie and Harry Fecitt, p. 235.

[81] Charles Chasie and Harry Fecitt, p. 234.

[82] Charles Chasie and Harry Fecitt, p. 234.

[83] The Indian National Army, also known as the Azad Hind Fauz, was an armed Indian nationalist movement led by Subhas Chandra Bose and a collaborator with Japan during the Second World War.

[84] Gyles Mackrell, *Mackrell Papers*. 1942. Centre for South Asian Studies, University of Cambridge, United Kingdom.

[85] *Mackrell Papers*.

order to curb daily absconsions and revolts, the mahouts were enticed with opium. Illustratively, the entry from 8 September states, 'Opium Mahouts, Raghoo and 2 others causing trouble. Opium left at Dapha but sent for some'.[86]

The non-complaint mood was also widespread among other workers in the operation. As early as on 19 June, the diary reported that 'Some of the porters have already absconded and only a small number of the total Capt. Wilson brought seem really useful'. On 9 July, Wilson himself reports, 'He [Mackrell] had sent word that another 4 porters had run away, we are now left with practically nothing except the Nepalis I picked up at Miao. The original from Lekhapani are either sick or absconded!' If they were not absconding, they cited illness or refused to work outright. The 9 July entry reports, 'Four of the Porters Wilson brought absconded last night on foot after saying they were too ill to do any work' and one from the 5 July elaborates, 'Porters are still hobbling about, except for 2. They go sick, some or most genuine. Others probably not [...] it is only by absolutely driving them, that one can even get a few odd camp duties done'.[87]

The ranks of the porters were made up mostly of indigenous peoples, such as the Adis[88] and the adivasi[89] labourers from tea plantations in Assam—all of whom are listed simply as 'porters' in the rescue operation diary. The indigenous Mishmis, uncommonly, are addressed specifically, owing to their indispensable and valued role in helping the rescue team navigate the territory. At the start of the operation, the diary indicates that they are cooperative and describes them appreciatively, although, unsurprisingly, they are also liberally supplied with opium. Despite this, the Mishmi guides unambiguously express that they 'want to go home',[90] perhaps even escalating their defiance as one entry hints apprehensively: 'Mishmis have refused to go over again to help. Their headman is sick [...] Mackrell thinks he has not the control he had when he was fit'. Friendship in wartime Indian Northeast was sought out, sought after, bought and extracted.

The notion of the 'enemy' too was continually subverted by the peoples of the region. In *The Wild Green Earth* (1946), Ferguson, who was a

[86] *Mackrell Papers.*

[87] *Mackrell Papers.*

[88] Listed as 'Abors' in colonial archives.

[89] 'Adivasi' refers to the tribal peoples from the Indian mainland who were brought in to work in tea plantations in Assam in the nineteenth century by the British.

[90] *Mackrell Papers*, 4 July.

part of the special wartime force called the Chindits, described an instance when some Naga tribes tried to manoeuvre the wartime conditions to gain political advantage against their territorial adversaries. On 'his arrival at the Chinese-garrisoned village of Hkalak Ga in Naga territory',[91] Ferguson writes:

> I liked what little we saw of the Nagas. They were much more jungly and primitive than those other Nagas who live near the Manipur Road [...] They would do nothing for money; opium was their only currency [...] and without it the Nagas would neither work nor give us information [...]
>
> The relations between the villages were strained [...] they would not band as a nation against us or [...] the [Japanese][92]. We had amusing confirmation of this [...] when we intercepted a runner carrying mail for the [Japanese] south of the Chindwin. Among his dispatches was one from a junior [Japanese] officer to a senior, complaining [...] for the umpteenth time that the Nagas wouldn't help the [Japanese] against us, but were continually asking the [Japanese] for help against each other.[93]

Allyship, similarly, found itself in fractured terrains. In his first-hand account of the hasty evacuation from Burma in 1941 in a group led by the American general Joseph Stilwell, the *Time* war correspondent Jack Belden reveals one side of the mutual reservations between the British and the Americans, usually considered the most natural of allies. Belden described the British Army in Burma as a 'colonial army of occupation' with 'no practical and varied experience of modern war', ill-equipped and unprepared 'to fight a war against a powerful invader' who had 'organised for total warfare, with supreme confidence [...] armed with an appeal to the native population to help them overthrow the British'.[94] The central contradiction of the war wherein the British 'tried to invest their cause with a moral purpose by calling it a "people's war," a "war for freedom," or a "war for our way of life"' but 'could not and would not lead the Asiatic masses in a genuine struggle for liberation against both democratic and Fascist tyranny' was also not lost on him.[95] Given this unresolvable

[91] Quoted in Charles Chasie and Harry Fecitt, p. 149.

[92] The original word used in the quote was a commonly used racial slur. I've replaced it with the neutral 'Japanese' to convey the intended referent here and in the rest of the chapter.

[93] Quoted in Charles Chasie and Harry Fecitt, p. 150.

[94] Jack Belden, *Retreat with Stilwell* (London: Cassell, 1943), p. 7.

[95] Jack Belden, p. 8.

contradiction, although the British 'had an intense belief in their cause at home, they were moved to no high patriotic pitch in defending the colonies abroad' and 'had to continue its police work at the same time that it opposed a *new* adversary'.[96] But even as Belden lambasted Britain's 'colonial bureaucracy' and their 'blind faith in the power of an empire that was crumbling beneath their feet', he betrayed the limits of his anti-British empire rhetoric.[97] The 'crumbling' British empire, he wrote, had nothing but 'a tradition of victory only over unruly natives', adversaries who were at best minor nuisances compared to the place that the Japanese occupied in the Western imagination during this time as the meteorically rising Asian threat.[98] Their perceived superior 'moral homogeneity' gave them an impetus that the British army with its 'racial mixture' of 'Indian, Karen and Kachin soldiers' who were 'fighting merely for pay and not for any great cause' could not muster.[99]

The British shared similar sentiments of scepticism towards the rise of the American empire and influence. In his 1936 essay about the intimate workings of colonialism in British Burma, George Orwell momentarily gestures towards the overlap between the two empires and his seemingly accidental presence in what would soon become a playground for the two allied but contending empires. He writes, 'I did not even know that the British Empire is dying, still less did I know that it is a great deal better than the younger empires that are going to supplant it'.[100] In particular, the British did not hold the US-led strategies in the region in high regard. Churchill famously described the construction of the Stilwell road as the primary supply route to China as 'an immense, laborious task, unlikely to be finished until the need for it has passed'.[101] Aside from internal friction between the top echelons of the US military and the British army, there was also fundamental difference in the perceived objectives of the war in the region. For the British, 'the retaking of Burma was a worthy end in

[96] Jack Belden, p. 7. Emphasis added.

[97] Jack Belden, p. 7.

[98] Jack Belden, p. 7.

[99] Jack Belden, p. 7.

[100] George Orwell, 'Shooting an Elephant' (*The Orwell Foundation*, 20 Oct. 2010), <https://www.orwellfoundation.com/the-orwell-foundation/orwell/essays-and-other-works/shooting-an-elephant/>.

[101] Francis Pike, *Hirohito's War: The Pacific War, 1941–1945* (New York : Bloomsbury Academic, an imprint of Bloomsbury Publishing PLC, 2016), p. 892.

itself,' but for the Americans, the objective was to assist China, making 'the reconquest of Burma [...] 'merely incidental'.[102]

At the same time, other parallel projects of allyship were also being forged. The INA, after collaborating with Japan in capturing British Burma, made their way to the Indian Northeast to fight alongside the Japanese army. Interestingly, assistive gestures were extended towards the Japanese-INA alliance by the peoples of the region. In fact, even before the Japanese army and the INA had reached the vicinity of the kingdom of Manipur, the Kukis of Manipur had already made contact with them by 1943.[103] Lambert, a Central Intelligence Officer, reported at the time that 'Tongkhotang, Chief of Haokip clan and son of Pache [...] was believed to have crossed Chindwin River [...] to contact the Japanese and have asked for 400 rifles to fight against the British'.[104] The Kukis went on to assist the Indo-Japanese army with 'materials, labour, intelligence services and campaigns [...] [and] men' throughout the war.[105] Moreover, the Kukis were not the only group to have forged an alliance with the Japanese. Angami Phizo, the leader of the Naga national movement, also cooperated with the Japanese by providing intelligence, 'with the hope that Bose and the Japanese would enable him to establish a separate Naga state'.[106] Notably, Tongkhothang, Chief of Haokip Kuki clan, had been one of the leaders of the 1917 Kuki rebellion against the British. The historian Jangkhomang Guite insightfully locates the assistance of the Kukis to the Indo-Japanese army—and by extension, Phizo's support to them—as 'separate resistance movement[s] embedded within the larger war'.[107] So, although geographically contained and arguably minor, these alliances were a part of the autonomous anticolonial projects in the region with their own distinct imaginations of indigenous postcolonial futures.

There were also indirect forms of anticolonial resistance that often gave way to friendly interactions with the Japanese. Desertion from the British

[102] Francis Pike, Chapter 11: "Limits of Empire: Doolittle and New Military Strategies"; Epub; 2016 publication; Source: https://www.bloomsbury.com/uk/.

[103] Jangkhomang Guite, 'Representing Local Participation in INA–Japanese Imphal Campaign: The Case of the Kukis in Manipur, 1943–45', *Indian Historical Review*, vol. 37, No. 2, Dec. 2010, pp. 291–309.

[104] Jangkhomang Guite, p. 292; quote from document in Assam State Archives.

[105] Jangkhomang Guite, p. 291.

[106] Namrata Goswami, Ed., 'Phizo: The Charismatic Protagonist of the Movement', *The Naga Ethnic Movement for a Separate Homeland: Stories from the Field* (Oxford: Oxford University Press, 2020), p. 19.

[107] Jangkhomang Guite, p. 291.

intelligence gathering unit, V Force, was frequent. Many joined the Japanese and INA camps after their desertion. For instance, one Hemkholet Touthang Kuki joined the V Force in 1943 only to desert them to join the INA Battalion, for which he was 'arrested and put in jail for six months'.[108] Given the fluid loyalties and compulsions for political self-determination, the indigenous communities were also subjected to constant suspicion and surveillance. At the end of the war, about 70 Kukis were 'charged with assisting the Indo-Japanese forces',[109] out of whom 8 were taken to trial. However, it is the extra-archival sources which are especially revealing of the dimensions of resistance which do not clearly present as either individual or organised action. In the Kuki Japan Gal La, or the orally transmitted 'Song(s) of the Japanese War', the historical sentiments of euphoria at the promise of the Japanese army's arrival remain fossilised securely:

> From the horizon of Burma plain;
> The Japanese plane floats its sweet note;
> Like the sweet note of the bamboo flute;
> Japanese plane hums in the sky above;
> When the Japanese plane floats its sweet note;
> The depressed farmer forgets the sadness...[110]

In contrast, the sentiments towards the British forces are far less genial: 'March on Britishers / Japanese soldiers are here to kill you / I knew Japanese bravery / British power wants to imitate them / But it will be impossible'.[111]

* * *

Until this moment, the hierarchical racial system of dominance and management of populations in the colonies[112] had served its purposes well. The war in the Indian Northeast foregrounded the malfunctioning of the simplistic categorisation of the colonised peoples along the

[108] Jangkhomang Guite, p. 294.
[109] Jangkhomang Guite, p. 296.
[110] Jangkhomang Guite, p. 300.
[111] Jangkhomang Guite, p. 301.
[112] See: Franz Fanon, *The Wretched of the Earth* (Harmondsworth: Penguin, 1971), and Tayyab Mahmud, 'Colonialism and Modern Constructions of Race: A Preliminary Inquiry', *University of Miami Law Review*, vol. 53, no. 4, 1999, pp. 1219–1246.

epidermal and physiognomic logic of colonial racial ideology in the face of the complexities presented by the political and epistemic liminality of the borderworld.[113] The ideological terrain demarcating the lines between 'enemy', 'friend', 'ally', and 'subject' dissolved and coagulated indeterminately in the region. In Allied veterans' accounts, which form a substantial body of writing on the war in the region, this malfunction becomes clear. A recurrent strand in these writings is the characterisation of the war in the region as distinctly gruesome and 'savage'. The 'savage' nature of the combat conditions included the challenges of navigating inhospitable jungles often described as 'green hell' and confronting the enemy at close quarters. This was compounded by their perceived engirdling by 'head-hunting' tribes, second only to cannibals in the colonial scale of savagery. These writings recast the perceived savagery of the location and its peoples into the savagery of the enemy. The war conditions relayed in many of the memoirs speak of the abrupt reduction of the self to bodies bereft of agency and function. Thirst, hunger and lack of sleep are described as the base reality of the battlefield, with the stench of human excretion, rotting carcasses of friends, enemies and mules, and the noise from the constant shelling and bombardment serving as its atmospherics. This debased state of lull punctuated the heightened periods of action. Street, who was briefly present at the 'Siege of Kohima' in April 1944, describes several instances of confronting the enemy at such close quarters that the confrontation frequently progressed to 'savage' hand-to-hand combat.

In one instance, he declares, 'You couldn't tell friend from foe, not until you were right next to them. Then it was too late, and hand-to-hand combat started'.[114] This takes on several meanings in the context of the region. The physical proximity to the enemy was aggravated by the confusion created due to the imperial tendency to group people based on epidermal and physiological categories of race. Street recalls one such moment of confusion, wherein the logic of the category of the 'mongoloid' race glitches:

[113] See: Mabel D. Gergan, and Sara H. Smith 2022, and Wouters, Jelle J. P. Wouter and Tanka B. Subba, 'The "Indian Face," India's Northeast, and "The Idea of India"', *Asian Anthropology*, vol. 12, no. 2, Dec. 2013, pp. 126–140.

[114] Raymond Street, p. 121.

> About three in the morning, three men I thought were non-combatants appeared. Somehow they had got through our front line positions and now were standing several yards away from me. I ordered them to halt. They huddled together, none of them reached for any weapons, and I had my rifle levelled at them. They had no means of identification, so I was uncertain what to do. Apart from the Royal West Kents at Kohima there were hundreds of *Asian* troops. Many didn't have British steel helmets and some had incomplete or no British uniforms [...] it was hard to say if someone unwashed and dressed in this or that was on our side or the enemy's. Many were unable to speak English, and the English of those that could was rather hit and miss. I was faced with a difficult decision: should I blast them away, only to find out I had killed or wounded three unarmed, innocent *local tribesmen* and caused the already exhausted troops to stand to unnecessarily — or should I let them pass?[115]

This racial confusion compounded by the lack of identification, and coupled with local anticolonial sentiments, was also mobilised by the Japanese army. Several accounts register the Japanese using racial camouflaging as a military tool. Asem Tomchou from Manipur recalls how '5-6 Japanese came in disguise as local people with the traditional baskets filled with fruits and vegetables (Shams) and did a reconnaissance of the British area at Oinam bazar' one time.[116] The following midnight, they launched an attack on the British position in the area, leading to 'many' British casualties, and after which, 'the area was filled with Japanese [... for] some time'.[117] The Japanese also used racial passing for survival and sustenance. One Hemabati remembers them going about in 'phanek (sarong) in the day time' and that 'it was easy to mistake them for local tribal women'.[118] She also recalls that one time 'some of them in sign language indicated they were hungry'; she 'boiled [some] eggs and offered them on plantain leaves [... which] they relished'.[119] Racial camouflaging also helped the Japanese in flitting in and out of the British-surveilled villages without detection. Pastor Kiezotou from Nagaland recollects, 'There were so

[115] Raymond Street, p. 111. Emphasis added.
[116] Hemant Singh Katoch, *The Battlefields of Imphal: The Second World War and North East India* (Abingdon: Routledge, 2016), p. 30.
[117] Hemant Singh Katoch, p. 30.
[118] Hemant Singh Katoch, p. 26.
[119] Hemant Singh Katoch, p. 26.

many Japanese soldiers in our village, but if the British came, they pretended to be Nagas and would ask our […] clothes so they could cover themselves […] and look like us'.[120]

The Japanese also used the ideas of pan-Asianism and racial affinity to garner support from the natives. They invoked a mythical connection with the Kuki people by claiming that they had 'descended from "chongji" brother of "chongthu" from whom the Kukis claimed their descent'.[121] They also tapped into the Kuki prophesy of 'Nongpokthong hangani'—which predicted that the British rule will end when the 'eastern gates will be opened'[122]—by claiming that they 'represented the eastern power'.[123] Similarly, they invoked racial affinities between themselves and the Nagas, asserting that both were of the 'Mongoloid stock' and the British, in contrast, were 'so different'.[124] They claimed that 'they had come to liberate the Nagas' and promised them 'independence from the British' and 'economic development after they won the war'.[125] The uptake of the notions of racial affinity was surprisingly successful. Many Nagas believed that 'the Japanese were […] [their] elder brothers'.[126] For instance, villages in the area of Meriema and Viswema 'had stories about being related to the Japanese' and the Indo-Japanese forces were received with enthusiasm when they first arrived.[127] Notably, the INA tried to invoke similar anticolonial sentiments against the British among the Nagas, which did not have the same success as 'the Nagas generally did not like the plains people and felt no affinity with them'.[128] In fact, accounts suggest that 'bad behaviour' by the INA soldiers turned the Nagas against them,[129] signalling a history of colonial racial configurations which has repercussions for the region to this day. In turn, ideas of racial difference from South Asia and affinity with the 'Mongoloid stock' were also cited by the Naga nationalist

[120] Charles Chasie and Harry Fecitt, p. 67.

[121] Hemant Singh Katoch, p. 36; quote from N. Lokendra Singh, *Manipur During World War II (1941—45): Socio-Economic Change and Local* Responses (Imphal: Manipur State Archives, 1993).

[122] Berenice Guyot-Rechard, 'When Legions Thunder Past: The Second World War and India's Northeastern Frontier', *War in History*, vol. 25, no. 3, July 2018, pp. 328–360.

[123] Hemant Singh Katoch, p. 36; quote from N. Lokendra Singh.

[124] Charles Chasie and Harry Fecitt, p. 64, p. 59.

[125] Charles Chasie and Harry Fecitt, p. 64, p. 59.

[126] Charles Chasie and Harry Fecitt, p. 64.

[127] Charles Chasie and Harry Fecitt, p. 64.

[128] Charles Chasie and Harry Fecitt, p. 64.

[129] Charles Chasie and Harry Fecitt, p. 64.

leader Phizo to stake claims for an independent Nagaland after the overthrowing of the British from the Naga hills. Naturally, the Northeastern peoples continue to invoke race, along with their indigenous and/or tribal status, in their expressions of self-fashioning and self-differentiation in relation to the Indian mainland.

Conclusion

Wars dictate and prefer unambiguous demarcation of the lines between the categories of enemy and friend, but in the wartime Indian Northeast, the two superstructures were both breached *and* reinstated. The specificity of the context—characterised by ongoing empire-building activities paradoxically in the final years of the British empire, the rising involvement of the USA in Asia and the production of the region as remote and existing in the interstices of India and Burma proper—evinced a whole host of complications. First, the Allies had to manage two levels of colonial anxiety and suspicion: one against the external enemy, Japan, and the other against the 'unruly' colonial subjects who were until recently the internal enemies. Second, the context rendered possible multiple formations and destabilisations of allyships. There were frictions between the Americans and the British regarding both strategy and ideology, as much as the war afforded opportunities for anticolonial formations such as the one between the Japanese, the INA and the Kukis of Manipur. The 'enemy' within and outside became an amorphous category, coalescing into different, porous formations of friends and anticolonial allies. The indigenous peoples from the region unwittingly caught in the war witnessed, cooperated, participated and subverted expectations of the two sides, all the while also partaking in the hierarchical and lateral conversations on race and making claims to independent, indigenous futures. Thirdly, the war laid bare the empire's racial logic *and* exposed its limits. The British grappled with racial confusion, often struggling to confidently distinguish between indigenous peoples of the region from the Japanese in battlefield conditions. The Japanese took advantage of this ideologically as well as strategically by invoking racial affinities and anticolonial Pan-Asian sentiments and through the military strategy of racial camouflaging.

At no point in time were the global processes of the racialisation of the region in both its hierarchical and lateral movements laid bare more transparently than during the Second World War. This is especially pertinent because the post-independent Indian nation refuses to acknowledge race as a factor in its engagement with the region. There is, in fact, a national and historical logic at play when the massacre at Oting in Nagaland is explained away as a case of 'mistaken identity'. Within this logic, it is entirely credible to confuse ordinary citizens from the Indian Northeast for internal enemies because they remain, in effect, a racialised category in a region apart, whose remoteness has been systemically sustained and reified by the post-independent Indian state.[130]

[130] I'd like to extend my deepest gratitude to the Cambridge Trust, Wolfson College, Centre for South Asian Studies and the Faculty of English—without the support of these nodes at the University of Cambridge and the wonderful people who make and sustain these nodes, this chapter would not have materialised.

Enemy Encounters in the Imperial War Museum's Second World War Galleries

Vikki Hawkins

In October 2021, new permanent Second World War Galleries and Holocaust Galleries were opened simultaneously at the Imperial War Museum in London. Spanning two floors and occupying over 3000 m^2, the £30.7 million project, which took five years to complete, put on display more than 3500 items and personal stories from more than 80 countries. To make the story of the Second World War more relevant to twenty-first-century visitors, the ambition for the new Second World War Galleries was to place the British experience of war into a wider, global context and deliver a narrative that explained how this global war continues to shape our modern world.

This was a move away from the previous Second World War Galleries (opened 1989), which were narratively Eurocentric, emphasising events within the British popular myth of the war. In those spaces, the footprint given over to showcases relating to the Blitz and the Battle of Britain was about equal to the total and entire campaign in the Pacific and the war in

V. Hawkins (✉)
British Museum, London, UK
e-mail: VHawkins@britishmuseum.org

H. Furneaux, M. Greig (eds.), *Enemy Encounters in Modern Warfare*, https://doi.org/10.1007/978-3-031-56748-3_10

eastern Europe. Visitors would have fairly concluded that within the strategic balance of the war, these events carried the same weight, or affected a comparable number of lives. Where events in Asia were discussed, they reflected British and Commonwealth experiences: the Burma Campaign and prisoners of war held by the Japanese were well documented, but Americans and Australians taking part in the island-hopping campaign across the Pacific, Chinese, and Japanese battles in China, or life on the non-British home fronts, were rarely mentioned.[1] The 'enemy' had an abstract presence. Although the leadership of Axis countries was mentioned and there was a sense that the enemy posed a constant threat, objects related to the enemy on display included battle souvenirs, or military equipment with no personal provenance. Individual people from Germany and Japan were not made visible other than through unnamed photographs of battles or scenes of atrocity. Their voices, perspectives, motivations, or responses to the conflict were de-prioritised. In a reflection on IWM's past exhibitions and practices, historian Roderick Bailey acknowledged that the museum has tried to move away from this approach, recognising that 'today, IWM's curators aim to confront established narratives, overturn myths, and encourage visitors to engage with what war is actually about'.[2] This is exactly what the Second World War curatorial team hoped to achieve. Confronting how the enemy was represented within the narrative and through the material culture displayed was vital in engaging visitors to think critically about the cause, course, and consequences of war.

Presenting the complex, global history of the Second World War in a limited space was challenging, and the curatorial team needed to identify interpretative approaches that would guide without overwhelming visitors through the retelling of the conflict. Consequently, the narrative of the galleries is woven around the theme of 'total war', defined as the unprecedented scale and excesses of modern warfare which affected almost everyone across the world. In the displays, transnational links between countries enable visitors to understand the concept of 'total war'. The extensive

[1] Vikki Hawkins, 'Displaying Marginalised and "Hidden" Histories at the Imperial War Museum London: The Second World War Gallery Regeneration Project', *War & Society*, 39.3 (2020), p. 210.

[2] Roderick Bailey, 'Commemoration, Veneration, Reflection: The Imperial War Museum and the Past', in *Gehören Ins Museum! Aber Wie? [Wars Belong in a Museum! But How?]* ed. by Elena Messner and Peter Pirker (Vienna: Edition Atelier, 2021), p. 163.

mobilisation of people, bombing of civilians, occupation, collaboration and resistance, imprisonment and detention, mass killing, forced labour, sexual violence, and separation and loss get to the heart of the Second World War experience, and the new galleries emphasise those features that distinguish it from other, previous conflicts. Bailey called for IWM to address 'offences like murder, sexual violence, and torture' which 'speak to a wealth of important themes and discussions relevant to IWM's focus, from power dynamics and behaviours within armed forces and conflict zones, to the suffering of victims and issues of accountability'.[3] Using 'total war' as an interpretive tool allowed for this: the case studies included in the galleries clearly show that the effects of 'total war' were not exclusive to one country and were instigated by both Axis and Allied countries. In some instances, the universality of 'total war' experiences blurred the lines between who fought on which side. However, as Stephan Jager correctly asserted, 'if the visitor merely sees images of suffering that seem similar, an exhibition lacks the historical specificity for the visitor to make connections between them. The visitor might just take away a general message of the universality of suffering'.[4] Therefore, the displays also provide conceptual and thematic examples of 'total war' within their historical context and alongside nation-specific examples, positing different readings, comparisons, and discussions for the visitor. A clear example of this can be seen through the galleries' representation of civilian bombing.

The Blitz looms large and familiar in the public imagination and IWM holds a large amount of material culture that records experiences of the Blitz—the aerial bombing campaign against civilians during the attacks on Britain in 1940 and 1941—including uniforms and equipment used by the civil defence services, personal diaries, and letters describing reactions to and lived experiences of the raids, and propaganda leaflets and posters advising on precautions including the building of shelters. 'These recognisable – even mundane – objects have the capacity to invoke memory with our visitors and encourage intergenerational discussions between them. To encourage that same level of engagement within a new global narrative structure, IWM uses the theme of total war to explore the experience of

[3] Bailey, p. 162.

[4] Stephan Jaeger, *The Second World War in the Twenty-First-Century Museum: From Narrative, Memory, and Experience to Experientiality* (Berlin/Boston: De Gruyter, 2020), p. 60.

aerial attacks elsewhere'.[5] On display for the first time is a Chinese shelter admission ticket facsimiled with the permission of the Three Gorges Museum in Chongqing. It highlights the plight of Chinese civilians in Chongqing who were forced to hide in caves built into the mountainous terrain to hide from the Japanese bomber aircraft. Rana Mitter, a historian of Chinese history, stated that 'most westerners have scarcely heard of the bombing of Chongqing' due to the fact that 'the events of the conflict took place far from American and European eyes and had little relevance for anyone other than the Chinese themselves'.[6] Mitter expanded that following the massive air raids of 3 and 4 May 1939, Chongqing became 'a symbol of resistance to people around the world, who were now certain that a global war could not be far off'.[7]

It is not just Britain and its ally China who were the victims of bombing raids (Fig. 1). A collection of material from the German home front including an Air Raid Precautions helmet, a poster warning civilians about the importance of a blackout, and a fireman's jacket, serve to both remind the visitor of objects they have encountered in the Blitz section of the galleries, and to clearly show that these objects were used by and belonged to German people. To emphasise the personal experience within the universal, the story of Dieter Classen is included. A teenager living in Essen, Classen was forced to transport the wounded and dead to a hospital after the bombing raids using a cart attached to the back of his motorbike. In a nearby showcase, a Japanese flag hangs over a display section related to the bombing of Japan from American bases in the Mariana Islands. Characters signifying that water is close by appear on the flag, most likely hung by air raid personnel for easy identification after bombing raids. Video footage of German and Japanese people lying wounded or dead in the streets or in hospitals shows the devastating effect of the Allied raids on civilians. To add further complexity to the narrative, pamphlets written by lobbyist pacifists outraged by the bombing, show how some British individuals questioned the morality of such tactics against enemy civilians. In her book *Seeds of Chaos*, on display in the galleries, Vera Brittain writes, 'no doubt there are many non-adult minds which find reason for satisfaction in the anguish that we have caused to the enemy. But others will reflect

[5] Hawkins, p. 213.

[6] Rana Mitter, *China's War with Japan 1927–1945: The Struggle for Survival* (London: Penguin Books, 2014), p. 10.

[7] Mitter, p. 10.

Fig. 1 IWM's new Second World War Galleries. These showcases feature objects related to the bombing of Chongqing including Chinese shelter admission tickets. Displayed in the background is a piece of the American battleship USS Arizona which was sunk at Pearl Harbor by the Japanese Navy Air Service. © IWM

more reasonably that each one of those million dead (to say nothing of the injured and several million homeless) have relatives and friends who will remember'.[8]

Despite having limited personal encounters with 'enemy' civilians, bomber crew did sometimes come face to face with the people of the countries they were targeting. The galleries highlight an example from 26 August 1944, two days after a B-24 Liberator bomber from the US Army Air Force was shot down over Hanover. Nine crew members were captured by Luftwaffe personnel and held as prisoners of war. The Americans were marched through the town of Rüsselsheim, damaged by an RAF bombing raid the previous night. German civilians attacked the POWs with rocks, shovels, and hammers, killing six of the men. The German civilians chose to show their anger at the devastation of homes and loss of

[8] IWM, LBY 80 / 2317, Vera Brittain, *Seeds of Chaos: what mass bombing really means* (London: New Vision Pub. Co, 1944), p. 33.

life inflicted by enemy aerial bombing by targeting individuals who represented that aerial threat the previous night. These are complex and complicated examples of civilian bombing during the Second World War. In contrast, the tendency in the past has been to focus on a simplified and sanitised depiction of warfare, especially bombing. The British perspective of the Blitz is dominated by narratives of people rallying together against the enemy, maintaining morale and fighting back with brave pilots and crew. By including objects related to the aerial bombing in both enemy and allied countries, personal experiences of people from multiple countries who feared the threat from the air, and different perspectives on the morality of such tactics, visitors are faced with a more complex history of events. In a total war, the bombing of civilians is a trans-national experience.

Another key principle in the representation of a global narrative was to bring encounters with individuals into focus, with a very strong emphasis on human stories and personal narratives. Viet Thanh Nguyen in his study on the memorialisation of war comments that 'things exist even when humans die, and thus museums often give more space to things than people'.[9] Yet in the IWM galleries, we wanted visitors to feel that they are not just seeing objects, but meeting people from the past, and from around the world. These individuals provide a window onto a wider cultural context of war, perhaps one very different from the visitors' own or one which challenges their preconceptions of enemy or ally. In the 1990s, the Imperial War Museum sought to inspire empathy with people in the past through immersive experiences including the popular 'Blitz experience' in which visitors would walk through the rubble of a London street, the sound of air raids overhead. Visitors had the opportunity to explore the spaces in an eerie darkness, listening to communal singing, the distant thud of bombs and the occasional cry from 'George' the cockney air raid warden. Visitors entered a shelter which shook, while smoke flooded the space, and a flash of light picked out images of London in flames. The 'Blitz experience' did little to illuminate for visitors the complex narratives and diverse personal experiences of the Blitz. There was no wider context about the mass burials that took place because of the inability of many families to pay for private burials or due to the lack of identification of the victims, the opportunistic crime, or the struggle for those who had been made homeless. Also excluded were objects or personal stories related to the bombing

[9] Viet Thanh Nguyen, *Nothing Ever Dies: Vietnam and the Memory of War* (Cambridge, MA: Harvard University Press, 2016), p. 164.

of civilians across Europe, specifically air raid victims in the enemy countries of Germany and Japan. In essence, the 'Blitz experience' was popular and entertaining, but it lacked integrity and did not reflect the diversity of the people around the world who had been affected by aerial bombing campaigns.

The galleries which opened in 2021 move away from this manufacturing of experiences, while still trying to present sensory moments and tangible interaction for the visitors. For example, in a gallery focused on the British Home Front, visitors encounter an Anderson shelter and a room from a deconstructed 1940s house. They act as set dressing for showcases displaying objects relating to the home front—gas masks, glow in the dark buttons used during the blackout, and documents relating to the safety of pets in an air raid. Overhead, digital screens play audio from air raid sirens and fighter aircraft set against indicative footage of an aerial battle, reminding visitors of the constant threat civilians faced from above. However, unlike 'the Blitz experience' of the previous iteration, the audio has been created to play on a long loop so as not to distract visitors as they explore the gallery. Crucially, the room of the 1940s house is not an exact recreation and doesn't claim to be. Instead, it is an interpretive device: the wallpaper, inspired by 1940s designs, was created using a repeating pattern of photographs from IWM's collection capturing scenes of evacuation and bombing. The mirrors and decorative cross stitch pictures on the walls are used to hold contemporaneous quotes from those living in the 1940s. In addition, the interior of the house includes objects, images, and film showing mass burials, the buying and selling of black-market goods, and the great losses that people endured. The people featured in the new galleries are not abstract, compound people of the past like 'George' who featured in the 'Blitz experience'. Instead, the stories of real individuals from across the world are included in the galleries, meaning visitors encounter authentically a variety of people, places, and situations created by this global conflict (Fig. 2).

There are over 100 people-centred stories in the galleries from over 30 different countries. There are soldiers and civilians, adults and children, and men and women. These people do not represent the beliefs and values of their entire country; instead, they represent a snapshot of ordinary people living through extraordinary times. These individuals had their own thoughts, feelings, and fears. Regardless of which side of the war or which country they were from, they were all humans who experienced joy and pain, who lost loved ones, who saw new places, learnt or discovered new

Fig. 2 In the centre of the home front galleries, visitors are introduced to the stories of civilian defence personnel who experienced the Blitz. © IWM

skills, fought hard for what they believed in and tried their best to survive through acts of bravery, resilience, and personal struggle. The breadth of the personal stories included allows visitors to find moments of emotional connection in the galleries—some visitors see aspects of themselves reflected in the thoughts, actions, and beliefs of those featured. For others, it inspires conversations about how their lives have been shaped by the actions of individuals in their own past. IWM's collection is vast and includes the personal papers, objects, and photographs of thousands of individuals. However, a significant proportion of these individuals are British, often white and male. During the development stage of the new galleries, identifying objects and stories associated with the enemy was challenging as IWM had not traditionally collected this material, other than through the donation of souvenirs by British troops. To rebalance the global perspectives of the conflict, the curatorial team actively sought out the stories and objects of the 'enemy', bringing their voices and experiences to the fore.

One book that had a profound impact on how stories of Japanese soldiers were presented in the galleries is entitled *Kike Wadatsumi no Koe* (*Listen to the Voices of the Lost Sea*). It is a collection of diaries, letters, and poems first published in 1949, compiled by the Japan Memorial Society for the Students Killed in the War—the Wadatsumi Society. The letters from these young Japanese soldiers are deeply moving, full of reflections on the beauty of life and the horrors of war, love felt for their families, loss of the futures they had planned, and the pain of coming to terms with the duty they felt towards dying for their country. The volume had an immediate and dramatic reception in post-war Japan. It was an instant success, with multiple reprints and revisions, but it was not until 1995 that Midori Yamanouchi and Joseph L. Quinn, S.J., translated the book into English 'to give a human face to each of those fallen college/university students' and to show that special attack forces members were not 'mindless, robot-like figures, who simply followed orders and died'.[10] Midori Yamanouchi lived through the war as a young boy, sheltering from air raids with his family and listening to the farewell speeches of the young special attack forces' members on the radio every evening. He reflected, 'As I look back at that time, hatred towards the enemy was not the primary emotion at all, it was almost absent – there was just the sense of fate, helplessness and sorrow'.[11] He hoped that 'after reading this [book], perhaps the peoples of the West may understand the Japanese a little better'.

The first individual featured in the book is Ryōji Uehara, an economics student at Keio University in Tokyo, who was called up to the Imperial Japanese Army in 1934. On 11 May 1945, he was ordered to undertake a suicide mission to fly directly into US ships at Katena Bay, Okinawa. Ryōji wrote several letters detailing his thoughts on the 'enemy'. In early musings, Ryōji quotes Suntzu, the Chinese military strategist of the Chun-Chiu Period, who stated that 'if you know your enemy as well as you know yourself, you could fight a hundred battles and not be in any crisis'. But Ryōji goes on to wonder 'in today's Japan, how many people really know the enemy, America? Anyone who wants to know anything about America, which is a free country, will first have to understand what liberalism is'.[12] He believes that Japanese political and military leaders 'lie to the ignorant

[10] Midori Yamanouchi and Joseph L. Quinn, S.J., *Listen to the Voice of the Lost Sea: Kike Wadatsumi no Koe* (Scranton, Pa.: University of Scranton Press, 2005), pp. vii–viii.

[11] Yamanouchi, p. vii.

[12] Ryōji Uehara, February 7 1945, in Yamanouchi, p. 235.

masses by using twisted eloquence to convince them of how America should be dealt with'.[13] As for Britain, Ryōji writes his 'ambitious hope was to have lived to see my beloved fatherland – Japan – develop into a great empire like Great Britain in the past'.[14] The night before Ryōji's final mission, he wrote a letter entitled 'My Thoughts'. Reflecting on his fate, he grappled with the 'honour involved in being designated a member of the Special Attack Corps', and his belief that a pilot is 'nothing more than that part of the machine which holds the plane's controls – endowed with no personal qualities, no emotions, certainly with no rationality – simply just an iron filament tucked inside a magnet, itself designed to be sucked into an enemy aircraft carrier'.[15] The inclusion of a copy of Ryōji's letter in IWM's new Second World War Galleries enables visitors to see these complex emotions and to encourage them to think critically about presumptions they might have had about the fanaticism and nihilism of kamikaze pilots. Ryōji's account is deeply human and sad, and despite his courage and acceptance of his fate, Ryōji was a victim of a terrible war. The original copies of his letters were kept by Ryōji's family. Academics at Keio University approached the family on IWM's behalf and asked for their permission to reproduce in facsimile form Ryōji's last letter, as well as photographs of him to display in the galleries. In practical terms, this was a long process which took over three years from the initial reading of Ryōji's words in *Kike Wadatsumi no Koe*, to receipt of the final digital files and images. Understandably, his family were apprehensive about why a national war museum in London would want to tell his story. To ensure sensitivity, the curatorial team gradually built trust by explaining the concept of the galleries and our intention to tell a global story through the lives of individuals. Translating the technical language of the copyright documents into Japanese was another hurdle with a limited budget, but due to the kindness and generosity of individuals in Japan who saw the potential in what we were trying to achieve, Ryōji's story can be shared with IWM's visitors. Ryōji's story is not intended to represent a universal experience of Japanese soldiering, but his powerful words might complicate and challenge people to think differently about reductive 'enemy' and 'ally' ascriptions, what each side thought of each other and what they

[13] Ryōji Uehara, February 7 1945, in Yamanouchi, p. 235.
[14] Ryōji Uehara, May 10 1945, in Yamanouchi, p. 8.
[15] Ryōji Uehara, May 10 1945, in Yamanouchi, p. 8.

thought of their own actions, even in the most extreme acts of killing (Figs. 3 and 4).

The global approach, personal perspectives, and trans-national 'total war' connections were tested and evaluated with members of the public from the outset of the redevelopment project. The responses and feedback received from these groups highlighted fascinating preconceptions about how IWM should display material and stories of 'the enemy'. The advisory committee of community members became known as 'The People's Forum' and included residents from the local Lambeth area, teenagers

Fig. 3 Ryoji Uehara's last letter and translation. © Vikki Hawkins

Fig. 4 Showcases dedicated to war in Asia and the Pacific are positioned over the areas of the map which their contents relate to; war in China, the Island Hopping Campaign across the Pacific Islands, and war in the Philippines and New Guinea. © IWM

who were taking part in IWM's 'Youth Ambassadors' programme, individuals who represented local charities, organisations, and archives including the Black Cultural Archives and the Holocaust Education Trust, as well as leaders of religious and cultural groups. We also ran Access and Inclusion workshops with people who had an interest in visiting museums and cultural heritage institutions, but experienced barriers dis-enabling them from doing so. The workshops and sessions with community members took many forms; sometimes meetings would be themed around specific ideas: the use of language, prototyping interactive displays, or discussing topics including Britain's relationship with its empire and the display of violent imagery. One particularly informative session was about the use of personal stories. The group immediately responded to objects that could be associated with an individual from the past, ranging from a tea cosy knitted by a British evacuee to a dead gerbil that was sent by a soldier serving in North Africa to his rather surprised little sister. The group were inspired by the power that material culture had to establish emotional connections to the people associated with the objects. Two particular personal stories caused some debate and disagreement: Heinz Wolfgang Schnaufer, a German night-fighter ace, and Frederick Cecil Harrold, a Royal Air Force (RAF) Hurricane Pilot.

Frederick took part in the Battle of Britain, the aerial struggle between German and British air forces, which occurred during the late summer and autumn of 1940. On his third operational flight, 23-year-old Frederick was killed in action when his single-seat fighter Hurricane was shot down by a Messerschmitt Bf 109 over Deal in Kent. Frederick's personal belongings were collected from the wreckage of the aircraft. A cigarette case, a religious medallion, and a bent door key were stark reminders that this young man would not be returning home. His ripped uniform and personal effects were donated to the Imperial War Museum.[16] More than half a century later, the curatorial team showed them to our People's Forum members as part of a discussion about the importance of personal stories. The People's Forum discussed their assumptions about the RAF: the men were young, brave, and suffered heavy casualties. They spoke about the names of aircraft that they recognised—Spitfires and Lancaster Bombers—of the nose art that aircrew would paint on their aircraft showing cartoon characters, beautiful women and the number of missions the crew had completed, or the tally of enemy aircraft they had shot down. People were familiar with the concept of aerial dog fights, when battles were conducted in the air at close range, but no mention was made of the pilots and crews in those opposing aircraft. Drawing attention to the people in the machines, we also showed our People's Forum the personal story and objects associated with Heinz Wolfgang Schnaufer. Heinz became one of Germany's leading night-fighter pilots, nicknamed 'the Ghost of St Trond' by the British after the location of his base in occupied Belgium and his ghost-like stealth in attacking bombers from below. On display in the gallery is the tail fin of Heinz's Messerschmitt Bf 110, which is covered in markings recording the destruction of 121 British bomber aircraft.[17] This equates to 850 airmen killed, wounded, or taken prisoner. Heinz survived the war but died in a motorbike accident shortly afterwards.

[16] See IWM EPH 10133-10138, Documents.17351 and UNI 13724.
[17] IWM, EPH 2961.

Both Frederick and Heinz took part in aerial battles to defend their respective countries and civilians from the threat of bombing and invasion. Although Frederick died very early in his career, his success as a fighter pilot would have been calculated in a similar way, by the number of enemy aircraft he was able to shoot down. Both men would have been subjected to the constant fatigue of flighting in dangerous conditions, with temperamental navigational and radio equipment, under the stress of seeing so many of their friends failing to return, being wounded or taken prisoner. It could be argued that these two men might have had comparable experiences, fears, anxieties, or thrilling stories to share. Their main point of difference then, is the fact of fighting on opposing sides. They were enemies. The curatorial team decided to write the personal story labels for these two men and share them with the People's Forum for feedback. People immediately responded with dislike to the decision to comment on how successful Heinz's career had been, that he had been decorated and met Hitler who congratulated him on the number of enemy aircraft he shot down. If Frederick had lived longer, perhaps he too would have been a fighter ace, praised for his skill. Why should the language used to describe the job that both men were required to do be any different depending on who it was they were killing? Our People's Forum members struggled with the notion that Heinz 'the enemy' and Frederick 'the ally' had comparable life experiences and skills. We encouraged them to interrogate why they considered one person to be brave and fighting for his country, and the other evil for killing civilians or allied men in air combat. When the members took away the ascriptions of enemy and ally, they concluded that while both men were likely influenced by the politics and culture of their country, they were fundamentally caught up in a conflict outside their control, that the conflict changed the trajectory of their lives and demanded actions of them, which we should not label in hindsight as 'good' or 'bad' depending only on their nationality. Community consultation and participatory practice can broaden and compliment narratives being developed within exhibitions and galleries and can be used to gather a range of perspectives, experiences and insights that sometimes shape the interpretation of the curator-led narrative. Such approaches are increasingly common in the museum-sector. However, in this example, there is also value in understanding the process of challenging the views of the group and individuals who were consulted, to develop an approach that could be used in the galleries to confront visitor assumptions and promote debate.

Personal perspectives bring diversity and complexity to a narrative, but the selection of objects, digital assets and language also played a significant role in making the galleries multivocal and inclusive of the 'enemy' point of view. One area of the galleries where the curatorial team found an opportunity to move away from a British-centric view of events and bring in new perspectives, was Japan's rapid conquests across Asia and the Pacific in 1941 and 1942. On 15 February 1942, British and Empire troops in Singapore were defeated by Japanese forces. The city was supposed to be a fortress, and it was a devastating and embarrassing turn of events when Lieutenant General Arthur Percival was forced to sign the largest surrender in British history. However, the predominant image that is usually used to portray these events is that of Percival defiantly marching with a Union Flag to negotiate the surrender. As an alternative telling of this narrative, when visitors to the Second World War Galleries interact with the screen that explains Japan's tactical attacks, they also see footage of Japanese women and children cheering and celebrating at the Yasukuni Shrine in Tokyo after hearing news of the fall of Singapore. The language has specifically been chosen to reflect Japan's military successes, so rather than British 'defeat' at Burma, it becomes Japanese 'victory' at Burma, reframing the event from the perspective of the enemy.

Key to this interactive is the inclusion of the reaction of those who were living in the countries invaded. Many Asian people were discriminated against under European or US rule. Their countries were rich in natural resources including tin, rubber, oil, and food that Britain exploited. To justify their invasion strategy, Japan promised these countries and their populations independence from the control of Western powers. Japan declared that it would create a self-sufficient bloc of Asian nations under the 'Greater East Asia Co-Prosperity Sphere'—although ultimately real power would remain with the Japanese, many civilians in these countries welcomed the invading Japanese troops. To reflect this perspective, the interactive includes film footage of civilians cheering the invading Japanese forces, providing water and fixing their bicycles which were being used to manoeuvre Japanese troops across the terrain. Encouraging visitors to think critically about predominantly British narratives and providing material culture that presents events from a different point of view is key to understanding the conflict and its legacy.

Within the galleries there are several stories of people who had daily encounters with the enemy because they were prisoners of war or civilian internees. In some cases, this meant daily brutality at the hands of their captors. IWM's collections hold an abundance of material recording the atrocities that were committed by Japanese personnel in the camp system, including the restriction of food and medicine, harsh punishments and forced labour. As S.P. MacKenzie explains, 'the extreme brutality with which enemy captives were treated by the Japanese has often been explained through reference to a Japanese military ethic based on a strict Bushido tradition that demanded absolute sacrifice and deemed surrender as unbearable dishonour'.[18] MacKenzie interrogates this claim, since in theory the warrior spirit of Bushido also included fair treatment for captured enemy soldiers and civilians. But 'by 1940, when a new Field Service Code was drafted, soldiers were told "Never live to experience shame as a prisoner"'.[19] Consequently, the large number of Allied troops who surrendered during the Pacific campaigns in 1941–1942 provoked contempt and disrespect in the minds of their captors, and POWs were treated badly. Yet examples of encounters between prisoner and captor show that despite the attempt by enemy soldiers to dehumanise allied soldiers through the restriction of food, medicine, clothing, and harsh working conditions, their encounters often resulted in small but powerful acts of resilience and defiance on the part of the allies, or occasionally through small acts of kindness on the part of the Japanese, forcing us to interrogate the power dynamics.

In February 1942, Dennis Houghton together with approximately 55,000 other British troops, became a prisoner of war in Singapore. Dennis and his friend Reg Bradley, a civil engineer, began to 'scrounge' carpentry and engineering tools for their camp at Changi. They wanted to create a camp workshop but hiding their tools whenever a guard appeared in their hut became dangerous. During encounters with the camp's Japanese interpreter, they asked to learn Japanese characters including those needed for the word 'workshop'. Having this information at their disposal, they boldly painted the relevant characters on a piece of board and 'after the next changing of the guard the sign was hung up in the officers' hut over our beds and our collection of tools arrange below it, so that the new guard assumed that we had the permission of the Camp

[18] S. P. MacKenzie, 'The Treatment of Prisoners of War in World War II', *The Journal of Modern History*, 66.3 (1994), p. 512.

[19] MacKenzie, quoted from Charlotte Carr-Gregg, *Japanese Prisoners of War in Revolt: The Outbreaks at Featherstone and Cowra during World War II* (St. Lucia, Queensland: University of Queensland Press, 1978), p. 25.

Authorities to run this workshop and never bothered us any more'.[20] Now with a place to work, Dennis was determined to build a flute to play in the camp's orchestra. By borrowing a flute from another prisoner, dissecting it to identify the parts he would need, Dennis used discarded metal, stirrup pumps, fire extinguishers, spokes from a bicycle wheel, a watch spring, gloves, and a packet of needles to build his own flute. Writing in the 1980s, Dennis commented, 'I feel the flute was a real lifeline in more senses than one'.[21]

Denis and Reg's trickery is not the only example in the galleries of prisoners of war defying their captors. F B Taffs, a POW working in a copper mine in Taiwan, took a great risk by smuggling out a piece of copper ore hidden in his bottom.[22] POWs working in the mines were thrashed with mining hammers, so stealing the ore was extremely dangerous. But to Taffs, it was an act of personal, human resistance against his captors. Interestingly, a more positive enemy encounter can be seen if visitors look closely at a jersey in the galleries that was worn by civilian internee Sargent Bulpin, which has signatures from his fellow inmates scrawled across the fabric. Towards the bottom at the back of the garment, the Japanese interpreter of the camp has also signed his name, next to which someone has drawn an arrow pointing to the name and added a comment: 'a good guy'.[23]

Civilians were interned in several countries, often being categorised as 'enemy aliens'. Following the Japanese attack on Pearl Harbor in December 1941, pre-existing racial prejudices against Japanese people in the USA intensified. Anti-Japanese propaganda fuelled suspicions that Japanese Americans were not loyal to the USA and should be considered the enemy. *LIFE* magazine published an article with portrait images of a Chinese man and a Japanese man. The article, which has been reproduced in the galleries, prejudicially suggests that readers will be able to identify whether someone is a Chinese ally or a Japanese enemy by their facial characteristics. On 9 February 1942, President Roosevelt issued Executive Order 9066, which required people with Japanese ancestry to report to assembly centres, leaving behind their homes, businesses, and possessions. Approximately 117,000 Japanese Americans were forced into internment

[20] IWM ACQI/EXH/H/572, Dennis Houghton, 'Another Lifeline', *Pan: The Journal of the British Flute Society* (1984), p. 32.

[21] IWM, ACQI/EXH/H/572.

[22] IWM, EPH 199.

[23] IWM, EPH 500.

camps located in California, Arizona, Colorado, Wyoming, Idaho, Utah, and Arkansas.[24]

Upon the declaration of war in Britain, German, and Austrian residents also became classed as enemy aliens. Internment tribunals were set up to examine every registered enemy alien over the age of 16. They were divided into three categories: Category A, to be interned; Category B, to be exempt from internment but subject to restrictions; and Category C, to be exempt from both internment and restrictions. The British authorities interned approximately 27,000 Germans and Austrians living in Britain. One of the largest internment camps was established on the Isle of Man, where Telkea Kuck was sent in 1940. Telkea, a German nurse and midwife, came to Britain in 1933 to work as a domestic servant. She had initially been placed in Category C, but a few months later she was recalled to another tribunal and interned. Her British partner, Hugh Shayler campaigned vigorously for her release, writing letters to the House of Commons trying to convince people of Telkea's anti-Nazi and pro-British feelings, claiming she 'would be only too glad to use her qualifications in any capacity which would further the national cause'.[25] Telkea wrote many letters of her own: to Hugh she frustratedly reported 'I want to get out' and to her friend Miss E. Rathbone, 'I am burning to take an active part in Britain's struggle for freedom'.[26]

Telkea was released a year later in November 1941. Her letters are displayed alongside material that shows that even for enemy aliens who had been spared internment, their presence in Britain was still met with prejudice and hostility. A Jewish aid committee produced a circular, warning refugees to 'avoid talking German in public'.[27] The British restaurant chain, food manufacturing, and hotel conglomerate J Lyons and Co. put up signs in their establishments proclaiming, 'No Alien Enemy whether naturalised or not, is employed by this Company'.[28] However, the galleries also highlight that some German people in Britain had a more positive experience. Walter Klemenz was a German prisoner of war who worked on the Duke family's farm in Kent. He became friends with the family and

[24] Kate Clements, Paul Cornish, and Vikki Hawkins, *Total War: A People's History of the Second World War* (London: Thames & Hudson, 2021), p. 139.

[25] IWM, Documents.14984/D 'Rushen Camp, Isle of Man. Appeal for release from internment. Submitted by Hugh Shayler'.

[26] IWM, Documents.14984/D.

[27] IWM, Documents.7300/B.

[28] IWM, Art.IWM PST 6366.

made a pull-along Dachshund toy dog for the three Duke children, Chris, Felicity (Phil) and Francis. Walter made the toy dog from an old apple crate, fixed together with pieces of scrap metal. He stayed in touch with the Duke family on his return to Germany after the war and thanked them for their kindness towards him.[29]

The enemy encounters discussed so far have been focused on men: as fighting forces, they were more likely to interact with each other. However, women also found themselves on the front line or facing threat from the enemy. One of the recurring 'total war' subjects in the galleries is gender and sexual violence. The total war theme label in the galleries states, 'no army in the Second World War was innocent of committing acts of sexual violence' which itself marks a big interpretive step change. Though material cultural evidence is scant, the galleries also include examples of German, Japanese and Soviet gendered harms and sexual crimes against women. Wartime propaganda often depicted the enemy as amoral defilers of women. The curatorial team considered how to better reflect and include the real complexity, and examples of British and American perpetration of conflict-related sexual violence as well. With little material culture available in IWM's collections to address these crimes, and limited space to address this aspect within the complex storytelling of the war, the extent to which the curatorial team has successfully redistributed this narrative within the galleries to include 'all armies' is limited. A fundamental interpretative choice for the galleries overall was to display only items that were made, used, or written during the war years and can therefore be interpreted for their significance at the time without historical hindsight or tendency towards teleological inevitability.[30] This means that post-war testimony, memoirs, or oral histories were not included in the galleries. Identifying objects that fit these parameters was often a challenge, and in this instance made adequately addressing allied instances of conflict related sexual violence even more difficult.

Included within the galleries is a photograph showing American troops uncovering the hiding place of a woman and her child on the island of Saipan in the Pacific. One propaganda message spread by Japanese troops to the civilian population was that they would be raped and killed if the Americans invaded and took control of the island. Many civilians, fearful for their lives in the hands of the enemy, hid in caves or leapt to their

[29] IWM, Documents.13564.

[30] Hawkins, p. 213.

deaths from the island's cliffs. Numerous testimonies of rape at the hands of American soldiers during the American invasion and occupation of Okinawa were recorded, including from Okinawan nurses and local patients who had been admitted to the US field hospital. However, the limited words available for a photo caption (usually to a museum standard of 15–20 words) perhaps prevent the nuances of this example of sexual and gender violence from being fully explored.

Direct exploration of British forces' perpetration of gender harms and sexual crime is still an omission in the galleries and remains a collections development gap which the IWM curatorial team are alive to. My hope is that the change in mode presented in the new galleries, to addressing it both directly (through physical display and in interpretation) and in a balanced way lays a new groundwork for more programmatic activity around this important storytelling for the future. One included object which enabled us to illustrate sexual violence more explicitly is a wooden Japanese sign that was initially thought to have come from a brothel in Burma. An estimated 100,000 to 200,000 women from Japanese-occupied territories, especially China and Korea, were forcibly held captive as *ianfu* (often translated as 'comfort women'), a Japanese euphemism for prostitutes. These women could be raped multiple times in a day by Japanese soldiers. Official Japanese military documents described these women as 'war supplies', counted along with provisions for food and weapons. After revisiting this through research and unprecedently open dialogue with Chinese and Japanese colleagues, stakeholders and museums, the curatorial team now propose that the sign is likely from a 'Comfort Station'. One side reads 'sold out', while the other says 'closed/temporary rest'.[31] With this new information, the sign has been displayed for the first time since it was acquired by the Museum in the 1980s.

To give a voice to the women who experienced these encounters of sexual violence, the story of Park Young-shim, a young woman from Japanese-ruled Nampo, Korea (now North Korea), who was forced by the Japanese military to become a sex slave is displayed close by. Young-shim was taken from her home along with countless other 'comfort women' to Japanese combat zones, first to Nanjing and then to Songshan in China, and later to Lashio in Burma. In 1944, Japanese troops in Burma retreated, leaving behind many of the women who were found by Allied soldiers. As the war drew to a close, women were put into situations where the threat

[31] IWM, EPH 5425.

of a new enemy emerged. On 16 August 1945, the day after Japan officially surrendered to the Allies, major railway stations in Tokyo were crowed with women desperate to catch trains out of the city, fearing that if they remained, they would be raped by US and other Allied troops who were expected to land in Japan.[32] This was also the case as the Soviet Red Army advanced into Berlin in April 1945. In the galleries, the inclusion of Marta Hillers' story highlights these events. She was a 34-year-old freelance journalist living in a Berlin apartment and kept a diary of the Soviet advance, including the multiple rapes that she suffered at the hands of the Soviet soldiers.[33]

There are examples in the galleries of women who found ways to fight against their enemies. Lyudmila Pavlichenko was a Soviet sniper in the Red Army who was credited with 309 deaths, making her the most successful female sniper in recorded history. The mobilisation of women in Britain in 1941 put some women on the front lines of combat, including those who joined the Auxiliary Territorial Service (ATS) and joined antiaircraft gun crews (Fig. 5). Lakshimi Sahgal, an Indian Nationalist, joined the Indian National Army (INA) when she was nursing prisoners of war in Singapore. A pair of INA epaulettes are on display.[34] The INA was an armed force formed by Indian freedom fighters and former POWs in collaboration with the Imperial Japanese Army. Its aim was to secure Indian independence from British rule. To achieve this, the INA fought alongside Japanese soldiers, mainly in the Southeast Asian theatre of the war. This created a unique situation where Indian soldiers in the British Army were fighting against Indian soldiers aligned to the Imperial Japanese Army. In some cases, they were fighting for the same perceived cause—freedom from British rule—but they were allied to different countries to achieve it. Lakshmi created a regiment of women within the INA called the Rani of Jhansi Regiment with the support of the Japanese Army. The female volunteers were trained in the same way as men, preparing to fight British and Empire troops at the Imphal Kohima border in Burma. However, just as Lakshmi's regiment approached the front line, the tide of the battle changed, and the Japanese troops were defeated. Lakshimi was arrested

[32] Tanaka Toshiyuki, Yuki Tanaka, *Japan's Comfort Women: Sexual Slavery and Prostitution During World War II*, (London: Routledge, 2003), pp. 112–113.

[33] Marta Hillers, *Eine Frau in Berlin* (*A Woman in Berlin*). The diary was first published anonymously in America in 1954 and in Germany in 1959.

[34] IWM, INS 4015.

Fig. 5 The mobilisation of women in Britain in 1941 is explained through the lives of six women who each embarked on a new role. © IWM

and imprisoned by the British. The increased representation in physical space of previously marginalised experiences, including those of women on all sides of the conflict, has marked a significant shift in how the 'enemy' is presented at the Imperial War Museum.

There are now showcases exploring the conflict between Japan and China, and between American, Chinese, and British Empire forces against Japanese troops in Burma, New Guinea, and the Pacific as well as the war between Germany and the Soviet Union. The curatorial team worked closely with colleagues at the Australian War Memorial, the Auckland Museum, the Museum of the War of Chinese People's Resistance Against Japanese Aggression, Chongqing's Three Gorges Museum, the National Museum of the Marine Corps, the National World War II Museum in New Orleans, the National Film Archives in Tokyo, and many more independent curators and historians who supported the interpretation of IWM's collection and material culture selected for display. This collaborative process has ensured that the practical process of curating the Second

World War Galleries is transnational and multi-vocal at its heart. These museums provided objects on loan, helped to identify individuals whose personal stories were included in the galleries, and crucially advised on national perspectives of the conflict.

In addition, the Imperial War Museum's curatorial team embarked on research trips to many of these museums to understand international perspectives and representations of the war. At the Museum of the War of Chinese People's Resistance Against Japanese Aggression in Beijing, an effort was made to align China with its allies across the globe through an extensive multi-gallery display of photographs showing people celebrating the end of war from countries across the world. The defeat of the enemy appeared to be inevitable within the context of this global alliance. In New Orleans, at the National World War II Museum, captivating personal stories from American servicemen and civilians populated the narrative. Encounters with the enemy were well documented through personal papers, oral histories, and photographs, but there was a lack of representation of the enemy's perspective on events or accounts from named individuals from Japan or Germany. The Yūshūkan Museum in Tokyo also had a fascinating collection of personal objects and accounts of the Japanese soldiers' war experiences, but allied viewpoints were rarely included. The Australian War Memorial in Canberra displayed a sign that Japanese soldiers had erected at the site of a battle in Malaya in 1942, which detailed a Japanese victory over Australian troops. The accompanying translation provided an interesting perspective from the enemy, but it was the only such object to give an alternative narrative of events. Although there are many excellent museums and exhibitions about the Second World War which are on a journey towards positioning national narratives into a global context, and others which utilise their collections to emphasise personal experiences, the distinction between friend and foe is always clear. The behaviour of the country within which the museum was located was rarely questioned, narratives still focused on good and evil, and the depiction of the enemy was usually one dimensional, disregarding the complex nature of individual agency in favour of strengthening national identity. In discussing the memorialistion of war in her chapter, Christine Sylvester makes use of Viet Thanh Nguyen's study of war memory and ethics to emphasise this point. Nguyen asserts that war memorialisation does not usually include the memories of the enemy, neither does it acknowledge the nation's own war atrocities but is often guilty of 'excessively remembering something else, such as the heroism of the nation's soldiers'.[35] In this way, IWM's new galleries are providing something different from other national and international war museums. A British national identity is still present within the narrative and selection of objects, but crucially the visitor is presented with multiple critical perspectives, and they are also given the opportunity to encounter the enemy—as individual, as state, as community, as agent, as victim of war—within the displays.

Writing in 2000, Sue Malvern reflected on the way the Imperial War Museum has updated and changed its galleries and displays, contemplating how each new 'extension will bring into question again who and what the museum represents and to whom'.[36] IWM's audiences have changed since the Second World War Galleries of the 1980s and 1990s. In the past, IWM relied on 'grandparent tour guides'—visitors who lived through the war and could introduce their own families to objects and stories that were familiar to them, creating those emotional and personal stepping stones to the past. The number of people within living memory of the conflict has significantly reduced in recent decades, and our changing audiences are now looking for a way to find contemporary relevance in the historical narrative and evidence of how the war has impacted their personal or cultural heritage. The curatorial choices about the representation of a global narrative, the prominence of personal stories, and the multi-vocal and transnational perspectives of 'enemy' and 'ally', all reflect the need to make the new Second World War Galleries relevant to twenty-first-century visitors who are part of a diverse international community. By bringing previously marginalised 'enemy' voices to the fore within the connecting theme of total war, visitors have the opportunity to perceive the objects and narrative of the war with additional context, instead of passively viewing a dominant British account which perpetuates prejudices and popular myths.

The physical footprint of the galleries echoes this theoretical rebalance. As Jay Winter commented, 'visual representations of the two world wars have evolved alongside changes in public perceptions of their character and consequences'.[37] This can be seen in the showcase space that has now been allocated to objects and content from Japan and Germany, the

[35] Nguyen, p. 17.

[36] Sue Malvern, 'Memory and Museums: Art and Artefact in the Imperial War Museum', *History Workshop Journal,* 49 (2000), p. 197.

[37] Jay Winter, 'Museums and the Representation of War', in *Does War Belong in Museums?: The Representation of Violence in Exhibitions,* ed. by Wolfgang Muchitsch (Bielefeld, Germany: Transcript Verlag, 2013), p. 31.

prominent faces of the people whose stories we tell, and the display of foreign language material. IWM's Second World War Galleries confront a more complex version—a retelling—of history, one that is not reduceable to a binary of 'good' and 'bad'. Running through the narrative are frequent and diverse examples of enemy encounters, each one challenging the visitor to think critically about the boundary between enemy and ally. This better reflects the divergent and multi-faceted nature of the conflict, protects against all-too-easy binaries and polarisation, and keeps twentieth-century conflict relevant and connected to our contemporary lives. Our hope is that the galleries set a benchmark of innovative and inclusive curatorial practice.

Acknowledgements I would like to thank my fellow curators of the Imperial War Museum's Second World War Galleries, Paul Cornish and Kate Clements, for their dedication to amplifying marginalised voices within the galleries, their impeccable research, and immense historical knowledge. It was a joy and privilege to curate the galleries together. My thanks also go to the wider project team at the Imperial War Museum who were essential in delivering the galleries: Ann Carter, Nigel Steel, Madeleine Brady, Nora Ni Dhomhnaill, Verity Bridge, Rob Lee, and Jenna Taylor. Thank you to Rebecca Newell for her thoughtful comments and feedback on my chapter.

‘Children Made to Carry Bombs’: Staging the Malayan ‘Emergency’ in Postwar Britain

Rio Creech-Nowagiel

First Glance

On 23 February 1952, the *Daily Mirror* published a front-page report with a headline that read ‘CHILDREN MADE TO CARRY BOMBS’.[1] It featured a photograph of a fully clothed man crouching down to the level of an unclothed child, whose sarong appears to have fallen to her ankles, revealing a grenade belt wrapped around her waist. At the time of publication, the *Daily Mirror* was one of the top daily newspapers in Britain, ahead of competitors like the *Daily Mail* and the *Daily Express*.[2] Those leafing through their copy of the newspaper might have reacted with horror:

[1] Donald Zec, ‘Children Made to Carry Bombs’, *Daily Mirror*, 23 February 1952, p. 14.

[2] Hobson John Waller and Harry Henry, *The Hulton Readership Survey: An Analysis for Advertisers of the Readers of National Semi-National and Local Morning and Evening Daily Newspapers Local Weekly and Bi-Weekly Newspapers National and Semi-National Sunday Newspapers Weekly Fortnightly and Monthly Magazines by Region Sex Age Social Class and Certain Other Characteristics* (Hulton Press: 1947).

R. Creech-Nowagiel (✉)
Imperial War Museums / Cardiff University, Cardiff, UK
e-mail: creech-nowagielr@cardiff.ac.uk

H. Furneaux, M. Greig (eds.), *Enemy Encounters in Modern Warfare*, https://doi.org/10.1007/978-3-031-56748-3_11

IMAGINARY READER Surely not the children?!
How low will the reds/commies go?!?

Those with less interest in foreign affairs might have reacted with indifference, scanning the page, instead, to read up on the latest episode in an ongoing dispute between two multi-millionaire industrialists over the price of steel. If pressed, they might have asked for a few pointers:

IMAGINARY READER Which one is Malaya?
Does this newspaper come with a pull-out map of the Empire?[3]

Suppose an informed passer-by had indicated the location of Malaya in Southeast Asia, which was then a set of colonial protectorate states and would later be reconstituted as the independent nation-states of Malaysia and Singapore. Our hypothetical reader might have hoped that Churchill's newly re-elected administration had good reason for spending taxpayers' money waging a war on the other side of the world. These concerns would have been well-founded, given that times were still tough for the majority of readers living in the metropole. The social and economic upheaval of the Second World War was far from a distant memory and access of working-class people living in postwar Britain to basic provisions like food and housing could not be taken for granted.[4] In this climate, the British establishment needed to make the expenditure of millions of public pounds on a protracted counterinsurgency campaign in a remote corner of the Empire seem like a good idea. Leaders in Britain knew all too well then, as they know now, that votes are won with appeals to emotion and patriotic ideals. As Wendy Webster and others have shown, feelings of safety, domesticity and their counterpart—the threat of an 'alien other'—were routinely weaponised in the gendered, racialised representations of colonial wars that circulated within the popular cultural domain in postwar Britain.[5] Stories that inspired fear were the kind that helped sell tickets at the box office and had newspapers flying off the shelves into the hands of

[3] 'Social Survey: Study of Public Knowledge of Colonial Affairs Conducted by Central Office', 1951, CO 875/2/3 (The National Archives, UK).

[4] Jordanna Bailkin, *Unsettled: Refugee Camps and the Making of Multicultural Britain* (Oxford Scholarship Online, 2018), pp. 1–15.

[5] Wendy Webster, '"There'll Always Be an England": Representations of Colonial Wars and Immigration, 1948–1968', *Journal of British Studies*, 40.4 (2001), pp. 557–584.

consumers. And what better way to make a distant war feel urgent and relevant to voters in the metropole than to appeal to their sensibilities by calling upon an age-old genre of storytelling: an imperial morality tale, set against an imagined Orient, starring heroes, villains and innocent 'natives'?[6]

Now imagine a flurry of activity as editorial staff at the *Daily Mirror* London offices made efforts to decipher between newsworthy updates and non-items before a looming print deadline. Amongst the staff members is Donald Zec, the journalist responsible for penning the report in focus, who was, by the 1950s, gaining an international reputation as an entertainment and film columnist.[7] A newsworthy story needed punchy copy: 'A police officer just back from Malaya has told me that the bandits are now dragging little children into the war'.[8] It also needed a sympathetic, attention-grabbing headline: 'CHILDREN MADE TO CARRY BOMBS'.[9]

What would help disguise the shaky foundations of a rumour and dress it up as a reliable news update? An accompanying photograph illustrating Zec's claim could be that catalyst—ideally, an image of a child carrying bombs. What luck that the paper had this to hand!

A Closer Look

Would the knowledge that this photograph was a behind-the-scenes reconstruction of a rumoured event, featuring a child actor and a film crew—likely taken many miles from the site of the insurgency—have bothered any member of our imagined readership?[10] The *Daily Mirror* openly acknowledged the status of this image as 'fiction'. Nothing about this seemed inconsistent with its use in this report as proof of a 'fact' about enemy barbarity in Malaya. The caption reads:

> In the film "White Blood", now being made in Malaya, Lotus Wong is caught with "planted" bombs. For Lotus this is fiction – but for some children this is fact.

[6] Caroline Elkins, *Legacy of Violence: A History of the British Empire* (London: Penguin, 2022), p. 473.

[7] Geoffrey Goodman and Matthew Engel, 'Donald Zec Obituary', https://www.theguardian.com/media/2021/sep/09/donald-zec-obituary [accessed 22 September 2022].

[8] Zec, 'Children Made to Carry Bombs', p. 14.

[9] Zec, 'Children Made to Carry Bombs', p. 14.

[10] Zec, 'Children Made to Carry Bombs', p. 14.

The *Daily Mirror*'s decision to feature a staged photograph, rather than a 'live-action' shot, tells us something about the quantity and type of counterinsurgency imagery that was available to editors in postwar Britain.[11] Its inclusion offers us some insight into the kinds of professional norms and epistemological biases that informed what was deemed publishable.[12] While this decision might have seemed insignificant to those involved in the publication of this report or to those who encountered it when it appeared briefly in print on 23 February 1952, I want to advocate here for the significance of this photograph as a point of entry into the ways in which the so-called Malayan 'Emergency' was scripted, rehearsed and presented to mass audiences in Britain. I also want to advocate for the care of those held within the colonial archive and, where possible, resist perpetuating the patterns of violence that shape how imagery of this kind was originally produced and the terms of its subsequent reproduction. Here, I use a traced version of the photograph in focus over a replica, a choice I discuss in detail in the concluding section of this chapter. We only need to zoom out a little to discover a surprising overlap between representations of the enemy in Malaya as enacted on screen and those that featured as stories in newspaper columns. From this vantage point, we can begin to tease apart the entangled threads between imagined enemy encounters and 'the real thing' when theatres of colonial war collided with the metropolitan stage.

Setting the Scene

By 1952, four years had passed since Britain's declaration of a state of 'Emergency' in Malaya, and there was little indication that the 'exceptional' measures enforced in order to suppress anti-colonial resistance activity would be withdrawn anytime soon.[13] While the killing of three plantation managers by resistance fighters in Perak in 1948 is often cited

[11] Susan L. Carruthers, *Winning Hearts and Minds: British Governments, the Media and Colonial Counter-Insurgency 1944–60* (Leicester: Leicester University Press, 1995), pp. 72–73.

[12] Erik Linstrum, 'Facts about Atrocity: Reporting Colonial Violence in Postwar Britain', *History Workshop Journal*, 84 (2017), pp. 108–127.

[13] John Reynolds, 'The Long Shadow of Colonialism: The Origins of the Doctrine of Emergency in International Human Rights Law', *Comparative Research in Law & Political Economy*, Research Paper No. 19/2010 (2010) < http://digitalcommons.osgoode.yorku.ca/clpe/86>.

as the primary trigger for this declaration, the British government had other reasons for defending its role as an imperial power that extended beyond the patriotic retribution and defence of white settlers. Malaya was, after all, one of the most profitable territories in the Empire at the time. A region rich in natural resources such as rubber and tin, Britain had spent decades as indirect rulers profiting from an extractive plantation system they had put in place through the passing of exploitative land and labour laws. In the postwar years, independence movements in prized territories across the British Indian subcontinent and elsewhere had been fought and won through the resistance of those who had been enslaved and colonised.[14] The covert tactics of the Malayan National Liberation Army (MNLA), the communist guerrilla army at the forefront of the independence struggle in Malaya, sought to disrupt Britain's sense of territorial control and its status as an economic power. As Souchou You and Michelle Gordon have shown, the myth of British exceptionalism, which suggests that Britain's retreat from Malaya was somehow smoother or more graceful than that of neighbouring European nations from their colonial territories, glosses over the dynamics of routine and extreme violence involved as Britain attempted to seize back control of Malaya after the fall of Japan.[15] Local demands for self-determination and improved working conditions were met with a host of racialised counterinsurgency measures designed to police, intimidate and divide Malaya's ethnically diverse civilian population including curfews, identity checks, mass arrests and forced resettlement.[16] Attempting to suppress the flow of information locally, the colonial administration imposed censorship restrictions on press and correspondence, scrutinising letters, books and newspapers for anti-imperialist sentiment.[17] These were tried-and-tested governance methods adapted from previous counterinsurgency campaigns waged across the British Empire, most recently in Mandatory Palestine.[18]

[14] Priyamvada Gopal, *Insurgent Empire: Anticolonial Resistance and British Dissent* (London: Verso Books, 2020), p. 13.

[15] Souchou Yao, *The Malayan Emergency: Essays on a Small, Distant War* (Copenhagen: NIAS Press, 2016); Michelle Gordon, *Extreme Violence and the 'British Way': Colonial Warfare in Perak, Sierra Leone and Sudan* (London, New York: Bloomsbury Academic, 2021).

[16] Huw Bennett, '"A Very Salutary Effect": The Counter-Terror Strategy in the Early Malayan Emergency, June 1948 to December 1949', *Journal of Strategic Studies*, 32.3 (2009), pp. 415–444.

[17] Rachel Leow, *Taming Babel: Language in the Making of Malaysia* (Cambridge: Cambridge University Press, 2016), pp. 135–175.

[18] Elkins, p. 461.

In Malaya, racial stereotypes and hierarchies embedded under indirect colonial rule informed the direction and intensity of Britain's campaign, shaping how these measures were used against armed and unarmed groups in ways that were regionally specific. Top-level British officials—including Sir Henry Gurney, who held the role of High Commissioner until 1951—openly expressed their hatred and paranoia towards Chinese people, who they believed were naturally conspiratorial, and made a concerted effort to pit them against Malaya's ethnically Malay population.[19] The colonial administration targeted Malayan-Chinese rural-dwelling communities, whom the authorities deemed 'alien Chinese squatters' and considered more likely than others to sympathise with the goals of communist fighters and offer them assistance.[20] Drawing support from indigenous Dyak hunters, swathes of British and Commonwealth troops with varying levels of skill and combat experience were sent on protracted ambushes tasked with 'weeding out' insurgents hidden in the dense tropical rainforest.[21] Aiming to weaken the morale of local populations and cut off supply lines between guerrilla forces and civilians, they operated under a policy of 'search and destroy' whenever they encountered sites deemed to be potential enemy hideouts.[22] Strategies aiming at the seizure and destruction of ancestral land, among others, formed the basis of a blueprint for counterinsurgencies waged across the Empire, including the 'Emergency' campaign declared in Kenya in 1952.[23]

Smokescreen

The British government and colonial administration's attempts to eliminate the threat posed by insurgents in Malaya were by no means limited to local measures to seek out and punish guerrilla fighters. Imperial officials enlisted support from all corners of the mainstream press in Britain, Malaya and further afield in their efforts to control how the conflict was viewed

[19] Elkins, p. 475.

[20] Tan Teng Phee, 'Oral History and People's Memory of the Malayan Emergency (1948–60): The Case of the Pulai', *Sojourn: Journal of Social Issues in Southeast Asia*, 27.1 (2012), pp. 84–119.

[21] Maureen Sioh, 'An Ecology of Postcoloniality: Disciplining Nature and Society in Malaya, 1948–1957', *Journal of Historical Geography*, 30 (2004), pp. 729–749.

[22] Bennett, p. 430.

[23] Karl Hack, 'The Malayan Emergency as Counter-Insurgency Paradigm', *Journal of Strategic Studies*, 32.3 (2009), pp. 383–414.

internationally and produce what Peter Taylor termed a 'synthesis of the political and public perception of the state's enemy'.[24] In 1948, the Foreign Office had established the Information Research Department (IRD), a secret propaganda unit charged with managing public opinion around the plethora of communist and independence movements that were rippling through the Empire.[25] This work required consensus on the contentious matter of naming the enemy. Initially, the official line that British leaders had taken was that any resistance activity in Malaya was the result of a lone group of 'bandits' rather than an organised anti-colonial struggle. The IRD had been hesitant to stress the affiliations of these groups with communism, for fear that others would 'jump on the bandwagon' if this detail were publicised.[26] This shifted in 1952, when it was written into official policy that all enemy actors be referred to as 'communist terrorists' or 'CTs'.[27] British leaders hoped that this terminology shift would win them favour in the United States by positioning them as defenders of the 'West' and their campaign as part of an ideological battle against international communism.[28] The effect of combining accusations of terrorism on the other side with positive propaganda around their own activities was akin to a smokescreen. This strategy enabled British authorities in Malaya to publicly point the finger at insurgents while ploughing on with their own reign of terror. This included the removal of hundreds of thousands of Malayan-Chinese civilians from their homes into 'New Village' barbed-wire camps under the 'Brigg's Plan'. In British-controlled propaganda, this was cast as an opportunity for rural-dwelling folks to modernise and seek enlightenment in the form of elections, medicine and schooling, in stark contrast to the harrowing accounts of everyday life on these camps offered by survivors of this system.[29]

Revelations about acts of brutality or atrocity committed by British troops in Malaya were usually shut down before they reached the columns

[24] Peter Taylor, 'The Semantics of Political Violence', in *Communicating Politics: Mass Communication and the Political Process* (Leicester: Leicester University Press, 1986), pp. 211–221.

[25] Susan Carruthers, 'A Red Under Every Bed? Anti-Communist Propaganda & Britain's Response to Colonial Insurgency', *Contemporary Record*, 9.2 (1995), pp. 294–318.

[26] Philip Deery, 'The Terminology of Terrorism: Malaya 1948–52', *Journal of Southeast Asian Studies*, 34.2 (2003), pp. 231–247.

[27] Carruthers, *Winning Hearts and Minds*, pp. 84–85.

[28] Carruthers, *Winning Hearts and Minds*, pp. 84–85.

[29] Teng Phee, p. 89.

of British newspapers. Imperial authorities routinely responded to critical reporting with bribes, personal favours or recourse to legal action.[30] On top of practical barriers, such as Malaya's remote geographical location, journalists sent overseas to relay news of colonial wars had to contend with an epistemological climate that Eric Linstrum described as an 'underworld of rumours, representations and allegations' while also conforming to shifting industry standards based on ideals of neutrality, factuality and restraint.[31] The introduction of legalistic frameworks for discerning what constituted 'truth'—and whose version of the truth was considered reliable—meant that news about violence in the colonies was often considered unpublishable.[32] In the rare case that reports of abuses carried out by the British military in the name of counterinsurgency made it into newsprint, there were institutional mechanisms in place for claiming plausible deniability. When, in April 1952, a series of leaked photographs published on front pages of the *Daily Worker* newspaper offered visual proof that British troops were engaging in long-outlawed headhunting practices, the government's first response was to throw the reliability of the imagery into doubt.[33] The mainstream British press largely ignored the scandal, despite the lengths that the left-wing *Daily Worker* newspaper went to in order to prove the authenticity of the photographs in question.[34] The British establishment's insistence on a fantasy—the honourable conduct of their troops—provided sufficient justification for denying the evidentiary value of these photographs as soon as they slipped from private hands and entered the public domain.

Filling in the Gaps: Rehearsals and Reconstructions

When it came to reporting on colonial wars, mainstream newspapers in postwar Britain tended to favour dramatic and sensationalist coverage over their commitments to tenuous concepts like 'neutrality' and 'truth'. Press

[30] Linstrum, p. 109.

[31] Linstrum, p. 112.

[32] Linstrum, p. 116.

[33] Rio Creech, 'All Too Graphic: Leaked Photographs of Colonial Atrocities during the Malayan "Emergency" Shocked Postwar Britain', *History Today*, (12 December 2021) <https://www.historytoday.com/archive/history-matters/all-too-graphic> [accessed 12 September 2022].

[34] Susan L. Carruthers, 'Why Can't We See Insurgents? Enmity, Invisibility, and Counterinsurgency in Iraq and Afghanistan', *Photography and Culture*, 8.2 (2015), pp. 191–211.

photographers sent to Malaya tasked with documenting Britain's counter-insurgency campaign drew heavily from a repertoire of exoticising, orientalising tropes produced by generations of colonial occupation. Imagined scenes of warfare in Malaya also had obvious appeal to editors in Britain because the style of warfare that British troops were engaged in was not generally conducive to the kind of spectacular live-action coverage that mainstream press outlets tended to favour.[35] Much of this coverage converged around the idea of an unseen or hidden enemy, which was, in part, a response to the covert tactics of guerrilla fighters, which included targeted assassinations and acts of sabotage. This idea was deeply unsettling to British authorities and threatened their self-image as an all-seeing, all-knowing military power. Like language, photography was a tool that counterinsurgents used to justify violence against insurgents and consolidate what, according to Kleinian thought, could be characterised as subconscious 'splitting' between themselves and their opponents.[36] Press photographers stepped temporarily into the role of casting directors, using enacted photographs to compliment the kind of stories their editors wanted to tell. Malayan troops were often positioned in a more ambivalent supporting role as loyal but ultimately subservient allies.

A report published in the *Illustrated London News* in February 1951 titled 'Two Kinds of Camouflage Used to Outwit and Defeat the Communist Bandits in the Jungles of Malaya' featured two such photographs (Fig. 1). The first choreographed scene depicts a 'locally recruited' RAF Regiment dressed in camouflage moving through the dense rainforest.[37] This imagery draws heavily upon a version of the Malayan rainforest conjured up and popularised by British travel writers during the period of colonial expansion, who imagined the forest as a 'jungle', a spatial 'other' and a dangerous, infectious space that required sanitation by white settlers.[38] Adorned in leafy camouflage against a leafy backdrop, the staging of 'locally recruited' troops as visually indistinguishable from the local landscape reinforces harmful stereotypes about the differences between

[35] Carruthers, *Winning Hearts and Minds*, p. 72.

[36] Julia Segal, *Melanie Klein*, Key Figures in Counselling and Psychotherapy (London: SAGE Publications Ltd, 2004), 2ND ED, p. 34.

[37] Illustrated London News Reporter, 'Two Kinds of Camouflage Used to Outwit and Defeat the Communist Bandits in the Jungles of Malaya', *Illustrated London News*, 10 February 1951.

[38] Maureen Sioh, 'Authorizing the Malaysian Rainforest: Configuring Space, Contesting Claims and Conquering Imaginaries', *Ecumene*, SAGE, 5.2 (1998), pp. 144–166.

COUNTERING THE INVISIBLE FOE WITH AN INVISIBLE ANTI-BANDIT PATROL: A WONDERFUL EXAMPLE OF JUNGLE CAMOUFLAGE BY A SMALL GROUP OF MEN OF THE LOCALLY-RECRUITED R.A.F. REGIMENT (MALAYA), MOVING THROUGH THE FOREST AGAINST THE COMMUNIST BANDITS IN MALAYA.

CARRYING ITS OWN CLOAK OF INVISIBILITY IN THE MALAYAN JUNGLE: A JEEP OF THE GROUND SECURITY FORCES, FITTED WITH SMOKE-BOMBS WHICH CAN BE IGNITED BY THE DRIVER IF THE VEHICLE RUNS INTO AN AMBUSH. THE SCREEN GIVES AN IMMENSE TACTICAL ADVANTAGE AND HAS BEEN USED WITH GREAT SUCCESS.

TWO KINDS OF CAMOUFLAGE USED TO OUTWIT AND DEFEAT THE COMMUNIST BANDITS IN THE JUNGLES OF MALAYA.

Throughout the period of Communist bandit activity in Malaya, one of the great advantages resting with the bandits has been their invisibility and so the power to disperse and reassemble in the jungle, choosing their own time and place for operations and ambushes. The two interesting photographs which we reproduce above show how the forces of law and order, the Ground Security Forces—which is the generic term covering Army, R.A.F. Regiment and Police—are invoking the same cloak of invisibility for their own counter-strokes. The upper picture shows five men of the R.A.F. Regiment (Malaya) giving an excellent example of camouflage as they move through the jungle on patrol. The R.A.F. Regiment (Malaya) is locally recruited but has a majority of British officers and N.C.O.s, and although it is designed primarily for airfield defence it takes its turn in jungle operations and has scored several notable successes. The lower picture shows a very interesting tactical device. By fitting a number of smoke-bombs to jeeps in such a way that they can be ignited by the driver, this maid-of-all-work vehicle is converted into a sort of "warship of the jungle." Smoke clears slowly in the sheltered jungle tracks and behind its screen an ambushed jeep can be diverted, reversed or temporarily abandoned invisibly: and the ambushed can in their turn ambush the ambushers.

Fig. 1 'Two Kinds of Camouflage Used to Outwit and Defeat the Communist Bandits in the Jungles of Malaya', Illustrated London News, 10 February 1951. © Mary Evans Library 'civilised' troops and those recruited from 'native' populations in arenas of irregular warfare.[39] Predictably, senior officers and white combatants were usually cast as main characters in rehearsed jungle warfare scenes, performing military superiority in an environment they perceived as hostile. The second scene in this *Illustrated London News* report sees members of the British Security Forces with their Jeep caught in the theatrical act of igniting a smoke bomb, which is described in the image caption as a 'cloak of invisibility'.[40] Counterinsurgent forces here picture themselves, from the safety of their vehicle, roleplaying a successful version of a fantasy colonial combat scenario. In this simulated scene, they gain mastery over opponents perceived as 'invisible' by weaponising invisibility and using it against them. These are anticipatory photographs, inflected with what Elissa Marder called 'the future tense', akin to 'the structure of anxiety itself'.[41]

While the presence of anxiety is rarely an indicator of whether the source of danger is real or perceived, there were occasional moments during the 'Emergency' period when British fears of ambush were realised. Despite their attempts to closely monitor flows of information 'on the ground', there were details about the MNLA's activities that slipped through the net and took British intelligence forces by surprise. The ambush and assassination of Sir Henry Gurney by insurgents in his car en route to Fraser's Hill in October 1951, for example, sent shockwaves through the British establishment and hit the headlines of newspapers internationally. This was viewed as a significant win for the MNLA, despite an admittance in hindsight by those responsible for the attack that the targets had been random and their high-profile status entirely unknown to the guerrilla fighters.[42] In a report on the incident, assumed to be deliberate, by the *Daily Mail*, Gurney was portrayed as a tragic figure caught in the line of fire in the heroic act of saving his wife, who survived the

[39] Patrick Porter, *Military Orientalism: Eastern War through Western Eyes* (London: Hurst, 2009).

[40] Illustrated London News Reporter, 'Two Kinds of Camouflage Used to Outwit and Defeat the Communist Bandits in the Jungles of Malaya', *Illustrated London News*, 10 February 1951.

[41] Elissa Marder, *The Mother in the Age of Mechanical Reproduction: Psychoanalysis, Photography, Deconstruction* (United States: Fordham University Press, 2012), p. 94.

[42] A. J. Stockwell, 'Chin Peng and the Struggle for Malaya', *Journal of the Royal Asiatic Society*, 16.3 (2006), pp. 279–297.

Fig. 2 'Bandits Lay in Wait 36 Hours', Daily Mail, 4 October 1951. © DMG Media Licensing

assault.[43] The report featured a birds-eye-view photograph, which was taken from 'a camouflaged rock', that is used to reconstruct the scene of the ambush with the addition of the newspaper's own speculative annotations such as: 'TERRORISTS FIRED FROM HERE' (Fig. 2).[44] The newspaper's annotations function like stage directions, dictating the position of the protagonists in this reconstructed scene: a villain ('bandits'), a heroic victim (Sir Henry Gurney) and an innocent bystander (Lady Gurney). The application of a generous dose of creative license to banal

[43] Daily Mail Reporter, 'Bandits Lay in Wait 36 Hours', *Daily Mail*, 8 October 1951, p. 4.
[44] Daily Mail Reporter, 'Bandits', p. 4.

imagery—in this case, an otherwise unexceptional photograph of the dense Malayan jungle—enabled newspaper editors in Britain to compensate for a deficit of live-action press imagery and relay speculative news about the war in Malaya in a way that would provoke fear and stir the emotions of metropolitan readers.

Speculative reporting drew upon the first-hand testimony of those deemed credible in the eyes of the British establishment. Extracts from an interview with an eyewitness called 'Mr Staples' who claimed to have been sitting beside the driver of the Rolls-Royce, featured in the same report by the *Daily Mail*. Mr Staples notes that the driver 'lost control' and recounts his experience unsuccessfully attempting to apply the brakes to the vehicle during the attack.[45] Rhetoric of this kind, whereby events are evaluated in terms of gains and losses of control (over land, resources, populations or the spread of ideas) recurs frequently in colonial and neocolonial counter-insurgency discourses, particularly those circulated in the pages of tabloid press outlets. In this report, the imagery of a vehicle under siege evokes the panic of white settlers over the prospect of losing Britain's grip over Malaya. The loss of control of Gurney's car touched a nerve because it played into collective fears amongst British observers and commentators about the derailing of the whole imperial enterprise.

Staging the 'Emergency'

Depictions of staged encounters between British troops and hidden enemy insurgents were by no means the sole reserve of illustrated and tabloid newspapers circulating within postwar Britain. Encounters of this kind also had commercial appeal and could be easily adapted for the big screen, making their way into the scripts of cinematic portrayals of Britain's campaign in Malaya. The best-known example of such a portrayal is Ken Annakin's *The Planter's Wife*, which hit British box offices in 1952 and was re-titled *Outpost in Malaya* for release in the United States that same year.[46] Initially titled *White Blood*, which was the colloquial term for the milky white latex fluid that runs out of the bark of a rubber tree once it is tapped, it is likely that the title was amended before the release of the film because of its racist overtones. Adapted from a British novel by S. C. George, who had been living in Malaya when the 'Emergency' was declared, the

[45] Daily Mail Reporter, 'Bandits', p. 4.

[46] *The Planter's Wife*, dir. by Ken Annakin (Rank's General Film Distributors, 1952).

film follows the domestic struggles of Jim and Liz Frazer, a planter and his wife, as they make efforts to defend their estate from attack by communist insurgents. The producers hoped that strategic casting choices would lead to success on both sides of the Atlantic. They cast Jack Hawkins as the hero, a British actor who was known for his portrayal of military men, and Claudette Colbert as the heroine, an American actress with mass appeal in Hollywood. Screenings of the film in Britain, the United States and France were relatively well attended but largely failed to win over critics.[47] Some took issue with holes in the plot while others were outraged by the film's heavy-handed imperialist message. The matter of how honestly or realistically *The Planter's Wife* had presented the conflict in Malaya was a source of huge contention amongst critics. Unsurprisingly, commentary that appeared in right-wing and centre-right publications tended to treat the film as an authentic account of the war and sang its praises as a triumph of realism. Donald Zec—the entertainment journalist who would later proclaim that children in Malaya were being 'made to carry bombs'—promised readers of the *Singapore Standard* in 1951 that the film would offer 'an account of the quiet courage of British women in the most treacherous area in the world'.[48] Commentary that featured in newspapers with more left-leaning affiliations, on the other hand, made persuasive cases for its dismissal. In his review for the *Daily Worker*, Thomas Spencer went so far as to describe *The Planter's Wife* as 'the most viciously dishonest war propaganda film yet made in Britain'.[49]

The extent of the film's fictionalisation of the war in Malaya was a meaningful topic for debate, given the likelihood that audiences in post-war Britain would treat *The Planter's Wife* as a source of information as well as a form of popular entertainment. The distributors deliberately left this element open to interpretation, promoting *The Planter's Wife* as an example of 'film journalism', which they defined in convoluted terms as 'topical fiction based on fact and telling a story that could and does happen even if its characters and actual incidents are imaginative'.[50] The

[47] Jon Cowans, 'The British Empire and Decolonisation', in *Empire Films and the Crisis of Colonialism, 1946–1959* (Baltimore: Johns Hopkins University Press, 2015), pp. 139–181.

[48] Donald Zec, 'Malaya's "Forgotten Women" Will Not Be Forgotten', *Singapore Standard*, 26 May 1951, p. 10.

[49] Thomas Spencer, 'Glorifies a Dirty and Insane War', *The Daily Worker*, 17 September 1952, p. 3.

[50] *The Planter's Wife: A Pinnacle Production* (Falstaff Press, 1952), SCR-14259, BFI Special Collections, Berkhamsted, UK.

on-screen casting decisions that informed the production of *The Planter's Wife* offer a clear indication of whose imagination mattered when it came to calling the shots off-screen. Through the characterisation of the 'plucky' Liz Frazer played by Colbert, the film centres an idealised version of white femininity, which is called upon to act as a final defence of the home and a nation under siege.[51] *The Planter's Wife* is littered with anti-Chinese stereotypes and references, which places it within a lineage of racist portrayals of 'yellow peril' produced by the West that can be traced back to the nineteenth century, when the entry of a diligent Chinese labour force into Europe triggered panic about the threat of invasion.[52] In the film's opening sequence, the camera switches between a birds-eye-view shot of two snakes travelling across the rainforest floor and another that follows two camouflaged guerrilla fighters as they crawl across the same floor.[53] As if to ensure the comparison between serpent and enemy combatant is not lost on viewers, the camera then switches to a close-up of a snake, which fades into a close-up of a partially disguised Chinese guerrilla fighter. This allusion reproduces primitivist and essentialising notions about people from the 'Orient', conjuring up the motif of the snake in Christian mythology as an inherently deceptive or cunning creature. These highly stereotypical and racist ways of imagining a 'native enemy' informed how colonial wars were conducted, setting the parameters for enemy encounters long before they occurred, as Tom Menger has shown in his contribution to this volume.[54]

The motif of the snake also serves a wider symbolic purpose within the context of the plot of *The Planter's Wife*, where the course of the conflict is framed as a fight to death between two animals—the snake and the mongoose. While the battle between the mongoose and the snake has its roots in ancient Indian religious scripture, it was taken up and made the subject of a short story by Rudyard Kipling in his 1894 *The Jungle Book* anthology. Kipling's colonial morality tale follows 'Rikki-Tikki-Tavi', the pet mongoose of a white settler family residing in British India, who fends off a cobra in his efforts to protect the child of the family.[55] The echoes of

[51] Webster, p. 567.

[52] Wendy Gan, 'Mr. Wu and the Rearticulation of "The Yellow Peril"', *English Literature in Transition, 1880–1920*, 55.4 (2012), pp. 441–459.

[53] *The Planter's Wife*, dir. by Ken Annakin (Rank's General Film Distributors, 1952).

[54] Tom Menger, 'Racial Difference and the Limits of Fellow Feeling: British, German, and Dutch Soldiers Encounter the Enemy in Colonial Wars, c. 1900', this volume.

[55] 'Rikki-Tikki Tavi', in *The Jungle Book* by Rudyard Kipling (Cleveland: Duke Classics, 2012), pp. 165–192.

Kipling's tale are palpable in *The Planter's Wife*, during a scene in which a snake enters the Frazer Bungalow while their small son, Mike, is playing. In the final scene, the snake's defeat is apparent when the character playing a Tamil houseboy holds up a mongoose in triumph, while the reunited Mr and Mrs Frazer wave off their son (who is sent off to boarding school in England). By invoking the story of the mongoose and the snake, *The Planter's Wife* positions the white settler family at the heart of the narrative as the institution under threat by an alien 'other' and reliant upon the loyalty of colonised groups—who are seen to possess innately tribalistic, warrior-like traits—for protection.[56]

The Planter's Wife featured a varied cast of child actors, most of whom remain uncredited. One such actor played the role of an infant assumed to be assisting insurgents with the smuggling of weapons, whose sarong drops as she enters the scene, revealing a grenade belt wrapped around her waist. Does this scene sound familiar? In a front-page report issued by the *Daily Mirror* on 23 February 1952, a photograph of the same child featured alongside an article claiming that insurgents in Malaya were resorting to new tactics in their armed struggle against British rule (Fig. 3).[57] The child was known to directors as 'Lotus Wong' and this is the name that Donald Zec used to refer to her in the image caption accompanying his *Daily Mirror* article, but we can make an educated guess that this title was the product of a white man's fantasy too. In a previous chapter in this volume, Holly Furneaux explored how histories of cultural appropriation complicate the task of correctly naming children 'adopted' by white British guardians under imperialism.[58] Indeed, it is likely that the name 'Lotus Wong' was inspired by the pioneering Chinese-American actress, Anna May Wong, who played the role of 'Lotus Flower' in her first major on-screen appearance at the age of seventeen in *The Toll of the Sea*, released in 1922.[59] As an East Asian woman working in white, male-dominated Hollywood, Wong was usually confined to playing one-dimensional roles as either villainess or victim, an Orientalist paradox that film scholars refer to as the 'Lotus Butterfly/Dragon Lady' paradigm.[60] As Lotus in *The Toll*

[56] For discussion of 'Martial Races' theory see Porter, p. 41.

[57] Zec, *Daily Mirror*, p. 14.

[58] Holly Furneaux, '"I Was a Stranger and ye Took Me In": Adopting the Enemy', this volume.

[59] Krystle Doromal, 'From Yellow Peril to Yellow Fever: The Representations of Asians from Ana May Wong to Lucy Liu', *Offscreen*, 11.5 (2007), pp. 1–7.

[60] Doromal, p. 2.

Fig. 3 Traced rendering of the photograph featured in 'Children Made to Carry Bombs', Daily Mirror, 23 February 1952

of the Sea, Wong plays a tragic figure who leaves her child in the care of a white couple before drowning herself.[61] The casting of Lotus in *The Planter's Wife* conforms to the same Hollywood trope of East Asian girl as

[61] Jean-François Staszak, 'Performing Race and Gender: The Exoticization of Josephine Baker and Anna May Wong', *Gender, Place & Culture*, 22.5 (2015), pp. 626–643.

docile victim whose destruction demands a white saviour. The paradox could be that, by virtue of her youth, she is an innocent 'native', but if the white characters stand back, she has the potential to wreak havoc and destruction by assisting enemy insurgents. This is the primitivist subtext informing the publication of the image of the half-naked child in *The Daily Mirror*, which was a plea for sympathy and support from their readers.

Circulating photographs of mistreated children to illustrate enemy barbarity and the injustice of war is long-standing and common practice amongst campaigners seeking to appeal to the emotions of viewers and win them over.[62] In some cases, the true status of a photograph assumed to have been captured spontaneously is only revealed after its initial publication. This was the case with George Lipman's iconic photograph of an Aboriginal Australian child and a white Austrialian child holding hands, who he asked to pose for the camera in exchange for money.[63] If the editors of the *Sydney Morning Herald* had known this when they decided to publish it in 1967, they were likely to have considered this detail an unnecessary and potentially damaging addition to the newspaper feature. What is striking about the photograph of 'Lotus Wong' is that its status as a staged image seemed not to matter to editors at the *Daily Mirror* and certainly didn't prevent them from using it as a token to illustrate the 'fact' that this treatment of children was commonplace in Malaya.[64] This was a 'fact' confirmed by a police officer, a source that the British establishment deemed credible.[65] From what we know about the MNLA's tactics, it is unlikely that they were regularly, if at all, involved in the recruitment of young children to smuggle weapons. If, indeed, the only children being made to carry bombs were those recruited to pose in costume and perform for the camera, we should interrogate the kinds of labour being demanded from the young person in this *Daily Mirror* photograph, whose real name I hope to learn but might never know.[66] The demands placed on her mutate, but do not diminish, as her image moves from cinema screens

[62] Susie Linfield, *The Cruel Radiance: Photography and Political Violence* (Chicago: University of Chicago Press, 2010).

[63] *Photography and Its Publics*, ed. by Melissa Miles and Edward Welch (London, New York: Bloomsbury Publishing USA, 2020), p. 118.

[64] Zec, *Daily Mirror*, p. 14.

[65] Zec, *Daily Mirror*, p. 14.

[66] Temi Odumosu poses this question in response to the image of a child 'unwillingly forced into cultural labour' in 'The Crying Child: On Colonial Archives, Digitization, and Ethics of Care in the Cultural Commons', *Current Anthropology*, 61.S22 (2020), p. 293.

to the pages of newspapers, buried in archives and revived for publication in academic volumes. These are issues to consider in the concluding paragraphs of this chapter, where I take Daniel Foliard's lead and refuse the lens of 'a retreating dolly' gently distancing the viewer from a 'comforting picture at the end of a Hollywood movie'.[67]

Repeat, Replay or Look Away?

The extent of fictionalisation of the Malayan 'Emergency' in British tabloids and other mainstream press outlets remains an open question and a meaningful topic of enquiry because it has the potential to enrich our understanding about how knowledge about wars fought in the colonies reached those living in the metropole. Like film producers, journalists navigated or abandoned their commitments to truthfulness, using staged imagery to compensate for gaps in intelligence and live-action imagery. They borrowed readily from the realm of fantasy in their efforts to entertain and inform readers. Studying enacted photographs that were circulated in postwar Britain during this time helps to expose areas of slippage between the 'rehearsal' and the 'real thing', revealing just how entangled these concepts were when it came to depictions of colonial war. This endeavour also carries with it certain responsibilities. Those 'calling the shots' when it came to choreographing and captioning imagery of this kind usually positioned actors according to the racialised, gendered hierarchies and stereotypes reinforced under colonialism. These dynamics of power and violence were at play at the time of production and are continually repeated, even when the content of the image in question is assumed, at face value, to be benign. They show up in the actions of white scholars, like me, who salvage these orphaned photographs from institutional archives and recontextualise them, hoping, without certainty, that a gesture of recovery might do something to add complexity to those tokenised and denied agency under imperialism.

I am inclined to agree with Susan Crane's argument that, in some cases, 'choosing not to look' means we can reduce the likelihood of assuming the perpetrator's gaze and re-terrorising those photographed.[68] I also see

[67] Daniel Foliard, *The Violence of Colonial Photography* (Manchester University Press, 2022), p. 270.

[68] Susan Crane, 'Choosing Not to Look: Representation, Repatriation, and Holocaust Atrocity Photography', *History & Theory*, 47.3 (2008), pp. 309–330.

good reason to believe in Daniel Foliard's claim that photographs have the power to 'destabilise' and 'turn the archive upside down', provided that they are not seen simply as illustrations.[69] Some of the choice paralysis I have felt around what to do with the *Daily Mirror* photograph of 'Lotus Wong' is linked to the way it defies conventional genres and subject categories. It sits somewhat ambivalently between a documentary image and a film still. It isn't what we would typically consider an example of a live-action war photograph or an atrocity image, even though we know it was taken against the backdrop of an 'Emergency' regime and a protracted anti-colonial struggle.[70] It is a propagandistic image, with proximity to celebrity culture, but hasn't to date achieved any kind of iconic status.[71] It isn't even a typical humanitarian image of a suffering child victim, designed to rouse sympathy in viewers, because the weapons attached to her signal that 'Lotus' is potentially destructive and cannot be trusted.

While writing this chapter, I deliberated over whether it was ethical or necessary to reproduce the photograph in focus. It is, after all, already accessible to those equipped with funds to pay for a subscription to a digital image library or a trip to a newspaper archive in Britain. What harm could a modestly circulated academic publication possibly do? In the end, I used a traced copy of the *Daily Mirror* photograph in question, which makes explicit my embodiment and the role that my own flawed hand has played in researching and writing about this image. This act of substitution heeds to calls by activist scholars for more self-reflexive approaches to handling and reproducing photographs produced under imperialism.[72] It tries to defy the logic of the coloniser, which dictates that we need to offer 'proof' of colonial violence in order to speak about it. In the opening section of this chapter, I invited you to suspend any disbelief and step into the shoes of a *Daily Mirror* reader living in postwar Britain. This experiment

[69] Foliard, p. 271.

[70] *Picturing Atrocity: Photography in Crisis*, ed. by Geoffrey Batchen, Mick Gidley, Nancy K Miller and Jay Prosser (London: Reaktion Books, 2012).

[71] Robert Hariman and John Louis Lucaites, *No Caption Needed: Iconic Photographs, Public Culture, and Liberal Democracy* (Chicago: University of Chicago Press, 2007).

[72] See Preface: Ariella Aisha Azoulay, *Potential History: Unlearning Imperialism* (London, New York: Verso Books, 2019); Sadiya Hartman, *Wayward Lives, Beautiful Experiments: Intimate Histories of Riotous Black Girls, Troublesome Women and Queer Radicals* (London: CPI Group (UK) Ltd, 2019).

was another small act of refusal of enlightenment ideals that state that academic scholarship should be limited to hard 'facts'. It was also an invitation to fellow scholars and readers to arrive at representations of the 'enemy' with our imaginative faculties engaged. Given the title of this volume, why wouldn't we use this as an opportunity to reflect deeply and meaningfully on the nature of our own enemy encounters?

Trusting Your Enemy: American Encounters with the Kit Carson Scouts During the Vietnam War, 1966–1973

Stefan Aguirre Quiroga

In early 1966, a rumor began to spread among the inhabitants of Thanh Son, Dien Bau District, Quang Nam Province in South Vietnam. Soldiers of the People's Armed Liberation Forces (PLAF) told the villagers that Ngo Van Bay, a PLAF soldier who had recently defected to the South Vietnamese government, had been tortured and executed by the US Marines that he had surrendered to. While the horrific rumor was still in the minds of the villagers, a sudden turn in the story took place. Like a ghost returning to the land of the living, Ngo Van Bay and two other defectors returned to Thanh Son to counter the rumor that was being spread by the PLAF. The villagers of Thanh Son unwittingly found themselves in the midst of a psychological war between the PLAF and the United States that was raging in hamlets across South Vietnam. The unexpected appearance of Ngo Van Bay resulted in a small victory for the

S. Aguirre Quiroga (✉)
Department of Historical Studies, University of Gothenburg, Göteborg, Sweden

H. Furneaux, M. Greig (eds.), *Enemy Encounters in Modern Warfare*, https://doi.org/10.1007/978-3-031-56748-3_12

United States and South Vietnam in an operation that American sources would later deem extremely successful.[1]

The path to Ngo Van Bay's return to Thanh Son began with his defection. In 1963, the South Vietnamese government inaugurated a program with the primary purpose of encouraging men and women who were in active rebellion against the South Vietnamese government to voluntarily defect without risking punishment. The program was commonly referred to in its shortened Vietnamese form, *Chieu Hoi* (*Chieu*, 'to appeal', *Hoi*, 'to return', with individual defectors known as *Hoi Chanh*) and was translated into English as the 'Open Arms' program. The program enabled approximately 250,000 PLAF soldiers, and later 2000 North Vietnamese soldiers of the People's Army of Vietnam, to be granted amnesty by the South Vietnamese government and be allowed to return to their families and civilian life. Ngo Van Bay defected to US Marines in April 1966 and was sent to a *Chieu Hoi* center in Da Nang shortly afterwards, a place defectors from the nearby province would have been sent to in order to undergo interrogation and political re-indoctrination, as well as the opportunity to attend vocational training to facilitate their re-entry into civilian life. Bay did not get that far. Once the exaggerated news of his demise reached the Da Nang *Chieu Hoi* center, US Army Chieu Hoi advisor Major James Kelley saw the opportunity for an operation to strike back against PLAF propaganda and arranged for the return of the supposedly deceased Bay.[2]

From this small and obscure operation grew the idea of using defectors in military operations alongside American forces. A month-long trial program for the use of defectors as auxiliaries alongside US Marines from the 1st Marine Division was launched in August 1966, during which time the auxiliaries were primarily used for psychological operations. By November 1966, by which time the program was officially adopted by the United States, the *Hoi Chanhs* were directly employed as combatants. Major General Herman Nickerson Jr., Commanding General of the 1st Marine Division, gave the program its name: The Kit Carson Scout program, named after the famous nineteenth-century American frontiersman and army scout, Christopher 'Kit' Carson (1809–1868). By the end of 1966, only the United States Marine Corps were employing the Kit Carson

[1] The Vietnam Center and Archive (VCA), Texas Tech University (TTU), The Kit Carson Scout Program, 1966–1968 (18 January 1969), pp. 1–2; J.A. Koch, *The Chieu Hoi Program in South Vietnam, 1963–1971* (Santa Monica: RAND, 1973), pp. 101–102.

[2] Koch, pp. viii, 13, 20–21, 66, 78–83. The Kit Carson Scout Program, pp. 1–2.

Scouts but full authorization for the program to be used by all American units in South Vietnam was given in the spring of 1967 and extended over the following years to include South Vietnamese, Australian, New Zealand, South Korean, and Thai forces. The program began with only six scouts, all of whom except one would be killed in action, to over 1500 scouts by the end of 1968. At the height of the program in 1970, 2500 Kit Carson Scouts were actively employed throughout South Vietnam.[3]

Defectors made the choice to join the Kit Carson Scouts towards the end of their six-week-long stay in the Chieu Hoi center. The program only accepted thoroughly vetted volunteers and preferred defectors with proven military experience. The defectors in turn volunteered for the Kit Carson Scout program for a variety of pragmatic and personal reasons that made it possible for them to continue their lives as combatants on their own terms. This included deeply personal reasons such as the desire for revenge and pragmatic reasons such as the need for a livable wage with which they could support their family or better their living conditions. Benefits for scouts included generous pay, a deferment from being drafted into the Army of the Republic of Vietnam (ARVN), access to American medical and dental care in South Vietnam, as well as guaranteed leave each month to visit their family.[4] In return, the scouts were to serve as guides, interpreters, and combatants alongside American soldiers in order to assist them in finding the enemy and to protect American soldiers from enemy ambushes and traps. The scouts were to be dressed in American uniforms, wield American weapons, use American tactics, and even eat American rations. Upon their acceptance as scouts, defectors were sent to designated Kit Carson Scout schools where they were given their first introduction to life in American employment and underwent weapons training (including both small arms and explosives), tactical training, and English lessons—all the while being instructed in American drill and conditioning exercises. After two to four weeks of training, the newly graduated Kit Carson Scouts were sent out to their designated units that were working in areas that the scout was familiar with from their previous life as a PLAF or PAVN soldier.[5]

[3] In the 9th Infantry Division, they were known as 'Tiger Scouts'. Koch, pp. 58, 102–103.

[4] For an extensive discussion, see Stefan Aguirre Quiroga, 'Phan Chot's Choice: Agency and Motivation among the Kit Carson Scouts during the Vietnam War, 1966–1973', *War & Society*, 39:2 (2020), pp. 126–143.

[5] Koch, pp. 103–104; The Kit Carson Scout Program, pp. 2–3; Tal Tovy, 'From Foe to Friend: The Kit Carson Scout Program in the Vietnam War', *Armed Forces & Society*, 33:1 (2006), pp. 80–84. For an example of the courses, see Gary Rausch, 'Kit Carson Scouts', *The Hurricane*, April 1969, pp. 10–13.

This put the Kit Carson Scouts in contact with the American soldiers they were expected to cooperate with. What did these encounters look like? How did this professional relationship function? What personal and cultural negotiations took place as a consequence of this unexpected encounter? In order to explore the encounter between American soldiers and the Kit Carson Scouts, this chapter will primarily make use of military memoirs, interviews, and articles from contemporary divisional newspapers written and published in South Vietnam during the war. This English-language source material is predominantly American-centric. The Kit Carson Scout perspective is very difficult for researchers to reach due to the dearth of source material produced by them directly. There are no interviews, memoirs, or other forms of narratives produced before or after the fall of Saigon in 1975 where they have had the space to tell their own story in their own voice. This has created a serious imbalance when writing about the Kit Carson Scouts. The few times they are given space to speak, their voice is mediated through American interpreters. Despite this imbalance, it is possible to extract their voices, albeit through an American filter. While this will leave gaps in our knowledge of the scout's experience due to the scarcity of source material, it would be a disservice to the scout's perspective not to attempt to include their voice wherever possible.

First Impressions and the Question of Trust

The status of the Kit Carson Scouts as former enemy combatants caused immediate tension with American soldiers upon their arrival at their designated unit. Initial American reactions could range from curiosity and disbelief to suspicion and outright hostility. American soldier Don C. Hall's platoon had a mixed reaction upon being introduced to two North Vietnamese scouts who were joining their platoon. 'Everyone's jaw in the hootch came undone at the same time', Hall wrote as he described the initial reaction.[6] Once dismissed by the lieutenant who had explained the concept of the program, the individual soldiers of Hall's platoon began to share their immediate thoughts with each other. 'Walter "Smoky" Shreiner ran toward us, angry and yelling' an explicit-laden rant about the scout's past in the PAVN, asking in shock: 'Jesus, what kind of fucked-up program are we in now anyway?' Another soldier was in disbelief, shaking his head

[6] Don C. Hall, and Annette R. Hall, *I Served* (Victoria: Trafford Publishing, 2001), p. 125.

at the concept: 'One gook's name is Zin, and get this, the other's [sic] is Rei. What would the real Kit Carson say to that?' Others understood that their new allies had their reasons to be there. 'It's their war', one soldier explained before he walked up to the scouts to welcome them to the platoon.[7] Some soldiers encountered the scouts for the first time in the field without prior introduction, something that could cause confusion:

> As we began to reconsolidate with the other platoons, I noticed what I thought was a VC, talking to the platoon commander. To my disbelief, this Vietnamese had a M16 rifle slung across his shoulder. For the life of me, I could not figure out what was going on. Then Hawkeye explained to me that he was what they called a Kit Carson Scout.[8]

For scouts who had never before served in an American unit, their initial introduction to American soldiers differed greatly from the period they had spent in a *Chieu Hoi* center and the military training they had gone through where they had first encountered American and Vietnamese instructors. Now they had to endure the disapproving, confused, and curious stares of American soldiers. The two North Vietnamese scouts that Hall encountered had different responses to being intensely watched by their new colleagues. While the older scout 'stood firm with his eyes glaring back at each one of us', the younger scout was nervous, cracking half a smile at the Americans. 'They have a hard time adjusting to Americans', explained one Kit Carson Scout instructor, 'they are really bothered by G.I.s the first time. Yet they realize that the G.I.s are the ones who need them'.[9]

But did the Americans want them? Efforts were made by officers when introducing the concept of the program to explain the importance of the scouts and to reassure the men about the scouts' new loyalties. Divisional newspapers that were written and printed in South Vietnam and aimed specifically at American soldiers recurringly featured positive articles about the Kit Carson Scouts that chronicled their successes in local operations. Articles such as the one written by Dick Arnold for the *Sea Tiger*, the

[7] Ibid.

[8] Frederick Fenwick, *Lasting Visions: With the 7th Marines in Vietnam, 1970* (Hoosick Falls: Merriam Press, 2019), p. 116.

[9] Hall, 119; Conrad M. Leighton, *War Stories: A GI Reporter in Vietnam, 1970–1971* (Jefferson: McFarland & Company), p. 32.

divisional newspaper for US Marines active in the I Corps area in northern South Vietnam, encouraged American trust in the Kit Carson Scouts by describing their role as scouts, their successful track record, efforts made to combat any possible infiltration by enemy elements, and that there 'has not been a single instance in which a returnee turned against a friendly unit in any way or attempted to lead it into an ambush'.[10] Other articles functioned as human interest pieces where an American soldier narrated a story of a particularly successful and loyal scout, occasionally featuring different testimonials from American soldiers that all enthusiastically described their confidence in the scout's abilities.[11]

Despite these publicized efforts, the positive tone present in these articles can contrast heavily against the more negative assessment made by American soldiers in private. Some American soldiers were unable to get past the scouts' former lives as their enemies and never learned to trust them. In a social military context where personal loyalty to each other in combat situations was ingrained in the psyche of American soldiers, the disloyalty of the scouts against their former comrades played an important role in creating distrust among American soldiers. If they had once betrayed their comrades-in-arms, what is to say that they would not do it again? In the words of one American soldier, 'Kit Carson Scouts were traitors and were always considered to be traitors. It was hard to put your complete trust in someone who tried his best to kill you'.[12] Scouts were threatened with being shot by their American colleagues if they did anything that could have been considered suspicious. Having to share sleeping quarters with a Kit Carson Scout in the platoon command post bunker, the skeptical Dale Farnham made certain to get his point across. 'I told Dau at night, 'Let me tell you, Mr. Dau, this here's the line. You cross it and I'll shoot you'. I just didn't trust him. You get that gut feeling'.[13] Others shared that gut feeling. In the high-stress environment of an active combat zone in conjunction with the pre-existing mistrust towards the scouts as traitors, some Americans saw them as fifth columnists who were

[10] Dick Arnold, 'Kit Carson Scouts', *Sea Tiger*, 22 March 1967, p. 3.

[11] For examples, see 'Former VC are saving lives of 5–60 infantry patrols', *The Old Reliable*, July 26, 1967, p. 3 and William Ludescher, 'Valuable Kit Carson', *The American Traveler*, 2 February 1970, p. 8.

[12] Robert J. Gouge, *"These Are My Credentials": The 199th Light Infantry Brigade in the Republic of Vietnam, 1966–1970* (Bloomington: AuthorHouse, 2004), p. 20.

[13] William L. Buchanan, *Full Circle: A Marine Rifle Company in Vietnam* (Mill Valley: Baylaurel Press, 2003), pp. 70–71.

actively trying to sabotage American operations and trying to get them killed. During the battle of Fire Support Base Ripcord in 1970, one soldier claimed to have seen a scout making signals with a flashlight in the direction of the enemy. 'If you see a Kit Carson shining a flashlight out there', the soldier stated, 'shoot the sonofabitch'. Lieutenant Colonel Andre C. Lucas subsequently made the decision to detain all the scouts present on the firebase in a bunker for the duration of the battle.[14] Moving in an area thought to be filled with enemies, James T. Gilliam and his fellow soldiers were increasingly suspicious of the scout accompanying them, former PAVN soldier Lum. In Gilliam's view, the scout was acting very suspiciously and displayed a lack of competence: 'We also quietly passed the word from man to man to watch Lum's every move, especially if a firefight started'. That night, Gilliam was awakened by the platoon's scout dog growling. In Gilliam's narrative, he claims to have seen Lum moving within the platoon's perimeter illuminated by the moonlight, picking up weapons, and fumbling with them. Gilliam brought up his rifle and called out to Lum who stopped in his tracks. As the men around him awakened as a result of the dog's barking,

> Lum was standing there with another man's rifle in his hand. The first weapon Lum had picked up belonged to Stony Burks. Burks woke up when the dog started barking, grabbed his weapon and the magazine fell out because it was not locked into the magazine as well as it should have been. The chamber was empty too. The conclusion for me was clear. Lum was disabling weapons at one of our positions in preparation for an attack. I think if he could have gotten all three weapons disabled, he would have dispatched the guard and signaled an attack.

In Lum's own defense, he claimed that 'he was just checking weapons to make sure they were loaded'.[15] The circumstances surrounding this incident, steeped in intense suspicion and hostility towards the North Vietnamese scout, make it difficult to draw the same conclusion as Gilliam so readily did. Gilliam's narrative falls into a broader collection of anecdotal stories told by American soldiers of scouts supposedly acting as infiltrators and being caught in the act. One story told by an American who

[14] Keith Nolan, *Ripcord: Screaming Eagles Under Siege, Vietnam 1970* (New York: Presidio Press, 2000), pp. 363–364.

[15] James T. Gilliam, *Life and Death in the Central Highlands: An American Sergeant in the Vietnam War, 1968–1970* (Denton: University of North Texas Press, 2010), pp. 178–179.

was part of the 'Brownwater Navy' has a scout on his barge caught making maps and plans for the enemy. It was speculated that 'since they couldn't seem to get us from the position we had mortars in, we imagined he probably planted some mines inside the barge itself', although no further proof is offered on this point.[16] The lack of corroborative proof unites these narratives. In a different incident that American soldier Robert Williams claimed to have happened 'in another platoon across there from me', a scout was caught in the act of leaving trail signs during a patrol, guiding the enemy in the direction that the American soldiers were walking. The scout was unceremoniously executed as a result.[17] Gilliam, writing decades after the supposed incident with Lum, expressed a similar desire to have had Lum killed. Lum's disappearance made Gilliam wonder 'if someone just put him on the express train to Hell like I should have done the night before'.[18] The lurid stories surrounding the scouts' supposed lethal disloyalty towards American soldiers speak to the extreme levels of distrust that existed among some Americans towards the North and South Vietnamese people and who in turn projected it onto the scouts. American soldiers often learned to hate their Vietnamese enemy during basic training and recruits were exposed to racist terms and nicknames to describe them. Suspicion of South Vietnamese loyalty only amplified once they were in South Vietnam due to the nature of the conflict and similar, lurid stories about South Vietnamese disloyalty. Suspicion was therefore something that was established well before US forces ever saw a Kit Carson Scout.[19] For these soldiers, the image of the Kit Carson Scouts would never change beyond the image of the enemy that already existed in their minds.

Learning to Trust Your Enemy

When scout Ba arrived at Company C of the 2nd Battalion, 12th Regiment, 1st Cavalry Division, he was deprived of his M-16. The nervous commander of Company C, Captain Bill Scudder, was not truly convinced that Ba had left his old life as a PLAF soldier behind and he feared the

[16] Rodney Lynch, 'Interview with Rodney Lynch' (interviewed by Jason Stewart for VCA Oral History Project, 23 September 2008).

[17] Robert Williams, 'Interview with Robert Williams' (interviewed by Louisa for the Vietnam War Oral History Collection, Morehead State University, 11 February 1984).

[18] Gilliam, p. 180.

[19] Christian G. Appy, *Working-Class War: American Combat Soldiers and Vietnam* (Chapel Hill: University of North Carolina Press, 1993), pp. 130–137.

possibility of being assassinated by a rogue scout. However, unlike the cases previously discussed, Captain Scudder's mistrust of Ba would change over time and Ba ultimately won him over. As a result, Scudder arranged for 'a rite of passage', a ceremony in which Ba was given an M-16. When Ba was wounded in action, Scudder met the man whom he had dubbed the 'best scout in the battalion' at an aid station to make certain that the scout received the best medical attention he could get. 'He was cold but still alive, so I tucked him in a blanket, the way you do a baby', Scudder remembered. Ba was put on the next available medevac to take him to a hospital ship offshore. Tragically, Ba never arrived. The helicopter he was on crashed into the sea before ever reaching the hospital ship.[20]

The distrust that defined the initial encounters between American soldiers and the scouts could disappear over time as the scouts had the chance to prove themselves to their new American colleagues. Witnessing the abilities and courage of their scouts was enough for some American soldiers to become true believers in the program. Many American lives were saved by the scouts and many more witnessed how the scouts put their lives on the line to fight their former comrades.

The most common role designated for a scout was to walk point, to be at the front of the patrol together with three or four other soldiers. To be a pointman was a critical role during a patrol and came with important tasks that included searching for mines and booby-traps, in addition to searching for signs of the enemy. This role came with high risk because a pointman was often the first target for the enemy in an ambush and the risk of stepping on a booby-trap was high. A scout who functioned as a pointman was therefore important in order to watch over American soldiers and to prevent them from being ambushed or stepping on a mine. Drawing on their past experiences, and working in territory where they had once served, a scout could often notice things in a way that American soldiers who lacked that knowledge could not. To be able to skillfully recognize the signs of a potential ambush, a hidden mine, or a hiding place for a sniper were all abilities that separated a scout from an experienced American pointman.[21] The example of scout Le Van Det showcases how

[20] Charles Krohn, *The Lost Battalion of Tet: The Breakout of 2/12th Cavalry at Hue* (Annapolis: Naval Institute Press, 2008), pp. 120–121.

[21] Rausch, pp. 10–13; Dave Furse, 'Converted Viet Cong prove worth as guides, fighters', *The Old Reliable*, July 24, 1968, p. 4; 'Scout Spots Ambush & Mines', *Tropic Lightning News*, 26 August 1968, p. 8.

the abilities of a scout could work in practice. The former PLAF soldier became known for being able to lead his colleagues to safety during firefights by taking advantage of his knowledge of the enemy's blind spots in their line of fire. During a different occasion, he prevented an enemy ambush by noticing enemy movement on the company flank and proceeding to engage the enemy on his own initiative. 'He saved my life on several occasions, particularly once when he pointed out the presence of a mine I was about to step on', one American soldier stated, adding that he had personally witnessed Det 'blow three live bunkers spraying automatic fire during a recent contact'.[22]

Scouts also served as 'tunnel rats', soldiers who were tasked with searching for the enemy and their hidden supplies in underground tunnels built by the PLAF, often descending into the claustrophobic space armed only with a flashlight and a 0.45 caliber pistol.[23] Tunnels, bunkers, base camps, and hidden supply caches used by the enemy to store food, weapons, ammunition, and equipment were prime targets for the scouts. One successful operation led by a scout resulted in the retrieval of '15 AK-47 assault rifles, one SKS carbine, one .30 cal. machinegun, 10 rounds for an 82mm mortar, two field telephones with an estimated 1200 feet of commo wire, one radio transmitter-receiver, and one Chinese 9mm pistol'.[24]

In the role of interpreter and intermediary between American soldiers and the civilian population of South Vietnam, the scouts played an important part in facilitating contact between the two parties and helping to extract critical intelligence. American soldiers could interrogate the civilian population on their own, but the language and cultural barriers that existed between them were difficult to overcome. The scouts were therefore a valuable asset who could discuss local concerns with civilians and understand their perspective. In the hunt for PLAF soldiers, the scouts also interrogated suspected soldiers who were hiding among the civilian population. In interrogating a suspect, scouts looked for the tell-tale signs that would give the person away as an enemy combatant, but that would have been difficult for an American soldier to spot, such as vocabulary or haircuts specific to the PLAF. The scouts also played an important role during interrogations with captured enemy soldiers in the field. Questions were either asked directly by a scout or by an American soldier whose

[22] 'Kit Carson Scout aids 1st Infantry', *The Army Reporter*, 18 November 1969, p. 16.

[23] Henry Zukowski, 'Kit Carson in the Hole', *Tropic Lightning News*, 13 April 1970, p. 8.

[24] Rich Egertsen, 'Infantrymen Led To Camp', *Southern Cross*, 7 November 1969, p. 1.

questions were translated by a scout. This was done either on the spot or over the radio if the American soldiers did not have immediate access to a scout.[25]

Kit Carson Scouts taught American soldiers to understand how vulnerable their camps and bases could be to an enemy assault, as well as not to underestimate the enemy's skill. Training demonstrations were held by scouts who had formerly been sappers in order to show just how easily they could penetrate a perimeter defense consisting of several rows of concertina wire, flares, and mines in the hope that this would make American soldiers more alert when they were on guard duty and to show them that the experienced sappers could be stopped. Wearing only shorts, a scout would crawl their way through these defenses, often for a large audience. During one such display put on for the defenders of LZ Stinson southwest of Chu Lai, former PLAF sapper Nguyen Trong Thu in only three minutes 'worked his way through three strands of wire, disarmed a claymore mine and avoided numerous trip flares to breach the perimeter's most concentrated defense'.[26] Another way for scouts to demonstrate abilities that they had acquired in the PLAF and PAVN was a demonstration in camouflage. Two scouts, Liang and Nguyen Lao, were participants in one such demonstration for aspiring snipers in the United States Marine Corps. Marine John Culbertson narrates how his entire class failed to spot the men they had been instructed to search for in a nearby hill. When the instructor raised his arm and asked Liang and Lao to step forward, Culbertson and his fellow students could not believe their eyes. 'Each Marine looked embarrassed, and for ten thousand dollars, there wasn't one of us who could even remember what part of the slope they had risen from', Culbertson wrote, humbled by the experience. 'It was just amazing that every time I felt like I was getting pretty salty, someone like Sergeant Liang or Nguyen Lao would show up and make me look dumb!'[27] Many other American soldiers likely shared Culbertson's sentiments, if not his humility, in the face of such demonstrations.

Most impressive for the Americans was the scouts' willingness to take the fight to the enemy and their bravery in combat. Several scouts were

[25] VCA, TTU, Hugh Lunn, 'VC Defectors Used as Scouts', January 1968 Roger Hayes, *On Point* (New York: Presidio Press, 2000), p. 205.

[26] Thomas Maus and Bill Eftink, 'Infiltration Tactics in Vietnam Now Being Taught By Enemy Experts', *Army Digest*, 1 January 1970, pp. 38–39.

[27] John Culbertson, *A Sniper in the Arizona: 2nd Battalion, 5th Marines in the Arizona Territory, 1967* (New York: Presidio Press, 1999), pp. 109–110.

awarded the Bronze or Silver Star for valor and many American soldiers had stories to tell about having their lives saved by a scout, sometimes on several occasions. One of the most celebrated scouts was 35-year-old Truong Kinh who served with the 5th Marine Division. During intense fighting over one day in the early stages of Operation Union II in June 1967, Kinh single-handedly killed 31 PAVN combatants, although by his own estimation he killed approximately 53 combatants, and led several wounded Marines to safety. Kinh's actions were widely publicized not only in divisional newspapers in South Vietnam but also in newspapers in the United States, sometimes with an accompanying photograph showing the scout wearing American uniform. One Marine admitted after the battle that there had been a 'queasy feeling among some of the guys when Kinh first joined up with us. We didn't much like working with guys who had been shooting at Marines. But after Union Two everyone was more than convinced that he was okay'.[28]

Becoming Friends with Your Enemy

What grew out of the work relationship between the Kit Carson Scouts and American soldiers was not only soldierly camaraderie and loyalty but also genuine friendships. These personal transnational encounters are one of the most overlooked aspects of American-Vietnamese relations during the Vietnam War, providing an entirely different image than the prevailing image of hostility and mistrust, anchored in popular depictions of the war.

One exceptional case is the friendship between 22-year-old Timothy W. Walker and 36-year-old Ngo Van Nam of the 9th Infantry Division. The two men first met each other at a poker game during which Walker slipped the South Vietnamese scout a $20 bill when it was clear that he was losing. After that, they became an inseparable duo who 'ate, slept, gambled and fought together'. Nam saved Walker's life on several occasions. On November 24, 1968, Walker accidentally triggered a booby trap while on patrol. Nam pushed Walker out of the way and took the brunt of the explosion. 'He got thirty hits and I only got three. That's about the only time we weren't together—when one of us was in the hospital'. To show his gratitude, Walker invited his friend to travel with him back home to the United States. After much wrangling with the American and South

[28] John T. Wheeler, 'Kit Carson Great Help To US Marines', *Clarion-Ledger*, 22 October 1967, p. 30.

Vietnamese embassy, Walker managed to acquire a 30-day visitor's visa. In the ensuing month-long trip in 1969, the two men explored the United States together, from California to New York, and visited landmarks, shopping centers, army bases, and old colleagues from the 3rd Battalion, 47th Infantry Regiment. The most personal encounter took place in Walker's hometown of Dayton, Ohio, where Nam met Walker's friends and family. In an extraordinary photograph accompanying an article about his visit published in the *Dayton Daily News*, a smiling Nam can be seen sitting together with Walker's four-year-old daughter Crystal who looks to be happy around the visitor. For Nam, traveling to the United States was a way to quench his personal curiosity about the country, knowing little of its realities beyond what the PLAF had first told him and later what his American colleagues had shared. He became amazed with the country's infrastructure, enthralled by escalators, and particularly fascinated with comparing family life in Dayton, Ohio, with his own home in the Cai Lay district of Dinh Tuong province, finding much to think about of 'the way houses were furnished, the way people acted within the homes, whether they were eating, sleeping, or just sitting around talking'. Walker paid for the majority of the trip out of his own pocket, with soldiers from the 3rd Battalion chipping in with money in order to send Nam to the United States.[29]

Although the Walker and Nam friendship was highly publicized because of the unique gesture on Walker's behalf, similar friendships and connections appeared throughout American units as American soldiers became acquainted with men from South and North Vietnam on a personal level. David M. Aldridge was a close friend with the scouts assigned to his company, something that was attributed to his own personal interest in Vietnamese culture. The scouts came to call him *anh hai* (eldest brother), and he became particularly good friends with Nguyen Van Luong, a former PAVN soldier who preferred to save the lives of his former colleagues instead of killing them. 'Whenever we corner the enemy in a bunker, or anywhere, I've promised Luong that he will get to talk to them before we

[29] Robert W. Engelhardt, 'R&R: USA', *Army Digest*, July 1969, pp. 48–50; Jack Jones, 'Guess Who's Coming To Dinner—A Former VC', *Dayton Daily News*, 23 March 1969, p. 4. Jones' article paraphrases Nam's words rather than quoting directly from him.

do anything', Aldridge explained, 'so they can 'Chieu Hoi' and not be killed'.[30]

American soldiers who belonged to racial minorities sometimes had a different experience with the scouts. Guy Dull Knife Jr., a Lakota man from the Pine Ridge reservation in South Dakota, became close friends with an older scout who taught him how to walk point. 'I got along with the Vietnamese real well the whole time I was there. We kind of looked alike in some ways and it seemed to make a difference. To me and them', Dull Knife explained.[31] Having a close relationship with the scouts was not always seen as something positive. Chamorro soldiers from Guam and other American territories in the Pacific were in some cases racially targeted for their close association with the scouts and passed up for promotion as a consequence.[32] Chamorro men were racially abused for resembling the enemy in the eyes of other American soldiers. Other soldiers were likewise racially targeted. Richard Chan, an American soldier of Chinese descent, was mistaken for a scout by a corporal who told Chan that he did not trust the scouts and that he should watch himself. Chan defiantly replied that he 'didn't particularly trust Vietnamese scouts either, or corporals who weren't aware of new replacements. That seemed to stump him. He walked off with this ignorant look on his face'.[33]

A 'buddy system' was introduced in some units to facilitate the transition from Kit Carson School to an American unit in the field and to help the scout feel at home. All American 'buddies' were volunteers and the system looked different depending on the unit. In the 1st Cavalry Division, the American buddy went to the Kit Carson Scout school to help transition the scout, while in the 101st Airborne Division, scouts were assigned a buddy upon arrival. One of the tasks that the American buddy was meant to assist with was teaching the scout English. Yet as Harry G. Enoch came to find out, this was easier said than done once he found out that his adopted scout, Kieu Van Ngoi, could not speak English. Attending classes

[30] James Clare, 'Scout and US Friend Make Successful Team', *The American Traveler*, 17 February 1969, pp. 1–2.

[31] Joe Starita, *The Dull Knifes of Pine Ridge: A Lakota Odyssey* (Lincoln: University of Nebraska Press, 2002), p. 279.

[32] Chalsa M. Loo, Salvador S. Ueda and Robert K. Morton, 'Group Treatment for Race-related Stresses among Minority Vietnam Veterans', *Transcultural Psychiatry*, 44:115 (2007), p. 123.

[33] Johnnie Clark, *Guns Up!: A Firsthand Account of the Vietnam War* (New York: Presidio Press, 2001), p. 23.

was 'kind of a farce without an interpreter. I've worked really hard with him, but it is difficult to tell if anything has sunk in. I know the English lessons haven't. The only word he knows is "airboy" (airborne)'.[34] The language barrier between the scouts and American soldiers remained a problem throughout the program's existence. Despite going through an obligatory English class, English proficiency among scouts was relatively low. There was no standardized curriculum and different Kit Carson Scout schools focused on different types of basic English to suit everyday communication, military, and intelligence purposes. To overcome the language barriers, the scouts used different methods. Some used improvised body and sign language to communicate while others used a combination of broken English and Vietnamese words that would have been familiar to American soldiers. A third language put to use was French. The word *beaucoup* (many/numerous) was used frequently by scouts in the field, especially to point that there were many enemies in the vicinity.[35] More fortunate scouts were paired up with an American soldier who had studied Vietnamese and who could help you communicate with the rest of your unit, as in the case of Nguyen Toa and his American colleague David A. Cutaia, who had attended Vietnamese language school in California. A similar example is Chau Asnarut and Kenneth P. LeGrice who communicated in fluent French that made it easier for Asnarut and the other scouts in the unit to quickly convey information to their American colleagues.[36] The longer time a scout spent around Americans, however, the more words they picked up from their improvised English teachers. Predictably, slang and colorful four-letter words quickly became part of the scout vocabulary. Visitors to Fire Support Base Oklahoma might have been surprised to hear scout Chau S'Manan call out 'get outta here' like a true Brooklyn native. 'No sweat' was another of the scout's favorite phrases.[37]

[34] William K. Block, Jr., 'Buddy Program Helping Scouts', *Cavalair*, 4 February 1970, p. 1; Thomas Jensen, 'Kit Carson Scouts', *Rendezvous With Destiny*, Spring 1970, pp. 30–32; Harry G. Enoch, *A Year In Vietnam With the 101st Airborne - April 1969–March 1970* (Morrisville: LuLu Press, 2015), pp. 67, 70.

[35] Rausch pp. 10–13; Gary W. McKillips, *UpTight*, 'A Lesson From History', Spring 1970, pp. 9–11; Ray Byrne, 'Scouts Finds Former Unit', *Tropic Lightning News*, 9 March 1970, p. 3.

[36] W.L. Christofferson, 'Drummer, Ex-Guerrilla Form Kit Carson Team', *Sea Tiger*, 30 June 1967, p. 8; David Coughlin, 'Scout's French Upsets VC', *The American Traveler*, 30 December 1968, p. 1.

[37] Thom Arno, 'S-T-R-E-T-C-H', *The American Traveler*, 8 December 1969, p. 6.

Stitching together a common language was not the only transnational cultural exchange that took place between American soldiers and the Kit Carson Scouts. The scouts were taught American sports, such as football, and introduced to the ever-present card games such as poker with its gambling aspects, but also to other games such as stretch (or mumblety-peg) where the 'goal of the game is to stick the knife into the ground beyond where the opponent can stretch his leg'.[38] Jokes and pranks were also commonplace. Truong Kinh, the scout who had famously killed over 30 PAVN soldiers in hours of intense fighting, even got in on the fun as one American soldier narrated:

> One way to enliven things was to put Kinh on the radio on the battalion frequency. Kinh went along with the joke. Because Kinh knew a North Vietnamese dialect, he preferred to use that, in case South Vietnamese were in the command bunker when the call came through. He would read grocery lists and talk about the weather, but one could only imagine the chaos in the command bunker back at battalion headquarters as they scrambled to either change frequencies or try and discover which of their units had been overrun by North Vietnamese.[39]

The exchange of food culture was another consequence of the close proximity between American soldiers and the scouts. From the moment the scouts arrived at the Kit Carson Scout school and throughout their time in an American unit, they were put on a steady diet of American rations. The scouts liked the food, although many complained that there was not enough rice and chose to cook more themselves. Kieu Van Ngoi was described as loving the food, sucking 'down C-rations like a vacuum cleaner'.[40] American soldiers sometimes offered the scouts different types of American snacks. Jay Taylor offered a scout edible pinyon seeds, which resulted in an unexpected outcome. 'He looked at them, and as he grabbed a handful he said to me, "Ahhhh, pinion seeds … Where you from, Texas?" I couldn't believe it. He had never been out of Vietnam, and he knew what pinion nuts were. He even knew what region of the States they came from'.[41] American soldiers in turn got to taste Vietnamese food as scouts on their own initiative prepared and cooked meals for their colleagues. 'It

[38] Arno, p. 6.
[39] 'Smile Sarge', *The Indianapolis News*, 7 June 1975, p. 18.
[40] Enoch, p. 78.
[41] Jay Taylor, *Point of Aim, Point of Impact* (Bloomington: AuthorHouse, 2009), p. 25.

was not uncommon for [Nguyen Dau] to step off the trail and rip some herb or root out of the ground, which he then would cook in a pot of water at the company compound. With a little rice, his vegetable potpourris were downright tasty', wrote William L. Buchanan, adding that after one meal, Dau 'broke out into song called Ve De Anh—a nice smooth melody in a clear tenor about patriotic love for the Ben Hai river, a symbol of the North Vietnam-South Vietnam partition. [...] As far as Sgt. Farnham and Sgt. Mihalcik were concerned, he was no threat to Pat Boone'.[42]

American soldiers also had to adjust to Vietnamese gestures of friendships that were completely foreign to them, such as holding hands. 'I felt his hand touching mine. He was trying to hold my right hand that was at my side, with his left hand', one American soldier described what he saw as a sudden intrusion into his personal space by a North Vietnamese scout. Despite admitting to feeling uncomfortable at the gesture, he stated that he understood the local custom and 'let it pass'. As the two men walked together hand in hand, the scout 'was swinging my right hand with his left like a father might do with a son on a walk'.[43] Gestures like these attest to the cultural negotiations entered into by Americans and scouts. The American influence was never truly hegemonic and required Americans to adjust to the South and North Vietnamese scouts, albeit within an American framework that privileged American culture. The encounter between American soldiers and the Kit Carson Scouts was sometimes the only opportunity that an American soldier had to get to know a South or North Vietnamese individual on an intimate and personal level. The scouts spoke of their difficult lives as a PLAF soldier or how they managed to survive the long trek down the Ho Chi Minh trail. Both sides shared details about their pre-war lives and their girlfriends, wives, and children. 'Do you have a wife?' Thomas Prater asked his scout, who answered in the affirmative. 'He pulled out her picture and showed it to me. It felt strange to look at his sweetheart that lived in North Vietnam. She seemed as real as Ruth. His love seemed to be as genuine as mine for Ruth'.[44]

To show their appreciation and friendship for the scouts in their units, American soldiers looked for ways to reward and help them. Pham Yung Xuan was rewarded a Honda motorcycle by his American colleagues for

[42] Buchanan, p. 74.

[43] Thomas Prater, *Echo Five Papa* (Victoria: Trafford Publishing, 2002), pp. 200–201.

[44] Prater, p. 201.

his loyalty and other forms of monetary gifts were common. Yet more serious circumstances brought a larger collective effort by American soldiers to assist their scouts. Soldiers from the 196th infantry brigade helped gather and transport lumber to build a new house for a scout whose home had been burnt down by the PAVN, leaving his family homeless. After losing both his legs in action, Nguyen Quan was gifted $1400 by his American colleagues in his battalion. In one extraordinary case, American soldiers even went on a mission to directly assist their scout. In 1970, US Navy SEALs put together an operation to rescue the kidnapped mother and brother of a scout. They were unable to locate the two victims during the mission south of Ca Mau, but they managed to rescue the scout's son who had managed to stay hidden.[45]

The death of a Kit Carson Scout brought an immense feeling of loss for those who had known him and felt protected by him. For South Vietnamese scouts who were killed in action, American soldiers sometimes collected money that was then handed over to the scout's family to provide economic assistance. When Ngo Dau was killed in action on December 9, 1968, he left behind a widow and five children. His unit raised money for their scout's family, but they also took the additional steps of assuring that Dau's widow would find employment to continue to support her family, arranging for her to work as a food handler in the battalion's rear mess. These tragic circumstances were often the only time that American soldiers got to meet their scout's family.[46] The body of a deceased South Vietnamese scout was sent to their home in order for them to be buried according to local customs, but this was not possible for North Vietnamese scouts. Their burials were improvised and sometimes personally arranged by their American friends. Dan Sutherland and his friends arranged a burial for their North Vietnamese scout, 23-year-old Dinh: 'He had no country as a deserter. Not one Vietnamese person in South Vietnam would take Dinh's body after he was killed. So we took up a collection, bought him a casket from a local dealer in An Loc and we buried Dinh in the An Loc cemetery'.[47] Bui Ngoc Phep's funeral was arranged by the

[45] 'Kit Carson Gets Honda', *Tropic Lightning News*, 2 March 1970, p. 3; Don Newton, 'Kit Carsons are lifesavers', *Southern Cross*, 4 September 1970, p. 3; '$1400 collected for Carson Scout', *Southern Cross*, 26 November 1971, p. 1; VCA, TTU, U.S. Naval Forces, Vietnam: Monthly Historical Summary, December 1970, pp. 13–14.

[46] 'Bobcats Aid Widow', *Tropic Lightning News*, 3 March 1969, p. 1.

[47] Dan Sutherland, *365 Days of Mental Siege: Within Two Seasons In Nam* (Bloomington: Xlibris, 2010), p. 345, 348.

entire 11th Cavalry Regiment and was markedly different than Dinh's anonymous burial, as Nguyen Cong Luan described it:

> His casket was put on an armored personnel carrier, followed by other M-113 APCs and an American platoon with white headbands, a Vietnamese mourning sign. The similar piece of white cloth was also wrapped around the lowered 90 mm canon of an M-48 leading the procession. Honor guards carrying the Eleventh Armored Cavalry Regiment flag beside the Vietnamese and American national colors, bugle sounding U.S. taps ... all made many Vietnamese spectators think Phep must have been a general.[48]

Despite the contrast between the two burials, on both occasions, American soldiers gave their money, time, and effort to honor their fallen comrade with a respectful burial in accordance with Vietnamese customs.

Conclusion

The transnational encounter between American soldiers and the Kit Carson Scouts is one of the more surprising and underexplored aspects of the Vietnam War. The Kit Carson Scout Program put enemies, who had as recently as months before fought each other, in the position of having to trust, live, and fight together for a common cause. For some Americans, trusting these defectors was too much to ask for. To abandon and later fight against your former colleagues, their supposed brothers-in-arms, was a violation of the concept of loyalty among soldiers, a concept that made American soldiers suspicious and sometimes hostile towards the scouts. This made American soldiers distance themselves from the scouts and in some cases make explicit threats to kill them. They were still the enemy.

American soldiers who overcame their prejudice or who learned to trust their former enemy found themselves as participants in an active transnational encounter that opened a window through which they could humanize the men they had once fought against. While their professional relationship gave Americans an insight into the enemy's mind and tactics, their personal friendships with the scouts gave them a deeper understanding of the greater South and North Vietnamese worlds that their friends came from. In the process, the pre-conceived image of the enemy was

[48] Nguyen Con Luan, *Nationalist in the Viet Nam Wars: Memoirs of a Victim Turned Soldier* (Bloomington: Indiana University Press, 2012), p. 569.

nuanced, causing conflicting emotions that could even lead to the reconceptualization of the soldier-self. In the words of one American soldier, 'It was strange getting to know a North Vietnamese man. We were a lot alike with the same dreams. [...] I hated and feared them at times. Now one of them was a friend. It seemed to me that my role as a grunt wasn't as clear-cut as when I arrived'.[49]

The enemy-turned-ally encounter investigated in this chapter complicates our understanding of American-Vietnamese interactions during the Vietnam War. The professional and personal relationships between American soldiers and the Kit Carson Scouts provide an intriguing perspective from below of a different type of encounter than the prevailingly negative interaction present in the dominant American memory of the war. Although negative aspects are still present, intact with hostility, prejudice, and distrust—these interactions and intimacies also introduced an entirely different dynamic filled with trust, respect, cultural exchange, and even grief. In order to gain a complete understanding of the experiences of American soldiers during the Vietnam War, with all of its contradictions, it will be an important task for future scholars to include the Kit Carson Scouts in that history. They deserve to be more than a footnote.

[49] Prater, p. 201.

'Islam is the religion of the sword not pacifism': *Strategic Nostalgia* and *Self-Othering* in Islamic State Propaganda

Jared Ahmad

As many of the chapters in this collection demonstrate, one of the most basic features of human communication is the construction of enemy images. Throughout history, nations have sought to define themselves by demonising, dehumanising and exaggerating the threat posed by 'enemy' populations, something that, as a consequence, has served to bolster recruitment efforts and mobilise citizens in support for wars and unpopular foreign policies.[1] Yet despite attempts to maintain clear boundaries

[1] See Richard K. Popp & Andrew L. Mendelson, '"X"-ing Out Enemies: Time Magazine, Visual Discourse and the War on Terror', *Journalism: Theory, Practice and Criticism*, 11. 2 (2010), pp. 203–207, for an overview.

See Al-Hayat Media, 'Islam is the Religion of the Sword Not Pacifism', *Dabiq*, Issue 7 (2015), pp. 20–24.

J. Ahmad (✉)
University of Sheffield, Sheffield, UK
e-mail: j.ahmad@sheffield.ac.uk

H. Furneaux, M. Greig (eds.), *Enemy Encounters in Modern Warfare*, https://doi.org/10.1007/978-3-031-56748-3_13

between adversaries, in truth they have always been unstable. Indeed, since the end of the Cold War these boundaries have become even more fluid and uncertain, with today's enemy images characterised by a growing ambiguity over *who* and *what* the enemy is. During the early years of the 'war on terror' era, for example, scholars called attention to the presence of reductive, Orientalist-inspired stereotypes and stock enmification tropes in media portrayals of the terror threat.[2] At the same time, however, over the course of this 'war' others called attention to a breakdown in distinctions between combatant and non-combatant, friend and enemy, with encounters with the terrorist-Other often appearing to be encounters with our-*selves* as enemies.[3] As if to further unsettle these distinctions, the very media through which we experience today's enemies has also undergone profound transformations. Distant acts of war and political violence are no longer confined to geographically bounded arenas of conflict, but, instead, seep out and become instantly accessible through widely available social media platforms such as YouTube, Facebook, Twitter and Telegram. Outside of online spaces, this content is further amplified via international news media, where user-generated media is often *re*-mediated into news stories in an attempt to retain and engage fleeting audiences. Within this new, hybridised media landscape, enemies appear closer than at any point in history, entering into our living rooms, bedrooms and intimate spaces via an increasing range of screens and devices.

It is at the intersection between debates concerning the enemy other, shifting conceptions of war and conflict, and new communications technologies that the current chapter is situated. Focusing on the Islamic State, a violent Salafi-Jihadi organisation which, at its height, ruled over 7.7 million people, the chapter offers a provocative take on the theme of enemy encounters by considering the way the group uses propaganda images to communicate with different sets of audiences. More specifically, the analysis shows how the group appropriate nostalgic and civilisational

[2] See Debra Merskin, 'The Construction of Arabs as Enemies: Post-September 11 Discourse of George W. Bush', *Mass Communication and* Society, 7.2 (2004), pp. 157–175; Justin Lewis, 'Terrorism and News Narratives', in Daya Kishan Thussu & Des Freedman (eds.), *Media & Terrorism: Global Perspectives.* London: SAGE (2012), pp. 257–270; Jared Ahmad, 'A Shifting Enemy: Analysing the BBC's Representations of "Al-Qaeda" in the Aftermath of the September 11th 2001 Attacks', *Critical Studies on Terrorism*, 9.3 (2016), pp. 568–590, for example.

[3] See Jared Ahmad, *The BBC, the War on Terror and the Discursive Construction of Terrorism: Representing 'Al-Qaeda'.* Cham: Palgrave Macmillan (2018).

imagery to portray the caliphate as both 'a place of romance, exotic beings, haunting memories and landscapes' *and* a space of fear, uncertainty and uncivilised danger.[4] These visual strategies are designed to exploit and amplify pre-existing psycho-social conditions within different audience collectives,[5] thus enabling Islamic State propagandists not only to utilise 'the weight of the media against the media' but also, to repurpose this phrase, use the weight of *culture against culture* to secure its goals.[6]

The chapter begins by discussing the long-standing relationship between terrorists and their (imagined) audiences, focusing in particular on the Islamic State's audience segmentation strategies and the way the group crafts distinct images for different audiences. Following this, the chapter outlines the methods and data, before moving to analyse two sets of recurrent images that appear across the group's communications output. Here, two core analytical concepts are introduced, *Strategic Nostalgia* and *Strategic Self-Othering*, which each help to shed light on the way the Islamic State's identity-based messaging appeals to both friends and enemies alike. The final concluding section points to the complex utility of enemy images in an era characterised by a loss of faith in political institutions and a questioning of the very notion of 'truth' itself.

The Imagined Audience and Islamic State's Identity-Based Messaging

Since the discipline of terrorism studies emerged in the late 1970s and early 1980s, one of the major areas of scholarly consensus has been the view that audiences are central to political violence.[7] From the Sicarii's high-profile assassination campaigns of early antiquity to Russian anarchists' 'propaganda of the deed' during the nineteenth century, terrorist actors throughout history have sought to communicate violent acts to idealised 'imagined audiences'. While the target audience may vary from act to act, terrorists use violence, or the threat of violence, primarily to

[4] Edward W. Said, *Orientalism*, p. 1.

[5] See Haroro Ingram, 'The Strategic Logic of State and Non-State Malign "Influence Activities": Polarizing Populations, Exploiting Democratic Recession', *The RUSI Journal*, 165.1 (2020), p. 13.

[6] Neville Bolt, *The Violent Image: Insurgent Propaganda and the New Revolutionaries*. Columbia: Columbia University Press (2012), p. 257.

[7] See Alex Schmid, 'Terrorism as Psychological Warfare', *Democracy & Security*, 1.2 (2005), pp. 137–146.

influence the behaviour of those watching. As Alex Schmid notes, 'by using violence against one victim, [terrorism] seeks to coerce and persuade others. The immediate victim is merely instrumental, the skin on a drum beaten to achieve a calculated impact on a wider audience'.[8] Though we should always keep in mind that audience members are not passive receivers of fixed messages, but rather active interpreters of such propaganda, terrorist actors have a keen understanding of their various target constituencies, and thus seek to tailor or segment their messages to specific groupings of friends and enemies.

For the Islamic State in particular these audience segmentation practices form a central pillar of its communications strategy. Indeed, since rebranding itself as *the* Islamic State in June 2014, the group's leaders and propagandists have sought to develop a powerful, 'identity-based' messaging strategy that divides the world into two imagined collectives; namely, 'believers' and '*dis*believers'.[9] These dichotomised social identities revolve around highly simplified and essentialised in- and out-group constructs that echo wider patterns of enmification seen in conflicts throughout history.[10] As former leader, Abu Bakr al-Baghdadi, makes clear: '[t]he world today has been divided into two camps and two trenches, with no third camp present: the camp of [true] Islam and faith, and the camp of kufr and hypocrisy'.[11] Purposefully echoing George W. Bush's reductive '[e]ither you are with us or the terrorists',[12] this stark, binary division of the world into friends and enemies serves to underpin the group's entire propaganda output and is a near-constant theme across its messaging.

These imagined audiences comprise both internal and external elements. For those viewed as 'believers', the internal in-group includes Sunni Muslims within the Islamic State's immediate territorial or

[8] Alex Schmid, 'Frameworks for Conceptualising Terrorism', *Terrorism and Political* Violence, 16.2 (2004), p. 207.

[9] See Alexandra Herfroy-Mischler & Andrew Barr, 'Jihadist Visual Communication Strategy: ISIL's Hostage Execution Video Production', *Visual Communication*, 18.4 (2018), p. 523.

[10] See Stephane Baele, Katharine Boyd & Travis Coan (eds.), *ISIS Propaganda: A Full-Spectrum Extremist Message*. Oxford: University of Oxford Press (2020), p. 103

[11] Abu Bakr al-Baghdadi, 'A Message to the Mujahidin and the Muslim Ummah in the Month of Ramadan', Al-Hayat Media Centre (July 4th 2014).

[12] George W. Bush, 'Address to a Joint Session of Congress and the American People', U.S. Government Archives (September 20th 2001). Available at: https://georgewbush-whitehouse.archives.gov/news/releases/2001/09/20010920-8.html (Accessed June 6th 2022).

geographic sphere of influence (e.g. Iraq, Syria, Libya, Sinai, West Africa and Afghanistan). Here, research consistently reveals how the vast majority of the group's propaganda is targeted towards internal audiences, due to the fact that it is produced in Arabic and focuses on local or regional themes.[13] In order to expand its territory and maximise recruitment efforts, however, there is an external dimension to the in-group, which encompasses Sunni Muslim supporters and sympathisers outside its direct territorial influence; a group traditionally characterised as the Ummah. Constituting a form of 'public diplomacy', much of this material is produced for foreign, non-Arabic-speaking audiences by the Islamic State's al-Hayat Media Center. Here, magazines such as *Dabiq*, *Rumiyah*, *Dar al-Islam*, *Istok* and *Konstantiniyye* produced, respectively, for English, French, Russian and Turkish-speakers, alongside a range of other formats such as videos, computer games, nasheeds and social media content, constitute dimensions of the Islamic State's 'full-spectrum' propaganda and thus help to reinforce its singular, totalising message.[14]

When seeking to communicate with members of this in-group, two of the most important strategic goals are the ***propagation*** of the Islamic State's ideology and the ***legitimisation*** of its actions and methods.[15] These goals concern the Islamic State's attempts to attract new recruits and expand the reach of their ideology, alongside the need to justify its actions and contextualise them within a wider religious and historical context. Thus, long reports discussing the importance of leadership[16] or issues of legality or permissibility within Islam,[17] films discussing the economy[18] and reports depicting governance and the community services overseen by the Islamic State[19] constitute a compelling and 'competitive system of meaning' designed to propagate the group's ideology, legitimise its actions

[13] See Aaron Zelin, 'Picture Or It Didn't Happen: A Snapshot of the Islamic State's Official Media Output', *Perspectives on Terrorism*, 9.4 (2015), pp. 85–97.

[14] See Stephane Baele, Katharine Boyd & Travis Coan (eds.), *ISIS Propaganda*, p. 88.

[15] Carsten Bockstette, *Jihadist Use of Strategic Communication Management Techniques*. George C. Marshall European Centre for Security Studies (2010). Available at: https://apps.dtic.mil/sti/pdfs/ADA512956.pdf (Accessed May 6th 2022), pp. 11–12.

[16] Al-Hayat Media, 'Imamah is from the Millah of Ibrahim', *Dabiq*, Issue 1 (June 2014), pp. 20–31.

[17] Al-Hayat Media, 'The Revival of Slavery Before the Hour', *Dabiq*, Issue 4 (October 2014), pp. 14–17.

[18] Al-Hayat Media, 'Rise of the Khilafah: Return of the Gold Dinar'. Dir/Prod Islamic State (August 2015).

[19] Al-Hayat Media, 'Healthcare in the Khilafah', *Dabiq*, Issue 9 (May 2015), pp. 24–26.

and reify the identities of the believer in-group.[20] As will be discussed in further detail below, a key factor in the promotion of such goals is the strategic use of nostalgia and its appeal to an idealised Islamic past.

The second imagined audience are those viewed as the Islamic State's enemies, or '*dis*believers'. Again, there are internal and external dimensions to this out-group. Traditionally labelled as the 'near enemy', the internal dimension consists of the 'apostate' or secular rulers of Muslim-majority nations within the group's immediate geographic sphere of influence.[21] Long before the 2011 'Arab Spring' uprisings, governments across North Africa and the Middle East held reputations as being undemocratic, deeply unpopular and mired in corruption. With the events of 2011, and subsequent destabilisation in countries such as Libya, Syria and Iraq, the resulting chaos created a breeding ground for further sectarian and political violence. Here, Islamic State propagandists utilise a range of names for this out-group: from 'apostates' and 'rejecters', to 'tyrants' and 'polytheists'. Taken together, they form a singular and composite enemy whose principal goal is to destroy the Islamic State's caliphate. In terms of the external dimension of the out-group, here the emphasis is on 'Western' nations such as the United States, France and Britain, alongside countries who form part of the global coalition against the Islamic State, such as Russia, Israel and India. Traditionally viewed as 'the far enemy',[22] the group use labels such as 'Crusaders', 'Christians', 'Jews' or the more general 'the West' to name and classify this out-group. As Brandon Colas has pointed out, much of the non-Arabic language publications referred to above contain materials and content specifically designed for external audiences, with articles by the captured British journalist John Cantlie, together with pieces written by prominent think tanks, intended for consumption by politicians, journalists and ordinary citizens.[23]

As Carsten Bockstette's work on Salafi-Jihadi strategic communication techniques suggests, a major objective when communicating with those viewed as disbelievers, whether they be internal or external, is the creation

[20] Haroro Ingram, 'An Analysis of Islamic State's *Dabiq* Magazine', *Australian Journal of Political Science*, 51.3 (2016), p. 458.

[21] See Fawas Gerges, *The Far Enemy: Why Jihad Went Global*. Cambridge: Cambridge University Press (2005).

[22] See Fawas Gerges, *The Far Enemy* (2005).

[23] Brandon Colas, 'What Does *Dabiq* Do? ISIS Hermeneutics and Organizational Fractures within *Dabiq* Magazine', *Studies in Conflict and Terrorism*, 40.3 (2017), pp. 178–179.

of fear and uncertainty.[24] This strategy has been explicitly spelled out in operational treatises by ideologues such as Abu Bakr al-Naji, who advocate for spectacular, transgressive and fear-invoking forms of propaganda.[25] Islamic State propagandists appear keenly aware of the rich history of stereotypes and negative portrayals of Muslims appearing within Western cultural texts and seek to appropriate them for their own ends. As Andres Ventsel et al. point out, the amplification of feelings of danger and the 'cultivation of a general atmosphere of fear and confusion are seen as key elements of present-day strategic communication' practices,[26] something that enables groups like the Islamic State to exploit and amplify favourable psychosocial conditions in enemy audiences, thus exaggerating threat perceptions (Table 1).[27]

Table 1 The Islamic State's imagined audiences

	Believers	*Disbelievers*
Internal	Supporters and sympathisers within the Islamic State's immediate sphere of influence (e.g. Iraq, Syria, Libya, Sinai, West/Central or East Africa and Afghanistan)	Secular or 'apostate' rulers of Muslim-majority nations and their populations, within the Islamic State's direct influence
External	Transnational supporters and sympathisers outside of the group's influence (e.g. Muslims in America, Europe, the Middle East and Asia). This grouping has traditionally been viewed by Muslims as 'the Ummah'	Secular rulers ('Crusaders') of non Muslim nations and their populations. This grouping has traditionally been described as 'the far enemy' by violent Salafi-Jihadi groups
Major communication goals	Propagation of the Islamic State's ideology and legitimisation of its methods and violent acts	Fear, intimidation and coercion of the enemy

This table is based on the work of Gerges (2005), Bockstette (2010) and Herfroy-Mischler & Barr (2014)

[24] Carsten Bockstette, *Jihadist Use of Strategic Communication Management Techniques*, p. 12.

[25] See Abu Bakr al-Naji, *The Management of Savagery: The Most Crucial Stage Through which the Islamic Nation Will Pass*. Trans. Will McCants. Cambridge: John M. Olin Institute for Strategic Studies (2004).

[26] Andreas Ventsel et al., 'Discourse of Fear in Strategic Narratives: The Case of Russia's Zapad War Games', *Media, War and Conflict*, 14.1 (2021), p. 22.

[27] Haroro Ingram, 'The Strategic Logic of State and Non-State Malign "Influence Activities"', p. 13.

Methods and Approach

While scholars have criticised studies that focus solely on textual analysis,[28] that is, analysis of the language and imagery of terrorist propaganda alone, such approaches are necessary to gain insight into the way the material was meant to be interpreted by audiences.[29] In order to interrogate these preferred meanings, the analysis below develops a form of visual discourse analysis[30] that draws upon Norman Fairclough's three-dimensional approach to critical discourse analysis.[31] Here, attention centres on the textual, intertextual and contextual dimensions of media representations to show how meaning is created over time, via layers of discourse, knowledge and power. The focus on images is important because visual modes of communication serve a central, performative role in propaganda narratives. Though we should be careful in overstating their 'power', images and visual representations are significant because they help trigger, often involuntary, cognitive and affective responses in audiences,[32] repressing the executive 'control' networks of the brain and stimulating the more primal, 'salience' networks.[33] Indeed, while language helps stabilise or 'anchor' the meaning of visuals, images are often more immediate and efficient than text alone, thus helping to conjure deep-rooted, often unspoken, cultural connotations and associations in audiences.[34]

The analysis centres around propaganda produced between 2014 and 2017—a period that coincided with large territorial and military gains for the Islamic State group in Syria, Iraq and beyond. Adopting an explicitly qualitative approach to the data collection, images were taken from the widely circulated propaganda magazines *Dabiq*, *Rumiyah* and *Dar al-Islam*, alongside films such as *Rise of Khilafah: Return of the Gold Dinar*

[28] Anne Aly, 'Brothers, Believers, Brave *Mujahideen*: Focusing Attention on the Audience of Jihadist Preachers', *Studies in Conflict and Terrorism*, 40.1 (2017), pp. 62–76.

[29] Stuart Hall, 'Encoding/Decoding', in Simon During (ed.), *The Cultural Studies Reader*. London: Routledge (1993), pp. 507–517.

[30] See Gillian Rose, *Visual Methodologies: An Introduction to Researching with Visual Materials*. 4th ed. London: SAGE (2016), pp. 186–219.

[31] See Norman Fairclough, *Media Discourse*. London: Arnold (1995).

[32] David Domke et al., 'The Primes of Our Times? An Examination of the "Power" of Visual Images', *Journalism: Theory, Practice and Criticism*, 3.2 (2002), pp. 131–159.

[33] Andrew Nierenberg, 'Why Does Propaganda Work? Fear-Induced Repression of the Executive Control Brain Network', *Psychiatric Annals*, 48.7 (2018), p. 315.

[34] Richard Popp & Andrew Mendelson, '"X"-ing Out Enemies', p. 205.

(2015). Though admittedly selective and subjective in nature, focusing on a smaller number of texts in close detail helps uncover symbolic and subtextual meanings that are often lost in large, quantitative studies of terrorist propaganda. In this regard, the next section seeks to outline two mutually conditioning messaging strategies that have been developed by the Islamic State and targeted towards the two interpretive communities identified above.

Strategic Nostalgia: Imagining a Place of 'romance, exotic beings, haunting memories, and landscapes'

As we have seen, Islamic State propagandists and ideologues have developed a simple, yet highly effective, messaging strategy that is designed to communicate with two imagined audience groups: namely, 'believers' and 'disbelievers'. When targeting the former, Islamic State media workers deliberately seek to exploit long-standing feelings of anger, humiliation and alienation thought to be felt by individuals within this (in)group. As Roxanne Euben has shown, violent Salafi-Jihadi organisations make continual references to the 'humiliation of Islam' in their propaganda, deploying 'the symbols and rhetoric of emasculation to conjure historical, cultural and political experiences of powerlessness' and to highlight the continued legacies of colonialism.[35] Such accounts seek to link the experience of Muslims during the Crusades with more recent historical events and processes, such as the abolition of the Ottoman caliphate, the drawing up of the Sykes-Picot Agreement, Palestinian dispossession and the outcomes of wars in Afghanistan and Iraq.[36] For Islamic State propagandists, however, the solution to such narratives of humiliation is an appeal to an imagined, somewhat romanticised and idealised history; something I refer to as *Strategic Nostalgia*.

According to Bryan S. Turner, nostalgia is characterised by a profound 'sense of historical decline and loss, involving a departure from some golden age of "homefullness"' or longing for 'personal wholeness and

[35] Roxanne Euben, 'Humiliation and the Political Mobilization of Masculinity', *Political Theory*, 43.4 (2015), p. 513.

[36] See Samantha Mahood & Halim Rane, 'Islamist Narratives in ISIS Recruitment Propaganda', *The Journal of International Communication*, 23.1 (2017), p. 20.

moral certainty'.[37] Here, nostalgic discourse and rhetoric serve a central role in the process of identity construction, whereby 'history becomes a useful instrument for people to understand who they are, where they come from and where they should be going'.[38] During periods of crisis or uncertainty, moreover, this 'backward-looking' perspective serves to further demarcate in/out-group identities, providing a screen on which to project idealised self-images and mobilise usable memories. For example, Myria Georgiou has shown how geographically displaced groups often draw on imagined, heavily idealised pasts as a means to counter the 'destabilizing effects of globalization', thus creating a sense of stability when faced with 'the loss of a home, land or loved ones'.[39] Crucially, for Georgiou, this turn to nostalgia constitutes an explicitly *political* or *strategic* process of identity construction, whereby romantic, historical and culturally resonant symbols are 'used to define a bounded self as distinctly different from other cultural groups'.[40]

For the Islamic State, this strategic use of nostalgia is most evident in the recurrent figure of the pre-modern, Islamic warrior or 'mujahid'. Appearing across the group's communications output, and typically pictured wearing a mixture of traditional, seventh-century clothing and modern military fatigues, such imagery conjures culturally constructed ideals of honour, bravery and heroism that have long been associated with the first generation of Muslims. Figure 1 provides an emblematic example of such imagery. Taken from the first edition of the Islamic State's English-language *Dabiq* magazine, and titled 'From Hijrah to Khilafah', the image presents readers with the striking, almost apocalyptic image of a lone rider, sat atop a black horse, and holding the black standard of the Islamic State, emblazoned with the testimony of faith: 'There is no deity but God, Muhammed is the messenger of God'. While the group's enemies are likely to view such imagery through colonial lenses, that is, as backward, exotic and uncivilised in nature, for those within the believer in-group

[37] Bryan S. Turner, 'A Note on Nostalgia', *Theory, Culture and Society*, 4.1 (1984), p. 147.

[38] Ezgi Elçi, 'Politics of Nostalgia and Populism: Evidence from Turkey', *British Journal of Political Science*, 52.2 (2022), p. 699.

[39] Myria Georgiou, '"Between *Strategic Nostalgia*" and "*Banal Nomadism*": Explorations of Transnational Subjectivity among Arab Audiences', *International Journal of Cultural Studies*, 16. 1 (2012), p. 25.

[40] Ibid.

Fig. 1 The figure of the pre/modern warrior mujahid (Dabiq, Issue 1, 2014: 34–35, © Al-Hayat Media)

they are likely to rouse heroic fantasies and exploit 'traditional Muslim male sensitivities around chivalry, honour, shame and sacrifice'.[41]

Setting aside the semiotic truism that figures pictured from below appear more imposing to viewers,[42] the layered symbolism surrounding the anonymous rider is explicit in its application of nostalgic themes. The presence of the horse, for example, serves to connote pre-colonial ideals of chivalry, bravery and honour, and thus helps to connect the Islamic State's contemporary struggle with the first generation of pious Muslims. According to Christopher Anzalone, horses were central to early Muslim territorial expansion and have since become venerated within Arab and

[41] Akil Awan, 'Religion, Youth and the Effects of ISIS Narratives on Radicalisation in the West', in *Panorama: Insights into Asian and European Affairs*, 1 (2015), pp. 57–70. Available at: https://counterideology2.files.wordpress.com/2015/07/from-desert-to-world-cities-new-terrorism.pdf (Accessed May 19th 2022), p. 67.

[42] See Gunther Kress & Theo van Leeuwen, *Reading Images: The Grammar of Visual Design*. London: Routledge (1996), p. 146.

Islamic cultural history.[43] In early Islamic poetry and visual art, notions of '[s]peed, prowess, strength and aggressiveness, arrogance and haughtiness, virility, pride and nobility' were conferred on both animal and rider, with horses strongly connected to the morals and principles of military jihad.[44] For the contemporary believer, such ideals serve to fulfil some of the existential motivations experienced by Islamic State recruits, offering an empowering, and much-needed, sense of meaning and purpose; specifically, the sense 'that one is an active participant in a cosmic battle to defend the sacred' against an all-powerful enemy.[45] As Akil Awan explains, such imagery constitutes 'an emphatic rejection of the banality and monotonous inanity of daily life, providing... a sense of being part of an elite group that compensates for the shortcomings of one's own trivial existence'.[46]

The presence of the black standard, moreover, evident in no less than 222 images in the group's *Dabiq* and *Rumiyah* magazines alone,[47] further serves to elicit nostalgic and historical associations, reflecting 'the battle standards... carried by Muslim forces in some of the earliest armed conflicts in Islamic history'.[48] Although for the group's enemies, the presence of this colour will conjure notions of protest, rebellion and, importantly, death,[49] for the believer in-group the colour draws further parallels between the Islamic State and the eighth-century Abbasid caliphate—a period widely considered by Muslim historians as the 'Golden Age' of Islam.[50] Used by a range of violent Salafi-Jihadi groups, according to Will McCants, Islamic State propagandists specifically employ black flags to promote a Manichean worldview, 'which permits no grey areas between

[43] Christopher Anzalone, 'Equestrianism in Jihadi Cultures'. Available at: https://ibnsiqilli.com/2019/10/25/equestrianism-in-jihadi-cultures/ (Accessed June 9th 2022).

[44] Jérémie Schiettecatte and Abbès Zouache, 'The Horse in Arabia and the Arabian Horse: Origins, Myths and Realities', *Arabian Humanities*, 8 (2017), p. 15.

[45] Simon Cottee & Keith Heyward, 'Terrorist (E)Motives: The Existential Attractions of Terrorism', *Studies in Conflict and Terrorism*, 34.12 (2011), pp. 963–986.

[46] Akil Awan, 'Religion, Youth and the Effects of ISIS Narratives on Radicalisation in the West', p. 66.

[47] See Stephane Baele et al. (eds.), *ISIS Propaganda*, p. 98.

[48] Afshon Ostovar, 'The Visual Culture of Jihad', in Thomas Hegghammer (ed.), *Jihadi Culture: The Art and Social Practices of Militant Islamists*. Cambridge: Cambridge University Press (2017), p. 88.

[49] Yuet See Monica Owyong, 'Clothing Semiotics and the Social Construction of Power Relations', *Social Semiotics*, 19.2 (2009), p. 202.

[50] Afshon Ostovar, "The Visual Culture of Jihad", p. 90.

the binaries of right and wrong, believer and unbeliever'.[51] There is also an important eschatological dimension to the use of this colour, moreover, as scholars of Islam have traditionally foretold that the arrival of the Mahdi, or 'rightly guided one', will be marked by the sight of 'Black Banners coming from Khorasan'.[52] In repeatedly using the colour black in its imagery, therefore, Islamic State propagandists are able to nostalgically construct the group as 'true' successors of the first generation of Muslims, knowingly invoking a 'golden age of heroic virtue, moral coherence and ethical certainty' that occurred prior to the rise of European colonial powers.[53]

An equally important, yet perhaps more subtle, aspect of *Strategic Nostalgia* can be seen in the settings and backdrops against which the black flag-carrying mujahid is situated. This figure is rarely depicted in the urban, metropolitan towns and cities across Iraq, Syria and beyond.[54] Rather, as can be seen in Fig. 2, he, for such imagery is always gendered, is most often pictured on horseback emerging out of a series of pristine, pre-colonial, and 'exotic' desert settings. Culturally and semiotically speaking, desert imagery constitutes an important marker of Arab and Islamic identity and is most often associated with antiquity and an idealised past.[55] In fact, for some, desert landscapes have long been considered to be the antithesis of 'Western civilisation',[56] where wild, untamed and primitive desert wastelands stand in opposition to modern, ordered and civilised metropolitan centres. Within violent Salafi-Jihadi propaganda, in particular, deserts not only serve to further cement the links with the first generation of Muslims[57] but also function to perpetuate the

[51] Will McCants, *The ISIS Apocalypse: The History, Strategy and Doomsday Vision of the Islamic State*. New York: St Martin's Press (2015), p. 21.

[52] See Samantha Mahood & Halim Rane, 'Islamist Narratives in ISIS Recruitment Propaganda', p. 26.

[53] Georg Stauth & Bryan Turner, *Nietzsche's Dance: Resentment, Reciprocity and Resistance in Social Life*. Oxford: Blackwell (1988), p. 47.

[54] See Al-Hayat Media, 'Reflections on the Final Crusade', *Dabiq*, Issue 4 (2014), p. 34, for a brief exception.

[55] Jala Makhzoumi, 'Landscape in the Middle East: An Inquiry', *Landscape Research*, 27.3 (2002), p. 219.

[56] Jessica Jacobs, 'Have Sex Will Travel: Romantic "Sex Tourism" and Women Negotiating Modernity in the Sinai', *Gender, Place and Culture*, 16.1 (2009), p. 47.

[57] See Combatting Terrorism Center Westpoint, 'Islamic Imagery Project: Visual Motifs in Jihadi Internet Propaganda'. Available at: https://ctc.westpoint.edu/wp-content/uploads/2010/06/Islamic-Imagery-Project.pdf (Accessed 6th June 2022), p. 27.

Fig. 2 The caliphate's romantic, far-away desert landscapes (Dabiq, Issue 10, 2015: 26; Rumiyah, Issue 3, 2016: 13; Rumiyah, Issue 12, 2017: 18, © Al-Hayat Media)

Orientalist-inspired myth that Islam materialised from the empty deserts, rather than urban spaces, of seventh-century Arabia.[58]

Here, however, the use of desert landscapes should not be seen simply as a passive visual backdrop against which the group's propaganda unfolds. Rather, these austere settings should instead be considered as constitutive dimensions of the Islamic State's explicitly male identity, functioning aesthetically and, crucially, nostalgically to reassert their hyper-masculine identity as fierce 'lions', 'knights' or 'warriors'. As Michael Leyshon and Catherine Brace note when discussing the relationship between fictional, diegetic space and identity: 'the desert is neither decentred to accommodate action and spectacle nor positioned as a mirror of their condition but is an *active agent* in the forging of new social relationships'.[59] In this sense, the Islamic State's repeated use of desert settings provides a stage on which the group's fighters perform nostalgic identities *and* a cultural referent for their lost power and masculinity. Whether it be via the figure of the lone warrior mujahid or images of loose bands of riders taking part in dramatic raids, the images identified here foreground nostalgic

[58] Frederick Denney, *Introduction to Islam*. 4th Edition. London: Routledge (2011), p. 69.

[59] Michael Leyshon & Catherine Brace, 'Men and the Desert: Contested Masculinities in *Ice Cold in Alex*', *Gender, Place and Culture*, 14.2 (2007), p. 168 (emphasis added).

conceptions of hegemonic masculinity that encompass the ideals of risk-taking, self-discipline, emotional control, physical toughness, aggression and violence.[60] These 'wild', pre-colonial settings and backdrops, therefore, help to subtly animate conceptions of rugged, desert masculinity, alongside a sense of struggle and physical mastery, that is often possessed by those who live in such environments. The implication here is that such landscapes require 'real' strength, fortitude and bravery, something that, according to Islamic State propagandists, appears to be lacking in more 'modern' and 'civilised' parts of the world. The near-total absence of women in the group's communications output, save for a small number of patronising articles offering 'advice to our sisters', can be said to further underscore this view.[61]

Alongside the use of black, as noted above, the heavy use of sepia tones and digital filters constitutes an additional dimension to the group's use of *Strategic Nostalgia*. While rooted in contemporary social media aesthetics, what Aaron Zelin describes as the 'Instagramification' of terrorist propaganda,[62] saturating these images in rich, sepia shades or adjusting their brightness, sharpness and colour saturation further underscores the group's nostalgic appeal to history. These techniques can be seen, for example, in the opening sequence to the propaganda film *Rise of the Khilafah: Return of the Gold Dinar* (2015), where raw footage of Islamic State fighters is juxtaposed alongside scenes from Ridley Scott's epic historical drama *Kingdom of Heaven* (2005) (see Fig. 3). Here, the temporal boundaries separating the Islamic State's contemporary caliphate and its idealised historical representation are made invisible, as the group's masked foreign fighters appear to battle alongside medieval Saracen warriors in a timeless war against the enemy Crusader.[63] Within this rapid fire opening sequence, all the elements of *Strategic Nostalgia* can be seen on display; from the representation of the hyper-masculine fighter and use of equestrian symbolism, to the repeated presence of black banners and the stark, empty desert landscapes. Thus, by invoking the figure of the pre-modern

[60] See Ramon Hinojosa, 'Doing Hegemony: Military, Men and Constructing a Hegemonic Masculinity', *Journal of Men's Studies*, 18.2 (2010), pp. 179–194.

[61] See Al-Hayat Media, 'A Brief Interview with Umm Basir al-Muhajirah', *Dabiq*, Issue 7 (February 2015), pp. 50–51.

[62] Aaron Zelin, 'Picture Or it Didn't Happen', p. 93.

[63] Al-Hayat Media, *Rise of the Khilafah: Return of the Gold Dinar*. Dir. Islamic State, Aleppo (September 2015).

Fig. 3 Juxtaposing scenes from Islamic State's contemporary war propaganda with scenes from mainstream Hollywood cinema (*Rise of Khilafah: Return of the Gold Dinar* 2015, © Al-Hayat Media)

mujahid, and calling on a range of subtle semiotic devices linked to equestrian mythology, colour and the vast, evocative desert landscapes of pre-colonial Arabia, Islamic State propagandists subtly draw upon nostalgic myths surrounding the first generation of Muslims whilst also presenting those within the believer in-group an idealised image of an exotic, far-away and, crucially, romantic contemporary caliphate. Importantly, this appeal to a nostalgic, idealised, pre-colonial past is significant because it enables the group to denounce appeals to borders, democracy and national identity that are frequently invoked by political elites across North Africa and the Middle East. In so doing, Islamic State propagandists can thus portray themselves as the 'true' inheritors of the religion. At the same time, however, in drawing upon this reservoir of useable memories and semiotic resources, the group knowingly mobilise an essentialised image of the self; one that, moreover, is often rooted in 'claims of an irreducible *otherness*' and linked to colonial encounters with the enemy Other.[64]

[64] Myria Georgiou, 'Between *Strategic Nostalgia* and *Banal Nomadism*', p. 25, emphasis added.

Strategic Self-Othering: Visualising the 'Demonic Hordes of Hated Barbarians'[65]

Appearing alongside such idealised, wistful and nostalgic images are another set of visual representations. While still rooted in an essentialised self-image, here the emphasis centres around the more reductive, stereotypical and negative aspects of the group's identity. Indeed, the intended audience of these propaganda images is not so much those viewed as believers, but rather the more diffuse category of disbelievers (e.g. the group's 'near' and 'far' enemies in the Middle East and beyond). These images appear to mobilise long-standing stereotypes surrounding Islam and the 'dangerous Orient' in a deliberate attempt to harness the fears, insecurities and repressed cultural memories that the out-group is said to associate with such people and places; a practice labelled here as *Strategic Self-Othering*.[66]

According to postcolonial theorist Edward Said, for centuries European artists, historians and politicians viewed 'the East', and Islam, through racist ideologies of superiority that reduced diverse peoples and cultures to a reductive, stereotypical mass of threatening figures and locales. This way of seeing and speaking, a process he termed 'Orientalism', helped divide the world into ontologically distinct spheres of 'the East' and 'the West'; a practice knowingly invoked in the Islamic State's own simplified, binary worldview. Over time, Said argued, portrayals of the East and Islam were reduced to a few archetypal Orientalist 'myths' or 'dogmas'. The most pervasive of these has been the portrayal of Islam as a violent, resurgent and conspiratorially spreading threat to the Occident.[67] Thus, from proto-Orientalist paintings depicting the violent expansion of Ottoman forces into Europe[68] and colonial representations of hordes of fanatical

[65] See Edward Said, *Orientalism*, p. 59.

[66] See Jared Ahmad, 'Picturing the "Hordes of Hated Barbarians": Islamic State Propaganda, (Self)Orientalism and Strategic Self-Othering', *International Journal of Communication*, 16 (2022), pp. 2935–2957, for an extended discussion of this concept.

[67] Edward Said, *Orientalism*; see also Deepa Kumar, *Islamophobia and the Politics of Empire: The Cultural Logic of Empire*. Chicago: Haymarket Books (2012), pp. 41–62.

[68] See Daniel Viktus, 'Early Modern Orientalism: Representations of Islam in Sixteenth and Seventeenth Century Europe', in David Blanks & Michael Frassetto (eds.), *Western Views of Islam in Medieval and Early Modern Europe: Perception of Other*. New York: St Martins (1999), pp. 207–230.

'savages'[69] to more recent news media portrayals of Middle Eastern and North African refugees as an undifferentiated mass of threatening others,[70] the image of Islam as a looming, massified danger has provided a near-constant backdrop to Western cultural and political encounters with Islam.[71]

For the Islamic State, these latent fears, insecurities and essentialised identities provide a deep repository of ideas, images and usable memories that can be leveraged for strategic purposes. In contrast to Said's one-dimensional view of Oriental agency, the group's propagandists and media workers can be seen as 'strategic plagiarists' who have a deep understanding of Western cultural forms and media logics.[72] In much the same way that the Islamic State skilfully appropriate nostalgic themes and pre-colonial symbols to communicate with the believer in-group, they also *recontextualise* Orientalist discourses and imagery in order to influence those viewed as disbelievers.[73] And, in so doing, the group are able to temporarily construct a collectivised subject position and instrumentalise stereotypical dimensions of its identity for political purposes, thus skilfully mobilising the weight of culture *against* culture.

Textually speaking, the group's description of its own fighters as 'soldiers of terror'[74] or 'bloodthirsty knights',[75] alongside the portrayal of its violence as a form of '*just* terrorism',[76] serves to demonstrate awareness of the notion of *Strategic Self-Othering*, as they represent a conscious attempt to portray the Islamic State's actions as barbaric and cruel, yet necessary and religiously legitimate. The repeated claim, moreover, that 'Islam is the religion of the sword not pacifism' further serves to support this view,[77] celebrating the Islamic State's otherness and its embrace of 'blood-red

[69] See John Springhall, '"Up Guards and at Them!" British Imperialism and Popular Art, 1880–1914', in John MacKenzie (eds.), *Imperialism and Popular Culture*. Manchester: University of Manchester Press (2017), pp. 49–72.

[70] See Lilie Chouliaraki & Tijana Stolic, 'Rethinking Media Responsibility in the Refugee "Crisis": A Visual Typology of European News', *Media, Culture and Society*, 39.8 (2017), pp. 1162–1177.

[71] See Elizabeth Poole & John E. Richardson, *Muslims and the News Media*. London: I. B. Tauris (2010).

[72] Haroro Ingram, 'An Analysis of Islamic State's *Dabiq* Magazine', p. 3.

[73] Lilie Chouliaraki & Andreas Kissas, 'The Communication of Horrorism: A Typology of ISIS Online Death Videos', *Critical Studies in Media and Communication*, 35.1 (2017), pp. 24–39.

[74] See Al-Hayat Media, 'Soldiers of Terror', *Dabiq*, Issue 8 (March 2015), pp. 17–19.

[75] See Al-Hayat Media, 'Forward', *Dabiq*, Issue 15 (July 2016), p. 7.

[76] See Al-Hayat Media, 'Just Terror', *Dabiq*, Issue 12 (November 2015), p. 3.

[77] See Al-Hayat Media, 'Islam is the Religion of the Sword Not Pacifism', *Dabiq*, Issue 7 (February 2016), pp. 20–24.

swords'.[78] More crucially, however, these statements are visually supported through the repeated use of images portraying its fighters wilfully indulging in 'uncivilised' forms of violence, such as beheadings, mass executions and religiously sanctioned punishments. In fact, for Islamic State media workers, this explicit and graphic 'barbarism' constitutes a clear rejection of the 'humane' or 'liberal' forms of violence carried out by Western nations,[79] which is often *invisible* to their own populations,[80] and a deliberate attempt to 'terrify the disbelievers in their own homes'.[81] The Islamic State shows what its enemies are often unable to see, and, in so doing, purposefully claim a rich tradition of cultural representations of 'Otherness'.[82]

Figure 4 provides several graphic illustrations of these *Strategic Self-Othering* practices, whereby disbeliever audiences are presented with a

Fig. 4 The Islamic State's Strategic Self-Othering Practices (Dabiq 2015: 55; Dar al-Islam 2015: 105; Rumiyah 2016: 1, © Al-Hayat Media)

[78] See Al-Hayat Media, 'From Our Sisters', *Dabiq*, Issue 9 (May 2015), p. 47.

[79] See Michael Dillon & Julian Reid, *The Liberal Way of War: Killing to Make Live*. London: Routledge (2009).

[80] See Simone Molin Friis, '"Beyond anything we have ever seen": Beheading Videos and the Visibility of Violence in the War Against ISIS', *International Affairs*, 91.1 (2015), pp. 725–746.

[81] Al-Hayat Media, 'Interview with Abu Sa'd at-Trinidadi', *Dabiq* (July 2016), p. 69.

[82] This position is explicitly outlined in an article by the Islamic State. See Al-Hayat Media, 'Why We Hate You and Why We Fight You?', *Dabiq*, Issue 15 (July 2016), pp. 30–33.

series of images depicting the bloody aftermath of violent beheadings by foreign fighters and volunteers. Despite constituting a relatively small percentage of the Islamic State's propaganda,[83] in selecting and foregrounding these images the group seeks to deliberately target disbelievers (both internal and external) in an attempt to promote fear, uncertainty and terror. In this sense, the use of images depicting Europeans, and other foreign fighters, has been specifically designed to maximise their psychosocial impact on audiences and exacerbate fears about the threat posed by the migrant Other within Western democracies. In fact, research shows that Islamic State propagandists push the more violent and transgressive aspects of its propaganda in the aftermath of attacks in European cities or in response to territorial losses in a deliberate attempt to scare disbelievers and exacerbate levels of uncertainty surrounding terror threat levels.[84]

While no doubt inspired by modern propagandists and ideologues, the strategic value of such images lies in the way they draw upon, and intensify, entrenched fears and discourses surrounding the 'dangerous Orient'. In particular, a core feature of these *Strategic Self-Othering* practices rests on the way they invoke the Orientalist-inspired myth that Muslims have a 'predisposition' to brutal, lascivious violence.[85] From the very first European encounters with the Orient, Islam offered a psychological screen on which to project simplified fears and fantasies about the Other, which, importantly, enabled artists, politicians and academics to construct a collectivised self-identity.[86] During these early encounters, cultural representations depicted Oriental subjects as invading 'hordes of hated barbarians' who were,[87] in part, 'fond of beheading and other cruel forms of punishment and torture'.[88] Here, European Orientalist painters such as Jean-Léon Gérôme and Henri Regnault revelled in graphic depictions of cruelty and despotism, whereby the Oriental Other functioned as a canvas on which

[83] Scenes of brutality and graphic violence make up around 13.8% of the images appearing within *Dabiq* magazine, for example. See Shahira Fahmy, 'The Age of Terrorism Media: The Visual Narratives of the Islamic State Group's *Dabiq* Magazine', *International Communication Gazette*, 82.3 (2020), p. 277.

[84] Ibid., p. 279. See also Jesse Wozniak et al., 'The Evolving Self-Presentation of the Islamic State, from *Dabiq* to *Rumiyah*', *The Social Science Journal* (Online first 2020), pp. 1–14.

[85] See Deepa Kumar, *Islamophobia and the Politics of Empire*, pp. 52–55.

[86] See Said, *Orientalism*.

[87] Ibid., p. 59.

[88] Daniel Viktus, 'Early Modern Orientalism', p. 220.

Fig. 5 Contrasting Henri Regnault's *Execution Without Judgement Under the Moorish Kings* (1870, © Wikimedia Commons) with the Islamic State's own 'art' (2014: 4, © Al-Hayat Media)

to project a range of anxieties. As can be seen in Fig. 5, the latter's macabre *Execution Without Judgement Under the Moorish Kings* (1870) offers an exemplary point of comparison. Depicting a physically imposing and indifferent executioner, the painting foreshadows themes invoked in the Islamic State's own propaganda 'art'. The transgressive nature of Regnault's work, a piece renowned for the level of detail the artist gave to the blood and its potential effects on viewers,[89] is echoed in the group's carefully staged violence and the way its photo-propagandists enhance the

[89] See Marc Gotlieb, *The Deaths of Henri Regnault*. Chicago: University of Chicago Press (2016).

saturation and vibrancy of images to make the gore appear more real.[90] In this regard, Islamic State propagandists demonstrate a keen awareness of the fact that, for most Western audiences, their only encounter with extreme violence is most often through the, equally enhanced, medium of cinema. Regnault's painting was also deliberately designed to place the viewer in a subordinate position, inviting them to simultaneously fear and fantasise about the muscular executioner.[91] The sheer size of this figure, in contrast to the bloodied, prostrate victim, moreover, is visually echoed in the way the fighters in Islamic State propaganda are often digitally enhanced to give audiences the impression that they tower over their bowed, submissive victims.[92] Nevertheless, while Regnault's Moor is pictured in orange robes, for the Islamic State it is its victims who are now clothed in such colours. Here, the use of the bright orange jump-suit serves to draw powerful visual parallels with the abuse of Muslim prisoners in places such as Guantanamo Bay and Abu Ghraib, thus serving as a timely 'reminder of the group's resistance to Western imperialism' and the ongoing legacy of the colonialisation of Muslim lands.[93] Despite these sartorial differences, the parallels between the cold, methodical and casual nature of such imagery are self-evident.

In fact, despite being targeted towards different imagined audiences, it is here where the parallels between *Strategic Nostalgia* and *Strategic Self-Othering* start to become apparent. This is because both practices are inspired by a rhetoric of humiliation, alongside associated feelings of hatred, vengeance and the desire for retaliation, in the way that they seek to restore balance and recuperate 'a lost sense of agency' for the group's followers.[94] The Islamic State's strategic use of nostalgia seeks to draw

[90] See for example Simon Copeland, 'Islamic State, Filters and Photoshop', *Crest Security Review* (Autumn 2021). Available at: https://crestresearch.ac.uk/comment/islamic-state-filters-and-photoshop/ (Accessed July 20th 2022.

[91] James Smalls, 'Menace at the Portal: Masculine Desire at the Homoerotics of Orientalism', in Joan DelPlato & Julie Codell (Eds.), *Orientalism, Eroticism and Modern Visuality in Global Cultures*. London: Routledge (2016), p. 34.

[92] See Al-Hayat Media, 'A Message Signed with Blood to the Nation of the Cross', Dir./Prod. Islamic State (February 15th 2015).

[93] Patrick Richey & Michaela Edwards, 'It's More than Orange: ISIS's Appropriation of Orange Prison Jumpsuits as Rhetorical Resistance', in Michael Krona & Rosemary Pennington (eds.), *The Media World of ISIS*. Bloomington: Indiana University Press (2019), pp. 177.

[94] Roxanne Euben, 'Humiliation and the Political Mobilization of Masculinity', p. 505.

explicit parallels between the first generation of pious Muslims, evoking identities rooted in an idealised, mythical past, thus restoring 'the proper hierarchical ordering of men and women, Muslim and non-Muslim, the dominant and submissive'.[95] At the same time, the self-Othering practices identified above serve to call upon deep-seated fears and imagined pasts that disbeliever audiences associate with the 'dark', 'uncivilised' and 'dangerous Orient'. In the context of today's post-truth environment, a period characterised by growing levels of social and political polarisation and disenchantment, Orientalist-inspired fears and insecurities dominate narratives put forward by populist and far-right political parties. In short, as Amanda E. Rogers notes, we see the Islamic State precisely how the group wants to be seen: to believer audiences it presents the image of a powerful, legitimate, romantic *mujahid* warrior, who is poised to bring dignity and self-determination to a historically humiliated in-group. To disbelievers, it offers the endless image of barbaric violence and uncivilised evil.[96] These encounters with the enemy allow the group not only to harness the power of media against media, but more precisely to use the force of culture against culture.

Conclusion

In much the same way that the boundaries separating war and terrorism have weakened in the years after the attacks of September 11, 2001, so too have the representations of the enemy. This chapter has sought to explore these shifting boundaries by focusing on the Islamic State's use of visual propaganda. As we have seen, the group's identity-based messaging serves to construct two imagined audiences—'believers' and 'disbelievers'—and two distinct messaging strategies; what have been termed, *Strategic Nostalgia* and *Strategic Self-Othering*. These strategies enable Islamic State propagandists to legitimise their violence to potential recruits from the in-group, providing them with an opportunity to 'remake themselves as heroic figures, belonging to an exalted elite',[97] while, at the same time,

[95] Ibid.

[96] Amanda Rogers, 'Evil™ - Islamic State, Conflict-Capitalism, and the Geopolitical Uncanny', *Critical Studies on Security*, 6.1 (2018), p. 120.

[97] Simon Cottee & Keith Heyward, 'Terrorist (E)Motives', p. 977.

leveraging Orientalist-inspired fears and insecurities that are thought to exist within the collective memories of its enemies.

Such practices have implications for wider debates surrounding contemporary encounters with the enemy and the weaponisation of new communication technologies. In a context characterised by growing levels of economic instability, the mainstreaming of anti-Muslim/immigrant sentiment, and rising mistrust towards democratic institutions, alongside a questioning of the very notion of 'Truth' itself, these strategic communication practices provide actors from a range of political and ideological backgrounds an opportunity to play on audiences' emotions and deeply held sense of identity and (ontological) security.

In fact, there is evidence of an emergent, strategic embrace between violent Salafi Jihadi and far-right extremist groups. Iconic propaganda that invokes nostalgic images of hyper-masculine, seventh-century mujahid warriors battling with powerful European Crusader knights helps to sustain a mutually legitimising relationship that sees violent Salafi-Jihadi and far-right extremist organisations increasingly reliant upon one another for ideological sustenance. As Joseba Zulaika and William Douglass note, in such a situation 'alleged enemies feed rhetorically into one another's interests as each side perceives political advantage in the very existence of the other'.[98] *Strategic Nostalgia* thus supports a wider narrative employed by the two groups that they are each involved in a collective, sacred and cosmic battle to defend the in-group against the threat posed by the out-group. The use of *Strategic Self-Othering* practices, moreover, further serves to reinforce this strange embrace, with the Islamic State's strategic use of violent, essentialised and stereotypical imagery functioning to legitimise exclusionary and racist rhetoric that highlights the incompatibility of Islamic cultural practices with a seemingly monolithic European identity. Though some scholars have called attention to the shared discourses and strategic interplay that exists between violent Salafi-Jihadi and far-right

[98] Joseba Zulaika & William Douglass, 'The Terrorist Subject: Terrorism Studies and the Absent Subjectivity', *Critical Studies on Terrorism*, 1.1 (2008), p. 29.

extremist groups,[99] much more research is needed into the visual and iconographic aspects of such a mutually sustaining relationship. Furthermore, given the increased exposure to online, extremist content that citizens around the world have experienced during recent Covid lock-downs, it is vital that we better understand the way visual images are used to help to construct 'realities' that appear more *real* or desirable than reality itself. This is because, as social theorists such as Slavoj Žižek[100] and Zygmunt Bauman remind us,[101] future wars are set to be characterised by an inability to distinguish between friends and enemies, allies and adversaries, so we need to find ways of understanding precisely where, and how, those realities intersect so that we can construct alternative, more peaceful images of, and encounters with, Others.

[99] See for example Donald Holbrook, 'Far Right and Islamist Discourses: Shifting Patterns of Enmity', in Max Taylor et al. (Eds.), *Extreme Right Wing Political Violence and Terrorism*. London: Bloomsbury (2013), pp. 215–238; Meili Criezis, 'Intersections of Extremisms: White Nationalist/Salafi-Jihadi Propaganda Overlaps and Essentialist Narratives about Muslims', *Journal of Education in Muslim Societies*, 2.1 (2020), pp. 88–96.

[100] Slavoj Žižek, *Welcome to the Desert of the Real*. London: Verso (2002).

[101] Zygmunt Bauman, *Liquid Fear*. London: Polity (2006).

Memorializing the Enemy Three Ways: Australia, Japan, the USA

Christine Sylvester

In 2020, the organizers of this collection invited me to present a keynote at their conference on Enemy Encounters. The invitation provided an opportunity to think about an element of war that had haunted every page of my recent book, *Curating and Re-curating the American Wars in Vietnam and Iraq*, without declaring itself openly: the 'enemy' people of Vietnam and Iraq who experienced the American wars on their home turfs. Surely there was ample space for enemy encounters to materialize across pages that commented on war curations at the Smithsonian National Museum of American History, the Vietnam Veterans Memorial (VVM), and Arlington National Cemetery, all places of war memory in and around Washington D.C. Such encounters did appear once the book took a turn toward war novels and memoirs written by Vietnamese, Iraqis, and Americans. Ordinary people were suddenly there, visible, real, and meaningful, even in fictionalized form. These same people were missing from many celebrated memorials that were erected to remember wars via the lens of military fighters, not them.

C. Sylvester (✉)
University of Connecticut, Storrs, CT, USA
e-mail: christine.sylvester@uconn.edu

H. Furneaux, M. Greig (eds.), *Enemy Encounters in Modern Warfare*, https://doi.org/10.1007/978-3-031-56748-3_14

Johanna Mannergren Selimovic (drawing on Tolia-Kelly, Waterton, and Watson) captures a similar moment of recognition at two exhibitions she visited in Sarajevo on the 1992–1995 siege. There, displayed for all to see, was 'a child's blue knitted sweater, a half-finished letter and a dented humanitarian tin…They moved me deeply…I experienced what Tolia-Kelly et al call a "visceral shudder."'[1] I recall having a similar experience at the Vietnam War display within the larger 'Price of Freedom: Americans at War' exhibit in the Smithsonian National Museum of American History. The trigger was a simple wooden bicycle of the kind the North Vietnamese used to transport war materiel along the Ho Chi Minh trail.[2] Bicycles like the one exhibited were loaded to the hilt and pushed, not ridden, up and down the long jungle trail from North Vietnam, through a bit of Cambodia, and into South Vietnam to support the war against the American invaders. Numerous Vietnamese women were involved in that war effort, but we do not see them there, nor do we see the bicycle as intended by its builders, stacked with goods. The USA tried time and again to neutralize that trail of war through thunderous B-52 bombing runs that missed bicyclists who could hear the planes coming and quickly sheltered in dense jungle cover. At the Smithsonian, the bicycle that won the war is naked, bundle-less, and without even a hint of its 'pilot'. And, it has been placed *under* the enormous tail of the advanced American vehicle that failed against those bicycles, those women, and the soldiers of the enemy North: a Huey helicopter.

This chapter focuses on the presence or not of ordinary people's experiences of war in three major war installations: the Australian War Memorial and attached museum, Japan's Hiroshima Peace Park and museum, and the American Vietnam Veterans Memorial. The sites differ in many ways but also share one key feature. Each commemorates *military* defeat—of a war or of a significant battle within a war—and does so by 'forgetting' to represent the ordinary occupants of the enemy state, the ones who

[1] Joanna Mannergren Selimovic, 'The Stuff from the Sieges. Transitional Justice and the Power of Everyday Objects in Museums,' *International Journal of Transitional Justice*, 16 (2022), pp. 1–15; Divya Tolia-Kelly, Emma Waterton, and Steve Watson, 'Introduction: Heritage, affect and emotion', in *Heritage, Affect and Emotion: Politics, Practices and Infrastructures*, ed. by Tolia-Kelly, Waterton, and Watson (London: Routledge, 2018), p. 6.

[2] Christine Sylvester, *Curating and Re-Curating the American Wars in Vietnam and Iraq* (London: Oxford University Press, 2019), p. 94.

shouldered the burdens of war in the shadows without recognition. The Australian National Memorial and museum in Canberra is part memorial to battles and soldiers lost early in WWI, and part a bombastic museum over-displaying elements of every overseas war Australia has joined. The Vietnam Veterans Memorial remembers the American war in Vietnam via granite panels showing the names of 58,000 Americans who died in a war the USA lost ignominiously and absolutely. Hiroshima Peace Park and museum, the third site, memorializes civilians who experienced the American atom bomb on Hiroshima city in 1945. Each site commemorates an armed conflict conducted by militaries of state with the goal Elaine Scarry emphasized in her 1985 book *The Body in Pain*, to 'out-injure the other', the recognizable enemy, to gain victory over it.[3]

Throughout the discussion I draw on Viet Thanh Nguyen's study of war memory and ethics called *Nothing Ever Dies: Vietnam and the Memory of War*.[4] Nguyen, whose family fled Vietnam in 1975 when he was very young, writes of 'industries of memory' that transform war efforts into national triumphs by emphasizing righteous causes, individual sacrifice by soldiers, military valor, and true patriotism. The industries involved include agencies of state (e.g. ministries of defense) and 'museums, archives, festivals, movies, history channels, interviews, that display partial, distorted stories of wars'.[5] Most war memorialization, he says, betrays what he calls 'just memory' by not including (encountering) enemy memories or acknowledging one's own war atrocities. To achieve just memory requires: 'constantly trying to recall what might be forgotten, accidentally or deliberately, through self-serving interests, the debilitating effects of trauma, or the distraction offered by excessively remembering something else, such as the heroism of the nation's soldiers'.[6] It requires what James E Young, a leading American scholar of war memorials, calls the collected memories of many that do not collapse into one narrative that subjugates other knowledges of a war—to use Foucauldian terms.[7] To put this in

[3] Elaine Scarry, *The Body in Pain: The Making and Unmaking of the World* (Oxford: Oxford University Press, 1985), pp. 88–89.

[4] Viet Thanh Nguyen, *Nothing Ever Dies: Vietnam and the Memory of War* (Cambridge: Harvard University Press, 2016).

[5] Ibid., p. 17.

[6] Ibid.

[7] James E Young, *Stages of Memory: Reflections on Memorial Art, Loss, and the Spaces Between* (University of Massachusetts Press, 2018).

terms Vikki Hawkins uses in her contribution to this collection, today it is necessary for museums 'to confront established narratives, overturn myths, and encourage visitors to engage with what war is actually about'.[8]

The Australian War Memorial, Canberra

When we think of Australia, the first characteristic that comes to mind is not likely to be international 'warrior state'. Yet its national war memorial shouts that theme, belying any associations one might make to a laid back, 'no worries' country (albeit with indigenous 'issues'). An oddity in mission and message, its war memorial opened in 1941 and consists of two structural and thematic sections. One is a sacred sanctuary in remembrance of Australia's early WWI failed effort to gain a key waterway at Gallipoli from Turkey to ease allied transport. Over a nine-month period, Turkish forces repeatedly overcame Australian forces and left 18,000 of them dead for a cause that had nothing to do with Australia per se. Some might call that extended campaign a war tragedy. For Australians, it is a triumph in two regards. It has come to enshrine a national determination to fight in the face of continuous loss as a matter both of stamina and honor. More tangibly, the Gallipoli campaign is said to have established Australian identity as a nation; for the first time, scattered populations across a large landmass would recognize themselves as Australian through a war they waged together far from home.

Two salient points. First, Turkey was not an enemy of fledgling Australia. This was a case of an atomized group of people associated with one empire fighting on its behalf against another empire rather than for its own good. In the end, despite the defeat at Gallipoli, Australia, would be on the winning side in the European Great War by virtue of its colonial relationship with Britain. It is also the case that enmity between the Diggers (soldiers from Australia) and Turkey or Turks has, for some today, been replaced by a mythology of wartime 'friendship'.[9] Second, those who

[8] Vikki Hawkins, 'Enemy Encounters in the Imperial War Museum's Second World War Galleries', this volume.

[9] See Mark McKenna and Stuart Ward, 'An Anzac Myth,' *The Monthly* (December 2015–January 2016), <https://www.themonthly.com.au/issue/2015/december/1448888400/mark-mckenna-and-stuart-ward/anzac-myth#mtr> [accessed 5 February 2023]. For other interpretations, see Christine Sylvester, 'National War Heritage at the Australian War Memorial and Hiroshima Peace Park,' *Millennium: Journal of International Studies*, 49.2 (2021), pp. 280–304.

encountered the enemy in the country's first foreign war were white males who grew to recognize their commonalities and define themselves and the nation through the Gallipoli campaign. Aboriginal communities were not in that fight, having lost their own sovereignty after 1788, the year the first fleet of British ships carried convicts into Sydney Cove. To many of them, that arrival enabled a foreign white nation to treat ordinary indigenous peoples as their real enemies, while basking in British victories abroad. The local enemies suffered a biopolitical panoply of white settler practices aimed to control the 'other'—forced religious conversion, state schools for stolen children, genocidal campaigns in Tasmania, and confinements on 'traditional' lands.[10] This is not openly acknowledged in the museum displays (Fig. 1).

The namesake structure of the memorial is the large Hall of Memory, designed by Richard Ure, winner of a competitive process. The Hall resembles a domed church with an inner sanctuary encircled by names of Gallipoli and other war dead carved on the walls. Large stained-glass panels depict figures who are meant to represent distinctly Australian national characteristics of resourcefulness, candor, devotion, curiosity, and independence. The Hall of Memory feels like a holy place where visitors whisper as they take it all in. The second structure of the Australian War Memorial is vastly different. Connecting doors swing open to a wing triple in size or more to the Memorial. It is filled to exuberant exhaustion with exhibits for every war Australia has participated in overseas from Gallipoli on. The museum wing takes hours to walk through, days to study. There are detailed battle dioramas, large war aircraft in galleries, and on the grounds, rooms of uniforms, guns and other weapons, even taxidermied animals (Fig. 2).

Within the museum, a Huey helicopter diorama immediately catches my eye. In an elaborately staged Vietnam War ground battle scene, this key American military vehicle, with an original US serial number scratched over, gets pride of place in the Australian Vietnam War exhibit. Whereas a similar staging of the Huey in the Smithsonian features three soldiers providing aid to a wounded colleague on the ground, Canberra's Huey is manned by six men who burst from it as the diorama comes startling to life with flashing lights and a soundtrack of shouting men and blasting assault rifles. The loud drama carries on for several minutes as the

[10] See *Genocide and Settler Society: Frontier Violence and Stolen Children in Australian History*, ed. by A. Dirk Moses (New York: Berghahn, 2005).

Fig. 1 Great Hall of Australian War Memorial (Weil)

epicenter of Australian encounters with the enemy in Vietnam. The prominence the museum gives this weapon breezily implies that the Huey is a distinctly Australian weapon, so decisive in that war it warrants special effects. But again, the Australians depicted here are fighting America's

Fig. 2 Grounds of Australian War Memorial (Sylvester)

war, not their own. It is Gallipoli redux. Why this Huey scene is at the center of the Australian war museum's imagination of the Vietnam War could not be answered by a curator I encountered (Fig. 3). Is this a case of just memory or an example of Australian militarized masculinity, of amped up 'breathing, insinuating death, industrial production, and orgasm'.[11]

Visitors to the museum wing carry on to scenes of Australian military action and weaponry in Afghanistan and Iraq, through rooms of military hardware and endless displays of eager warriors. It is an exhausting journey in an overwrought military-thrilled museum that seems to give each established department considerable space to shine. The Australian War Memorial is a national war museum with an ambitious expansion of the museum on the government agenda.[12] In the USA, elements of American war prowess are exhibited in the Smithsonian and in local and military branch museums, but Warrior USA does not have a national war museum.

[11] Nguyen, *Nothing Ever Dies*, p. 119.
[12] See Sylvester, 'National War Heritage'.

Fig. 3 Australian War Museum Huey (Weil)

The key take-away from Australia's bifurcated remembrance/celebration of war is hard to find in an institution in which one part is subdued and grateful and the much larger part is excited about going to war, despite the inevitable casualties that result. Could it be that Australia continually bucks up its founding identity by fighting faceless enemies for others in distant places? Does the Memorial and attendant museum refight Gallipoli to a win every day? What of local Australian 'enemies', those fledgling Australians slaughtered on the island of Tasmania between 1804 and 1830? At the time of this writing, visitors to the Australian War Memorial and museum would not see a range of dioramas featuring that long Black War, as it is called, nor have an opportunity to contemplate Turks fighting in Gallipoli. World War I experience sets Australia's military heritage on a course where its 'true' wars *only* engage militaries abroad, skipping over violent encounters within Australia.

The Vietnam Veterans Memorial, Washington DC

The American war in Vietnam, which raged from around 1965 to 1975, was a failure for the USA on all counts: America lost the war decisively, its military retreated in a panic as enemy Hanoi drove military tanks into Saigon April 30, 1975, and 58,000 Americans lives were lost along with 3–4 million Vietnamese, Cambodians, and Laotians (Fig. 4). This controversial, incompetently managed war faced strong public backlash, especially from 1968, when the Tet Offensive stunned American troops, to the formal negotiated treaty that only seemingly ended the war in 1973. Returning American war veterans, many of them unhappy draftees, met a disdainful public. No one thanked them for their service. There were no parades, no honors, no recognition at all, particularly in comparison to the thanks Americans would bestow on veterans of more recent failed American wars in Afghanistan and Iraq. For years, the country cringed at

Fig. 4 Vietnam Veterans Memorial (Weil)

the very thought of the Vietnam War and shied away from similar encounters with enemies, an avoidance pattern dubbed the Vietnam Syndrome.

In 1982, however, the Vietnam Veterans Memorial was erected on the National Mall in Washington D.C., the culmination of veteran pressure for something approaching a sign of just memory for Americans made to fight the war. That memorial ushered in a new period of US citizen respect for all its soldiers who die far from home in any war, let alone through the Vietnam debacle. The memorial effort was led by the Vietnam Veterans Memorial Fund (VVMF), an organization founded in 1979 that demanded the nation honor soldiers who fought as required by the state, even if the cause was never clear and the war took place in faraway jungles unknown to most Americans. Energetically, the VVMF raised funds, argued with Congress for a prime memorial location on the National Mall, and successfully ran an open design competition that attracted 1432 submissions. Congress imposed only two rules: the memorial must contain the names of the 58,000 known American soldiers who died or were missing in action in Vietnam, and the memorial could not be political in the sense of taking a direct or implied position on the war; that is, it could in no way boast of war achievements, lament the failure of the USA to prevail, or criticize the war and its opponents. The VVMF-appointed the jury determined whether each entry met that standard. The process for selecting a memorial design went relatively smoothly.[13]

The resulting decision, however, unleashed a storm of protest. The winning entry was simple, abstract, and cutting-edge in its unconventionality. It showed thin slabs of polished black granite angled down a small rise in the earth, joining at the middle to form a V. Year of death in the Vietnam War was the organizing principle for listing names on panels that ranged in height from very short for early and later years of the war to high above the heads of visitors in the deadly year of 1968. As visitors perambulated the wall, mirror images of themselves would stare back from the reflective polished surface. Despite its cutting-edge design, which challenged the history of war memorials of military men, many veterans were disappointed by its abstract elements. They insisted repeatedly that the VVMF supplement the plan for starkly listed names with a traditional image, like a statue depicting heroic American soldiers. After considerable debate threatened to derail the project entirely, a statue was planned at a remove from the main memorial. Designed by noted memorialist,

[13] See Sylvester, *Curating and Re-Curating*.

Frederick Hart, it depicts three soldiers emerging from a forest area looking haggard and defeated. Other design complaints that were not accommodated concerned the thematic black color of the main memorial, which struck some as a mark of disgrace or condemnation of the war. It also irked many veterans that the winner was a young woman, an Asian-American, and a 'mere' undergraduate student at Yale University rather than an established architect—likely to be a man at that time—or someone with a military background. It is instructive in this regard to note that the memorial designer, Maya Lin, all of 20 years old at the time her submission prevailed, is now considered one of America's foremost landscape and ecology memorialists.

In all of this, the enemy encountered in Vietnam is nowhere to be seen or acknowledged at the memorial; ditto within Vietnam for Republic of Vietnam force members who fought with the Americans against the Hanoi government. Those were all 'disappeared' from the memorial, along with America's war allies, such as the Australians, who were the last allies to depart Vietnam, and Korean forces subsidized by the American government to fight. Most civilians who suffered the war in Vietnam or lost a loved one, are also unmentioned at the Vietnam Veterans Memorial, as are American military that made it home and subsequently died of cancers associated with American use of the Agent Orange defoliant. People in war-adjacent Cambodia and Laos experienced the late period of the American war, but are also absent, as are war protesters in the USA and military deserters in Vietnam. A National Parks employee is often available at the memorial with a roster of names and where to find them on the memorial, but in keeping with the requirement to avoid politics, no information about the Vietnam War is presented there; an information booth is located at some distance away.

Notwithstanding the controversies that surrounded the winning design, as soon as the Vietnam Veterans Memorial was erected it attracted the public. Even today, fifty years after the war ended, this memorial is one of the most visited sites in the Washington area. It is constantly enlivened by pop-up object assemblages that family members and friends create and leave beneath the panel carrying the name they know. These small displays by ordinary curators change daily but also tend to be apolitical. Rarely do any of them condemn the war, the US government, the public protesters or the draftees who refused induction. They also do not tend to praise soldiering or the war. What they do is focus the mind on war as a personal experience of overwhelming loss. One of the most esteemed war memorials in the USA continuously beckons visitors despite an absence of usual military tropes, battle dioramas, statements of cause and defeat, and images of encounters with enemies. Unintentionally, it stands as a direct contrast to the Australian War Memorial.

Hiroshima Peace Park and Museum, Hiroshima

The Australian War Memorial appears to applaud the country's numerous war campaigns, including losses, as a mark of national willingness to join military conflicts that others start overseas. The Vietnam Veterans Memorial emphasizes military casualties in a failed American war conducted overseas. The Hiroshima Peace Park stands in contrast with both of those memorials in one major way: it highlights the loss of an estimated total of 30,000 *civilians* from one bomb dropped from an American military aircraft on one day in 1945. That bomb was extreme, novel, nonhuman, nonprivileging, and nonsparing of any person, landscape, animal, or architecture in its path. At the Peace Park, the weapon itself is the enemy encountered, not the Americans who deployed the bomb, not the aggressive Japanese military that waged a cruel war in Asia at the time, not the Japanese emperor or any other group of people. The enemy is The Bomb.

The dominant American narrative on the bombing of Hiroshima (and Nagasaki a few days later) explains away what could count as a war crime or atrocity as a laudable device that saved American military from certain death in an invasion of Japan. At the Hiroshima Peace Park, it is clear that the Japanese do not take a strategic view of the events of that August day. It is all about the encounter of ordinary people with a purposely planned, purposely targeted and 'successfully' launched deadly weapon on a civilian city without warning. Members of the American military who opened the bomb bay on the Enola Gay plane, as well as Japanese military fighting and dying in places distant from Hiroshima, are present through their absence. In a short story about aftermaths of the war in Vietnam, Nguyen asks: 'What does one say to a ghost, except to ask why he was here?'[14]

The atom bomb detonated at 8:15 on an August weekday morning as children were settling into school. 'Little Boy' joined them, exploding directly above the city's main hospital, as planned. Read John Hersey's 1946 deservedly famous *New Yorker* article '*Hiroshima*' for a six-person

[14] Viet Thanh Nguyen, 'Black-Eyed Women', in *The Refugees* (New York: Grove Press, 2017), pp. 1–22.

account of the moment which leaves little to the imagination.[15] The outcome was total devastation and ultimate injury to all. America arrived on the ground soon after the emperor capitulated to what he referred to as a defeat by science not by way of a bona fide fight. The USA occupied and administered Japan until 1952 but was slow to send food supplies to the desperate area. American researchers arrived quickly to study effects of atomic radiation on the body; in violation of medical ethics and international rules of war, they did not provide victims with medical treatment. The Americans believed the Japanese got what they deserved.[16]

Many years later, in 1996, the general area of the bomb was declared a UNESCO World Heritage Site, but well before then, on an area along Ota river system at the epicenter of the atomic blast, the Peace Park was being marked out. In May 1949, the architectural journal *Kenchiku zasshi* announced competition for a Hiroshima Peace Park. It did not specify any particular style, requiring only that the design show multiple park installations that suit the designated area. The jury considered 132 entries and awarded the design to Kenzo Tange, a noted architect and advocate of the superiority of the Japanese and their culture, who was an imperial collaborator during the war.[17]

The Peace Park is nearly as large as the National Mall in Washington, but unlike it, the Peace Park is filled with memorials to local and international victims of one war and one bombing. Its focus on civilian rather than military war losses comes across poignantly at one central monument: the Children's Peace Monument (Fig. 5).

Compared to the military-centered displays at the Australian War Memorial and military only names on the Vietnam Veterans Memorial, the very existence of a war memorial to the many children who died in a war disaster is remarkable. Yet this memorial does not dwell on death. Rather, it celebrates youthful perseverance, determination, and appreciation of Japanese traditions. It is a must-see attraction for classes of school children, who arrive in their school uniforms each day and respectfully gather at the memorial to sing, recite tributes, and add the paper cranes they have made to streamers of these birds flowing over the monument.

[15] John Hersey, 'Hiroshima,' *The New Yorker*, 23 August 1946, pp. 1–60.

[16] John Dower, *Embracing Defeat: Japan in the Wake of World War II* (New York: Norton, 2000), p. 24.

[17] Hyunjung Cho, 'Hiroshima Memorial Peace Park and the Making of Japanese Postwar Architecture', *Journal of Architectural Education*, 66.1 (2012), pp. 72–83.

Fig. 5 Children's Peace Monument (Sylvester)

The paper cranes reflect the story of Sadako Sasaki, a local girl who was two when the bomb exploded near her home. She was seemingly unharmed, healthy and strong for almost ten years. Then she developed leukemia, the atomic disease that attacked so many survivors of the bombing. Japanese folk belief says cranes can live 1000 years, so Sasaki set about to turn that tale in her favor by pledging to make 1000 origami-style paper cranes to ward off her own death. She had almost reached the lifesaving number by the time she died in 1955 at age 12. The story is sad, but there is much joy expressed at the Children's Monument inspired by Sasaki's hope in the future; it was her classmates who suggested the memorial and helped raise the funds to install it. The memorial fits with what Tange's architectural collaborator, Sachio Ōtani, claims was always the main aim of the Hiroshima park project: 'not to mourn the dead, which was a past-oriented and backward-looking act, but to commemorate peace, which was a future-oriented and forward-looking act'.[18] From what many consider the worst bombing atrocity of World War II emerges something historian Lisa Yoneyama terms Japan's 'obsession with the future' in narrating Hiroshima.[19]

There is also a museum within the park that takes the visitor through epochs of Japanese history to the development of the atom bomb in Europe and the USA. It includes a digital bomb's eye recreation of 'Little Boy' as an object falling, falling, then detonating and creating the mushroom cloud that huddled over Hiroshima city. The location of the blast was not serendipitous in any way. Hiroshima is surrounded by hills and mountains that would intensify the effect of the blast and radiation on the city. The Peace Park museum displays everyday objects burned, misshapen, or mostly destroyed by the blast. Some photos taken soon after show skin peeling off people's bodies as they try to leave the rubble. There is much to shudder at while encountering a display that engages visitor sensibilities so directly. The strongest visceral shudder I experienced there arose in front of a display case containing only one object—a nearly incinerated tricycle, empty, burned and alone. A three-year-old boy had been cycling on that bike a mere 1500 meters from the blast hypocenter. His name was Sinichi Tetsutani and that was his vehicle of encounter with the enemy bomb (Fig. 6).

Especially compared to the Australian War memorial/museum, displays in the Hiroshima Peace Park and museum are humble—burned shoes, tattered school uniforms, and buckled metals of civilian life. They show a descent into the ordinary for meaningful war memories, to use Veena Das' language, as do many prosaic items left at the VVM.[20] The Hiroshima Peace Park and its museum, as well as the Vietnam Veterans Memorial, also encourage visitors to ponder what they see without suggesting what

[18] Sachio Ōtani, interviewed by Terunobu Fujimori, 'Senji modanizumu kenchiku no kiseki, Tange KenzŌ no jidai 01', Shinkenchiku (1998), p. 86.

[19] Lisa Yoneyama, *Hiroshima Traces: Time, Space, and the Dialectics of Memory* (Berkeley: University of California Press, 1999), pp. 18–21.

[20] Veena Das, *Life and Words: Violence and the Descent into the Ordinary* (Berkeley: University of California Press, 2007).

Fig. 6 Child's Burned Bicycle (Sylvester)

to think and feel in encounters with war memory. By contrast, the Australian War Memorial and museum can leave the visitor confused and worn down by the seeming celebration of entering wars that are not their own.

Conclusions

Nguyen points out that 'weaponized memory becomes part of the war machine's arsenal, deployed in the struggle to control reality'.[21] The war memorials considered here offer three versions of weaponized memory. Hiroshima's memorialization of utter destruction is the most hopeful about ending future encounters with an enemy's latest bomb and with war as an accepted politics of engagement. The Vietnam Veterans Memorial quietly places viewers into war death encounters through self-reflecting granite. The Australian War Memorial and museum lead with regret but rapidly nix that with the joy of national discovery achieved as an enthusi-

[21] Nguyen, *Nothing Ever Dies*, p. 109.

astic joiner of wars others start. In each case, the human enemy shot at or citizen marching against war are outside a war memorial's purview. Memorials list no enemies and only a limited range of participants 'obedient to the generals and statesmen who continue speaking on their behalf, telling the story that the Fatherland remembers their sacrifice'.[22] Intentional absences lurk everywhere—the My Lai massacre by American soldiers in Vietnam, the Australian wars against native peoples at home, and the Korean Comfort women made to service the Japanese forces. There are so many omissions in war memorials that 'just memory', as Nguyen defines it, seems unreachable, incomprehensible even, owing to the myriad invisibilities created by remembering something else.

Which brings us to the present. The 6 January 2021 insurrection at the US Congress saw some armed citizens reimagining the enemy as elected members of government from opposing parties. The 'right' to weaponize the US Capitol against such enemies was proclaimed and violence paraded confidently, selfies abounding.[23] Meanwhile, a contemporary failed war abruptly ended, sort-of. The USA repeated its clumsy Vietnam finale fifty years later by simply up and leaving Afghanistan, providing space for new wars to flare in vacated territories. In today's war regions, enemies recycle, rebrand themselves, and memorialize their achievements through draconian take-overs and subsequent measures against women or ethnic and religious minorities. The big throwback war of today is the Russian invasion of Ukraine. We daily witness incomprehensibles produced by Russia targeting, with complete abandon, the people and culture of Ukraine for the old-fashioned goal of restoring its own empire. Along a trail of bodies, 'enemy' becomes multireferential, and as Hawkins sites it, 'global'.[24] It is all around us instead of frozen in type and in time on war memorials or in museum dioramas crafted by the industries of war memory Nguyen laments. Yet if one takes Hiroshima as the model of war in extremis, the war that destroys everything—the military, ordinary citizens, cities, the

[22] Ibid., p. 28.

[23] Dan Barry, Mike McIntire, and Matthew Rosenberg, '"Our President Wants Us Here": The Mob That Stormed the Capitol', *The New York Times*, 9 January 2021; Deena Zaru, 'The symbols of hate and far-right extremism on display in pro-Trump Capitol siege', *ABC News*, 19 January 2021.

[24] Hawkins, 'Enemy Encounters in the Imperial War Museum's Second World War Galleries', this volume.

economy, political institutions, the school you attended—an even more extreme omnibus enemy operates constantly, biding its time and building but parsing its power as a world fights other wars. It is climate change. No memorials will survive to represent the final viscerality and dented humanitarian tins that the climate enemy will throw everyone's way. And it most certainly cannot/will not be killed by any bomb. That is the end-game war shudder.

USSR 2.0 vs. The Fourth Reich: Depicting the Enemy in the Official Occupation Media of Kherson Oblast

Mykola Homanyuk
Translated by Valeria Khotsina

At the end of June 2022, in the Russian-occupied part of Kherson Oblast, the first issue of *Naddnepryanskaya Pravda* began to be distributed free of charge in places frequented by the local population. The newspaper's logo was completed with the Soviet Order of Honour and a double-headed eagle—Kherson's coat of arms from the time of the Russian Empire. A note at the bottom read that the newspaper was issued by the Kherson Oblast Military-Civilian Administration. The issue began with an address by the head of the administration, Vladimir Saldo, to the population of the Oblast. It started with the words: 'Those in power in Ukraine are trying in every possible way to disrupt a peaceful return to normal life by terrorising and intimidating the local populace, but they will not succeed. We will

M. Homanyuk (✉)
Kherson State University, Kherson, Ukraine

H. Furneaux, M. Greig (eds.), *Enemy Encounters in Modern Warfare*, https://doi.org/10.1007/978-3-031-56748-3_15

withstand these attempts and they will not break us'.[1] The newspaper contained a description of the enemy, and this aspect was the focus of all subsequent issues of *Naddnepryanskaya Pravda*.

The editors had a challenging responsibility: to create an enemy out of the state whose nationals the readers still remained, and to which they were still linked in a variety of ways: economic, informational, kinship, legal, and so on. This article aims to identify the negative portrayal of the enemy created by official publications of the occupation authorities in Kherson Oblast and how it was used to achieve practical goals.

The empirical basis of the research includes the official publications of the occupation authorities, i.e. *Naddnepryanskaya Pravda*, *Tochka Otscheta*, which was the newspaper of the Kherson City Military-Civilian Administration, and the official Telegram channel of the Kherson Oblast Administration (established on 18 March 2022 as the Telegram channel of the Rescue Committee for Peace and Order that later transformed into the Kherson Oblast Occupation Administration), as well as *Golopristansky Vestnik* and *Nasha Kakhovka*, the printed publications of other municipalities in Kherson Oblast. The study only covers the period from 18 March to 11 November 2022. This is until the Ukrainian army liberated the entire right bank of Kherson Oblast, which includes Kherson, its regional centre.

In addition, the content of the official Ukrainian media in Kherson Oblast was analysed. As there were no Ukrainian newspapers published in Kherson Oblast during the occupation, the official website of the Kherson Oblast State Administration (during the war—the Kherson Oblast Military Administration) was chosen as a point of comparison.

Communication and Media in the Occupied Kherson Oblast

From day one of the full-scale invasion, social media became the main source of information in Kherson Oblast. Telegram channels, Viber groups, and Facebook provided information about the most important thing—the war. And this was true across all channels and groups—both news channels and chats for school parents, colleagues, gym goers, badminton fans, etc. By the second or third day of invasion, it was clear that

[1] 'Address by Vladimir Saldo, Head of the Kherson Military-Civilian Administration, to the residents of Kherson Oblast'. *Naddnepryanskaya Pravda*, no. 1, p. 1.

something was going wrong for the Russians. Videos of burning military equipment and photos of dead Russian soldiers flooded social media. These posts were shared with extraordinary enthusiasm. Pro-Ukrainian content completely dominated the information space. Here is a vivid example: As early as 27 February, a report appeared online that Roma people in Kherson Oblast had stolen a tank from the Russians, followed by a seven-second video clip showing a tractor towing an armoured personnel carrier, with people running after it. The clip showed an armoured vehicle instead of a tank; it was not clear whose tractor it was, where it was going, or who was chasing it, but few people questioned the legitimacy of the tank story.[2] The initiative in the information war was taken immediately.

Until 30 May, mobile communications and Internet in Kherson Oblast were provided by Ukrainian operators with large mobile operators (Vodafone, KyivStar, Life) and local Internet providers offering cable Internet. There were some breaks in communication, but because of national roaming, it was possible to connect to another network for free, and registering a virtual SIM card (which gets a new number without the need of a physical SIM card) in Kherson also helped to keep communications running.

On 30 May, all Ukrainian communication was cut-off. The invaders announced over loudspeakers that the Ukrainian side was the initiator of the cuts. There was no communication in the city for several days. On the morning of 1 June, stalls started selling SIM cards at Svoboda Square. These SIM cards were made by an unknown company and featured a service package called 'Stay in touch'. The phone numbers began with +7 99 00. The company turned out to be Velcom Telecom. As time went on, cable Internet also appeared which was run via Crimea. For a while, Ukrainian political and news sites were inaccessible to the majority of the population. But this problem was quickly solved with the help of VPN. The Russians constantly tried to block popular VPN applications, so users had to change them every now and then. As a result, attempts to cut-off the population of the occupied territories from Ukrainian and international information sources proved futile. Ukrainian journalists worked remotely in the cities of Kherson Oblast, research was conducted, teachers held classes and exams, and cultural projects were implemented—there was

[2] Mykola Homanyuk and Janush Panchenko, 'From a Pilfered Nail to a Stolen Tank: The Role of a Media Event in the Consolidation of the Ukrainian Political Nation', *TRAFO – Blog for Transregional Research* (March, 2023) <https://trafo.hypotheses.org/46010>.

even an underground exhibition of works by Kherson artists for Ukraine's Independence Day, the opening of which was held on Zoom.

The main sources of operational information in the region have been the following Telegram channels: *Khuevyi Kherson* (467 thousand subscribers[3]), *Nikolaevsky Vanyok* (341 thousand subscribers), *Trukha Kherson* (129 thousand subscribers), *Kherson: voyna bez feykov* (69 thousand subscribers), *Konstantin Ryzhenko—Zhurnalist* (55 thousand subscribers), *Khersonska OVA* (22 thousand subscribers), and others. The analytical content was spread by YouTube vloggers. The occupation Telegram channels, such as *Kirill Stremousov* (65 thousand subscribers), *Administratsiya Khersonskoy Oblasti* (25 thousand subscribers), *Tavria* (16 thousand subscribers), and *Svobodnyi Kherson* (11 thousand subscribers) clearly underperformed in terms of reach. They tried to make up for their lag in electronic media with television and the print press. The TV centre was seized within the first days of the occupation and soon began broadcasting Russian channels, the TV channels of the Donetsk and Luhansk People's Republics and a local private *VTV* channel that cooperated with the occupiers. The *Krym-24* TV channel broadcast special news bulletins for the occupied territories under the name 'Kherson-24'. Sometime later, the Kherson news office of the *Suspilne TV* channel was used to create the occupation state TV and radio company *Tavria*, whose activities were monitored by the Russian special services. However, before the war, Ukraine had a broadcasting system that used digital signals, while Russian channels used analogue signals. This hampered the delivery of information, as the consumer needed a special set-top box to access it.

In the city where the Internet and television were constantly out of order, the occupiers tried to make the print press an important propaganda channel. However, these publications could only be called press with a stretch—the vast majority of articles were anonymous. Several newspapers were distributed in Kherson Oblast. The first newspaper was *Komsomolskaya Pravda (KP)*, or rather its regional editions. At first, it was a joint issue for Zaporizhzhia and Kherson Oblasts, *Komsomolskaya Pravda. Liberated Territories*, which was later supplemented by *KP. Kherson*. The newspaper *Naddnepryanskaya Pravda* had the largest number of copies distributed. It was named and styled after *Naddnipryanska Pravda*, the Kherson Oblast Council newspaper that had been published

[3] All data in this paper is provided as of 20 January 2022. Telegram channels and groups catalogue <https://uk.tgstat.com/>.

since 1927. In Russian, this newspaper was called *Pridneprovskaya Pravda*, because the word *Naddnepryanskaya* does not exist in the Russian language. So the new name sounded like market sellers offering 'Abidas' instead of 'Adidas'. Another local newspaper, *Tochka Otscheta*, typically had extended columns that talked about generals and naval commanders from the times of the Russian Empire. These columns were sometimes continued in subsequent issues. In other towns within Kherson Oblast, the invaders also initiated their print media, such as *Golopristansky Vestnik*, *Nasha Kakhovka*, *Chernomorets*, and others.

Throughout the occupation of Kherson, the occupiers made active use of outdoor advertising: billboards, posters, and stickers. The abundance of text in the advertisements was striking. It was just too much for billboards. Billboard advertisements are meant to quickly catch people's attention and make a lasting impression so that they think about the ad after passing by. Such ads are supposed to be read in a very short time. In our case, a significant proportion of the billboards counted 30 to 80 words of text. This means that one had to stop for a few minutes to read it. To put it simply, these billboards were similar to handwritten wall newspapers that were utilised during Soviet times to convey information.

Thus, the occupation media had to shape the image of the enemy through the newly created channels, while Ukrainian sources of information continued to have a major impact on the population. The only type of media that was naturally dominated by the occupiers was the print media, but it had several significant shortcomings. Firstly, the publications would come out very late, and secondly, their distribution did not run smoothly. Despite their declared circulation of up to 320,000 copies, newspapers were not easy to get hold of, even if you wanted them.

Finally, a few words about the main speakers of the occupation media. Both in the newspapers and on the administration's Telegram channel, most of the content was either anonymous or signed by unknown authors, most likely fictitious. But there are several familiar names. The first is Kirill Stremousov, initially head of the Rescue Committee for Peace and Order, and from 26 April 2022 deputy head of the Kherson Oblast Occupation Administration for Family, Youth, Sport and Information Work. Before Russia's full-scale invasion of Ukraine, Kirill Stremousov was known as a scandalous blogger, an anti-vaccination campaigner, the head of the Kherson organisation of the Ukrainian Socialist Party, the head of the Henichesk Fisheries Protection Inspectorate, a participant in the Orange Revolution and an organiser of pro-Russian rallies. Such a varied career made him more of a trickster than a statesman. Stremousov died in a car crash on the left bank of Kherson Oblast on 9 November 2022 under unclear circumstances. After his death, an icon was made in his likeness, which is used by the invaders in religious ceremonies. Stremousov was apparently the chief media curator of the Kherson Oblast Administration.

In addition to Stremousov, the occupation media regularly published appeals from the leaders of the occupation structures. One of them was Vladimir Saldo, the mayor of Kherson between 2002–2012 and a member of the Verkhovna Rada of Ukraine in 2012–2014. In 2015 he lost the elections for the mayor of Kherson and in 2019 the elections to the Verkhovna Rada. In 2016, Vladimir Saldo was detained in the Dominican Republic on suspicion of involvement in a kidnapping. The head of the new Kherson city administration, the Ukrainian national Aleksandr Kobets, and the head of the Kherson Oblast government, the Russian national Sergei Yeliseyev, both FSB officers, also addressed the people of Kherson.

Enemy No. 1. President Volodymyr Zelenskyy

First of all, it should be noted that despite the status of the official media, both the newspaper and especially the Telegram channel of the Kherson Oblast Administration do not differ much from the tabloid press in terms of content. A considerable number of articles and posts contain unverified allegations that lack evidence, unsupported assumptions, or fictitious claims that can be readily refuted. Moreover, the Telegram channel contains shared posts from private Telegram channels, such as the ones from the internationally-wanted former Ukrainian MP Ilya Kiva, as well as videos by top officials including the heads of the Kherson oblast and city administrations, and shared posts from official sources such as the Ministry of Defence of the Russian Federation, the Ministry of Foreign Affairs of the Luhansk People's Republic, and the head of the Kherson administration. The language of the articles and posts is often full of insults, criminal and other jargon, neologisms, tautologies *('new neo-Nazis')*, and even obscenities:

- *President and Commander-in-Chief of the Armed Forces of Ukraine (Wuhan)!!! Cocaine addict and buffoon Zelenskyy!!! Ukrainians,* *wake up, open your eyes, and regain your consciousness!!!! F*ck, don't you realise that you're all being prepared to get slaughtered???*[4]

- *An ingenious interpreter of the role of Napoleon in film and life, and a virtuoso of unconventional piano playing.*[5]

In writing about Volodymyr Zelenskyy, the occupation media have used a series of clichés aimed at creating an image of the Ukrainian president as an alcoholic and drug addict, a corrupt official, and a political puppet of the USA and the West in general. Special attention is paid to his incompetence as head of state, his ethnic origin and his professional background (actor and comic):

- *Zelenskyy, who never served in the army, probably does not understand most of what is going on and what he is doing to the state. This is all the result of drug addiction and the total incompetence of the last P-resident of Ukraine.*[6]
- *At the behest of his Washington bosses, Zelenskyy is leading people to the slaughter. It is no secret that Ukraine can never win. The statehood of the Ukrainian people has been traded for Biden's election campaign.*[7]
- *The Ukrainian president's wealth is estimated at $850 million, while his official salary in 2021 was just $12,000. A question arises: How did Zelenskyy make most of his fortune in office, and what does he spend it on? How cynical to ask a democratic president and a Jew such questions!*[8]
- *Why aren't the Ukrainian channels reporting on their Kherson defeat yesterday? This rhetorical question haunts many Ukrainians today, especially those whose sons, husbands and fathers were thrown into this useless crucible as cannon fodder at the personal behest of the 'comedian' Zelenskyy.*[9]

[4] Telegram post (*@VGA_Kherson*, forwarded from *@The_Kyva*, 14 April 2022), <https://t.me/VGA_Kherson/916>.

[5] Gennady Shkil, 'There is one Holy Russia', *Naddrepryanskaya Pravda*.

[6] Telegram post (*@VGA_Kherson*, 1 April 2022), <https://t.me/VGA_Kherson/689>.

[7] Telegram post (*@VGA_Kherson*, forwarded from *@Stremousov_Kirill*, 21 October 2022), <https://t.me/VGA_Kherson/4188>.

[8] Telegram post (*@VGA_Kherson*, forwarded from *@RadioStydoba*, 26 April 2022), <https://t.me/VGA_Kherson/1309>.

[9] Telegram post (*@VGA_Kherson*, forwarded from *@Stremousov_Kirill*, 30 August 2022), <https://t.me/VGA_Kherson/3064>.

Much attention has been paid in the occupation media to Zelenskyy's appearance. His facial expression ('stuck-up muzzle'[10]), his facial hair, his clothes, as well as the interiors in which he appears: 'It is not yet known exactly where Ukrainian President Volodymyr Zelenskyy is. He demonstrates that his previous experience as an actor helps him to slip into the role of a completely different president depending on the moment: caring, concerned, determined, dead tired. Sometimes, however, his "tired leader" role looks like a boozer who has lost all sense of proportion, stoned on illegal drugs mixed with alcohol'.[11]

The September issue of the newspaper carried a full-page article entitled 'Why Zelenskyy always wears the same T-shirt'. The anonymous author claims that the president's new image (moustache and beard, khaki clothes, refusal to wear an official business suit) was designed by 'Western stylists':

- *Either they wanted to turn Zelenskyy into Rambo, or they mistakenly offered him the image of a tramp.*
- *His untrimmed appearance makes him look like a partisan or a tramp.*[12]

There are contradictions in the Jewish component of Zelenskyy's image in the occupation media of Kherson Oblast. On the one hand, Zelenskyy is portrayed as a Zionist: a Telegram channel published a famous photo of Zelenskyy wearing a kippah at the Wailing Wall[13] with the caption that in the photo he is supposedly 'swearing an oath to Zion'.[14] On the other hand, it stresses that it is the Jews themselves who are criticising Zelenskyy: 'Look at the sarcastic reaction of Knesset member from the Religious Zionist Party, Simcha Rothman, after Zelenskyy's address to members of the Israeli parliament: 'President Zelenskyy asked us to help Ukraine now, just as the Ukrainians helped the Jews during the Nazi occupation. I want

[10] Telegram post (*@VGA_Kherson*, forwarded from *@godeye_ru*, 19 March 2022), <https://t.me/VGA_Kherson/50>.

[11] Telegram post (*@VGA_Kherson*, 24 March 2022), <https://t.me/VGA_Kherson/200>.

[12] 'Why Zelenskyy always wears the same T-shirt', *Naddnepryanskaya Pravda*, no. 12.

[13] 'Volodymyr Zelenskyy visited the Wailing Wall and joined the prayer for peace in Ukraine', website of the President of Ukraine Volodymyr Zelenskyy (23 January 2020), <https://www.president.gov.ua/ru/news/volodimir-zelenskij-vidvidav-stinu-plachu-ta-doluchivsya-do-59429>.

[14] Telegram post (*@VGA_Kherson*, 24 March 2022), <https://t.me/VGA_Kherson/206>.

to say right away that we must reject his request firmly. We are not monsters, we only want to bring good to all people'.[15]

In terms of foreign policy, he is portrayed as a puppet who is driving the country towards disintegration and the assimilation of different regions into other states: 'Poland has its eye on Ukrainian territories, but the Kyiv regime is clearly open to it. Zelenskyy is aware of the imminent disintegration of the country and has simply decided to use it to his advantage'.[16]

At the same time, the authors of the occupation media, drawing parallels with World War II, threaten Zelenskyy with Hitler's fate: 'Like Hitler, he will soon be driven out and torn to pieces by yesterday's "associates" and partners. After all, history is not a teacher, but a warden; naturally, it punishes people for not learning their lesson'.[17] (These are misquoted words of the Russian historian Vasily Klyuchevsky: 'History is not a teacher, but an admonisher, a mentor of life; it teaches nothing, but only punishes people for their ignorance of its lessons'. Among other things, the word 'admonisher' is replaced by the word 'warden'—a prison guard).

And finally, another characteristic of Zelenskyy that is often cited in the occupation media is his political toxicity. A lengthy article entitled 'Zelenskyy's virus is deadly and is transmitted through handwriting', written in the style of investigative journalism with an esoteric flavour, concludes that it was contact with Zelenskyy that led to the resignation of British Prime Minister Boris Johnson, Italian Prime Minister Mario Draghi, Estonian Prime Minister Kaja Kallas (who returned to head the government after her resignation), Bulgarian Prime Minister Kiril Petkov and even the death of Japanese Prime Minister Shinzo Abe in 2020.[18]

Enemy No. 2. The State of Ukraine and the Ukrainian Political Leadership

Addressing the residents of Kherson on the eve of the withdrawal of Russian troops from the city, the head of the city's military–civilian administration, Aleksandr Kobets, described the Ukrainian political leadership as

[15] Telegram post (*@VGA_Kherson*, forwarded from *@migdalsergei*, 20 March 2022), <https://t.me/VGA_Kherson/98>.

[16] 'Ukraine in one year', *Naddnepryanskaya Pravda*, no. 12, p. 8.

[17] Telegram post (*@VGA_Kherson*, forwarded from *@Stremousov_Kirill*, 30 August 2022), <https://t.me/VGA_Kherson/3064>.

[18] 'The Zelenskyy virus is deadly and transmitted through handwriting', *Naddnepryanskaya Pravda*, no. 9, p. 4.

follows: 'With all its might, the neo-Nazi regime in Kyiv is trying to prevent us from building a new life. The shelling of civilians, acts of terrorism and sabotage continue. But this will not help the Ukrainian authorities. All attempts at "counter-offensives" are being crushed by the resilience of our brave defenders. The puppet regime of Zelenskyy and those who support it are responsible for everything that is happening now'.[19] This message from the occupying government of Kherson Oblast is also characterised by its mention of the use of 'forbidden methods of warfare' by Ukraine: 'Due to the probability of the Ukrainian regime using forbidden methods of warfare and the information received about Kyiv's preparation for a terrorist attack on the Kakhovskaya HPP dam, we face the immediate danger of flooding on the territory of Kherson Oblast and the mass destruction of civil infrastructure'.[20] The head of the Kherson administration, Vladimir Saldo, calls the Ukrainian elites 'gentlemen who prevent us from building a peaceful life'.[21] Kirill Stremousov, who personally participated in the interrogation of Ukrainian citizens, used more vivid metaphors for the Ukrainian leadership: 'It is in the Nazis' blood to kill. These demons have always been interested in atrocities and abuse. These are the issues that people like Herashchenko [Ukraine's deputy interior minister in 2019–2021, a popular blogger] enjoy and praise. I don't even doubt that the Ukrainian Nazis would be prepared to rebuild concentration camps with gas chambers and torture chambers like Auschwitz and Buchenwald, but this is no longer realistic because the Russians are coming. We are going to rid the world of this Nazi plague, and this is already an undeniable fact. So Herashchenko, Avakov [Minister of Internal Affairs of Ukraine in 2014–2021], Turchinov [Secretary of the National Security and Defence Council of Ukraine in 2014–2019], Poroshenko [President of Ukraine in 2014–2019], Zelenskyy and the like should get ready for a military tribunal that will definitely take place in Kyiv. By the way, the West will bring these accomplices of fascism to the people's court itself, as it did in Nuremberg'.[22] In addition to the Nuremberg Trials, the occupying

[19] Telegram post (*@VGA_Kherson*, forwarded from *@prokherson_ru*, 10 October 2022), <https://t.me/VGA_Kherson/3957>.

[20] Telegram post (*@VGA_Kherson*, 31 October 2022), <https://t.me/VGA_Kherson/4424>.

[21] Telegram post (*@VGA_Kherson*, forwarded from *@SALDO_VGA*, 13 July 2022), <https://t.me/VGA_Kherson/2488>.

[22] Telegram post (*@VGA_Kherson*, forwarded from *@Stremousov_Kirill*, 9 August 2022), <https://t.me/VGA_Kherson/2835>.

media also threatened to organise 'their own Hague Trial' for the Ukrainian political leadership.[23]

Apart from terrorism and the betrayal of national interests, the Ukrainian political leaders face accusations of implementing an illegal language policy: 'The authorities, both state and local, have been mandated to use the Ukrainian language under the leadership of Zelenskyy. Proficiency in Ukrainian has been declared obligatory for individuals seeking public office and police service positions. Starting in the year 2030, all high school graduates will be required to have a command over the Ukrainian language. The idiocy is getting stronger'.[24] They are also accused of illegally trading human organs during the war: 'According to the doctor of Kherson orphanage, Kherson authorities prevented an attempt by the Ukrainian side to kidnap children from the orphanage in order for them to be taken to Europe for organ transplantation'.[25] The accusations include expansionism and, specifically, the intention to attack Crimea: 'Stocks of medals for the Capture of Crimea were discovered in the military recruitment offices of Kherson Oblast. It is important to note that the term "capturing" has been used instead of "liberating". The creators of the awards did not consider Crimea to be Ukrainian'.[26] A photo of the medal 'For the Capture of Crimea' with an award form was presented as evidence. The form contained several grammatical errors and misspelled the name of the President of Ukraine.

In the occupation media, Ukraine and its leadership are referred to as 'Ukroreich', 'Ukrainostan'—suggesting, on the one hand, a link with Nazi ideology, and on the other hand, Ukraine's backwardness, perhaps equating it with Afghanistan: 'Terrorist Biletsky [the founder of the Azov Battalion] has been quoted saying that his ideal society would resemble that of Afghanistan or Somalia, with no economic or cultural development but only universal armament and power being concentrated with warlords'.[27]

[23] Telegram post (*@VGA_Kherson*, 21 March 2022), <https://t.me/VGA_Kherson/99>.

[24] 'There is no Nazism in Ukraine, they said, because Zelenskyy is a Jew', *Naddnepryanskaya Pravda*, no. 5, p. 7.

[25] Telegram post (*@VGA_Kherson*, forwarded from *@HersonVestnik*, 26 April 2022), <https://t.me/VGA_Kherson/1361>; Telegram post (*@VGA_Kherson*, 27 March 2022), <https://t.me/VGA_Kherson/539>.

[26] Telegram post (*@VGA_Kherson*, forwarded from *@ukraina_ru*, 24 March 2022), <https://t.me/VGA_Kherson/199>.

[27] Telegram post (*@VGA_Kherson*, 20 March 2022), <https://t.me/VGA_Kherson/66>.

The purported influence of the US and the EU over the Ukrainian authorities is a frequent narrative in the occupation media. Even the official Telegram channel of a Russian Federation constituent reportedly shared the following pieces: 'The leaders of the country are just fantastic: chocolate baron,[28] drug addict, clown [presumably referring to Volodymyr Zelenskyy], faggot, and swindler [presumably referring to the head of the Office of the President, Andrii Yermak]. In the chaos of this clownish faggotry, it is the West and the root of faggotry, the EU, that is covering for them'.[29]

The fact that representatives of the Jewish elite, including Zelenskyy, the previous president Poroshenko, businessman Ihor Kolomoyskyi, and politician and businessman Hennadii Korban are in power demonstrates that Ukraine is only serving as an instrument for anti-Russian policies: 'In a speech given by Ihor Kolomoyskyi in Israel, he stated that Ukrainian *goyim* are no better than a herd of Russian schleppers. Blinded by the idea of "independence", they have turned into a biomass capable of slaughtering even their own mother in the name of independence of the Ukrainian state. Twenty years of propaganda have made them the ideal fighters against Russians in the form of their Slavic siblings – at present, any Ukrainian who has killed a Russian person will be acknowledged as a hero'.[30]

Apart from Volodymyr Zelenskyy, other representatives of the political leadership and media have also received significant personalised criticism from the occupation media, with special attention given to Oleksii Arestovych. Arestovych was an advisor for the Office of the President of Ukraine (2020–2023), a blogger (who holds a YouTube Gold Play Button), and is sometimes referred to as 'the main talking head of the regime'.[31] The Kherson occupation media depict him as an anarchist, esotericist, adventurer, transvestite, homosexual, paedophile, and provocateur. According to them, it is Arestovych who is the main source of misinformation for the residents of Kherson Oblast:

[28] Apparently referring to Petro Poroshenko, founder of the large Ukrainian confectionery corporation Roshen.

[29] Telegram post (*@VGA_Kherson*, forwarded from *@liferuss*, 15 April 2022), <https://t.me/VGA_Kherson/991>.

[30] Telegram post (*@VGA_Kherson*, 24 March 2022), <https://t.me/VGA_Kherson/202>.

[31] 'Counteroffensive: a meat grinder to account to the West', *Naddnepryanskaya Pravda*, no. 11, p. 4.

- *The messages of nationalism have lost touch with reality and are filled with Arestovych's imagined ideas... The Ukrainian side's announcement of Kherson soon coming under Kyiv's control seems like the ravings of a madman.*
- *Vladimir Saldo, the head of the administration of Kherson Oblast, stated that the claim made by Arestovych, a representative of the Kyiv authorities, that 'Russians in Kherson have mentally surrendered' is a false information attack. Kyiv propagandists use psychological methods of information warfare to zombify their own population, spreading false information and baseless allegations with a confident air.*[32]

It is worth noting that, in Ukraine, Arestovych has been consistently criticised, which eventually resulted in his resignation as an advisor to the Office of the President of Ukraine. Representatives of right-wing forces often accuse him of pro-Russian sympathies and of flirting with Ukraine's Russian-speaking electorate.

Interestingly, during the period of occupation, the occupation media did not publish any texts on the top military leadership, which includes Commander-in-Chief Valerii Zaluzhnyi, Commander-in-Chief of Land Forces Oleksandr Syrskyi, Minister of Defense Oleksii Reznikov, and commanders of the fronts.

When discussing the Ukrainian political leadership and their allies in the EU and USA, the authors of the invaders' Telegram channel accuse them of being gay and use derogatory words such as '*pidoras*' or '*pidor*' (rus. faggot). Using these words implies that they are immoral and wicked. It is important to mention that Ukrainian soldiers also refer to Russian soldiers as '*pidory*' in their everyday conversations. In Russian and Ukrainian, the word *pidor* can be used in two ways. The first and original meaning of the word is gay. The second commonly used meaning is that a *pidor* is a very bad and extremely mean person, who cheats and acts like a scoundrel.[33]

[32] Telegram post (*@VGA_Kherson*, forwarded from *@SALDO_VGA*, 15 July 2022), <https://t.me/VGA_Kherson/2500>.

[33] The Russian military refers to Ukrainian servicemen as '*ukropy*', '*khokhly*', '*vsuki*', and '*pidory*' (rus. faggots). The word '*ukropy*' has been used by separatists in eastern Ukraine since the start of the Russian-Ukrainian war in 2014. It literally means 'dill', but is essentially a distorted derivative of the word 'Ukrainian'. With time, the word '*ukrop*' in Ukraine became equivalent to 'patriot'. A political party with that name was even founded. The term '*khokhly*' has been utilised as a general or derogatory term for Ukrainian individuals since the era of the Russian Empire. Whereas the word '*vsuki*' only appeared in 2022. It is formed by merging *VSU* (rus. Armed Forces of Ukraine) and *suki* (rus. bitches). In addition to its primary meaning, in criminal slang, *suki* are prisoners who openly collaborate with the prison administration (snitches).

Enemy No. 3. Armed Forces of Ukraine (AFU)

When discussing the Ukrainian military, the occupation media means the Armed Forces of Ukraine or all military structures of the country, including the Territorial Defense Units and National Guard. Some sections of the Armed Forces of Ukraine, like the Azov regiment, are depicted as completely malevolent, while common soldiers are portrayed as being victimised by the Ukrainian government. However, exceptions are made under certain circumstances. These reports started emerging in August–September 2022 when Ukraine launched a counter-attack in Kherson Oblast.

The occupation media depicts the AFU as having poor technical equipment and low competence among officers and generals, resulting in heavy casualties and a division between soldiers, officers, and high command.

- *A captured soldier of the Ukrainian 58th motorized infantry brigade talks about the deplorable state of Ukrainian units: 'The leadership left us behind. We had to stay in a potato warehouse, locals brought us food. The unit had only two BRDMs*[34] *and two tanks, one wouldn't start, the other barely fired. They were hit almost immediately'.*[35]
- *AFU soldiers who discovered leaflets asking them to surrender and approached the DPR*[36] *troops, urged their fellow soldiers to follow suit. Many of them say that the Ukrainian military leaders do not assist them and even order them to attack military equipment with assault rifles.*[37]
- *Russian intelligence in Kherson found a group of injured Ukrainian soldiers hiding in the region. The Ukrainian servicemen requested to be included in the exchange lists. But the AFU command, in their usual way, discarded the soldiers who were sent to the front line as cannon fodder and abandoned without appropriate equipment.*[38]

[34] Armoured reconnaissance and patrol vehicle.

[35] Telegram post (*@VGA_Kherson*, 20 March 2022), <https://t.me/VGA_Kherson/80>.

[36] Donetsk People's Republic.

[37] Telegram post (*@VGA_Kherson*, forwarded from *@zvezdanews*, 27 March 2022), <https://t.me/VGA_Kherson/478>.

[38] Telegram post (*@VGA_Kherson*, 21 July 2022), <https://t.me/VGA_Kherson/2568>.

- *[Ukrainian soldiers] are being thrown uncounted into the trenches with only assault rifles to hinder Russian attacks with their bodies.*[39]
- *To hide the loss of soldiers, the AFU command gave instruction to bury them in mass graves without identification. Lime was poured over the corpses and the burials were classified.*[40]

One of the widespread claims is that Ukrainian Armed Forces servicemen 'use civilians as shields' to protect themselves during combat operations.[41] It is widely alleged that the Ukrainian army is attacking residential areas in Ukrainian cities: 'No one—not even Azerbaijanis with Armenians, nor Chechens—has ever attacked their own people. Here we are fighting subhumans who have deliberately set up fighting positions in residential areas and are firing Grad rockets at high-rise buildings. This is a form of pure fascism, where acts of inhumanity are taken to the ultimate extent'.[42]

According to Russian journalists or anonymous witnesses, the Ukrainian armed forces are targeting points where humanitarian aid is distributed,[43] staging provocations to falsely blame the Russian army for their crimes,[44] and using small arms to shoot civilians.[45]

As the Ukrainian counter-offensive against the Russian forces drew nearer and the evacuation of Russian troops and occupation structures from the western right bank of the Dnieper River was being arranged, the number of reports of Ukrainian military assaults on civilians increased considerably:

- *The so-called Ukrainian 'administration' of Kherson Oblast is calling on citizens to not comply with Vladimir Saldo's request to temporarily*

[39] 'Counteroffensive: a meat grinder to account to the West', *Naddnepryanskaya Pravda*, no. 11, p. 4.

[40] Telegram post (*@VGA_Kherson*, forwarded from *@The_Kyva*, 14 April 2022), <https://t.me/VGA_Kherson/918>.

[41] Telegram post (*@VGA_Kherson*, forwarded from *@sputniklive*, 25 March 2022), <https://t.me/VGA_Kherson/270>.

[42] Telegram post (*@VGA_Kherson*, forwarded from *@militaryreview*, 27 March 2022), <https://t.me/VGA_Kherson/432>.

[43] Telegram post (*@VGA_Kherson*, forwarded from *@izvestia*, 7 May 2022), <https://t.me/VGA_Kherson/1732>.

[44] Telegram post (*@VGA_Kherson*, forwarded from *@YuriBarbashov*, 8 April 2022), <https://t.me/VGA_Kherson/770>.

[45] Telegram post (*@VGA_Kherson*, forwarded from *@HersonVestnik*, 12 April 2022), <https://t.me/VGA_Kherson/837>.

move civilians to other areas in Russia. Ukrainian nationalist groups are planning to carry out attacks in Kherson Oblast with the intention of causing harm to civilians. They have openly stated their goal of maximizing harm to innocent people.[46]

- *Ukrainian armed formations have once again shown their bestial nature. Today's shelling is further proof of that. This is a deliberate plan to intimidate and kill ordinary people who want to protect themselves and cross to the left bank of the Dnieper River.*[47]
- *Kherson Oblast civilians became victims of serial maniacs. Their aims are straightforward: to disrupt infrastructure and cause chaos in daily life. This is a cannibalistic strategy: civilians are the targets.*[48]
- *In the Mykolayiv Region of Ukraine, territorial defence militants are compelling men to join them and shooting them on the spot if they refuse.*[49]
- *The Mariupol theatre bombing*[50] *is a provocation by the Ukrainian military, who are willing to sacrifice their fellow citizens.*[51]

In October, the AFU was accused of planning mass killings of civilians in Kherson Oblast:

- *The Ukrowehrmacht detachments have been ordered to kill all civilians in the area without distinction, which is genocide carried out by the Ukronazi government under Anglo-Saxon influence.*[52]
- *The neo-Nazis aim to kill and commit genocide against the peaceful citizens of Kherson Oblast. The soldiers confirmed that anyone who remained in the areas that were taken back would be physically elimi-*

[46] Telegram post (*@VGA_Kherson*, 19 October 2022), <https://t.me/VGA_Kherson/4158>.

[47] Telegram post (*@VGA_Kherson*, 21 October 2022), <https://t.me/VGA_Kherson/4178>.

[48] 'Ukrainian terrorist attacks swept through Kherson. What you need to do', *Naddnepryanskaya Pravda*, no. 3, p. 3.

[49] Telegram post (*@VGA_Kherson*, 22 March 2022), <https://t.me/VGA_Kherson/143>.

[50] On 16 March 2022, the Russian Armed Forces bombed the Donetsk Academic Regional Drama Theatre in Mariupol, Ukraine. It was used as an air raid shelter during the siege of Mariupol, sheltering a large number of civilians. The estimations of the number of deaths that occurred due to the bombing have varied, from at least 12 to 600 — *trans. note.*

[51] 'Torture, terror, genocide. The crimes of Kyiv that are falsely attributed to Moscow', *Naddnepryanskaya Pravda*, no. 8, p. 3.

[52] Telegram post (*@VGA_Kherson*, forwarded from *@Stremousov_Kirill*, 23 October 2022), <https://t.me/VGA_Kherson/4235>.

nated, following statements made by surrendered Ukrainian Armed Forces representatives.[53]

These reports were part of a campaign aimed at making the residents of Kherson vacate the right-bank part of the region. The invaders used harsh psychological tactics during this campaign, including intimidating and spreading false information. In the final two weeks of the occupation, the media consistently reported that the city was becoming dangerously unsafe. People were informed to leave by text message, even at night. Cars with loudspeakers drove through the streets. Civilians received voice messages from 'Ukrainian nationalists' promising to exterminate all remaining 'traitors' on the would-be recaptured territory.

The actions of the AFU allegedly include shelling civilian objects and churches, as well as burning fields. Killing and terrorizing civilians were said to be the main objectives of these military operations: 'It's becoming clear that Kyiv's campaign launched on August 29 did not have global objectives of capturing significant territory. The goal was to make it hard to return to normal life in these areas. This was to prevent the area from becoming an example for people in the rest of Ukraine, in particular by psychologically influencing their population'.[54]

The AFU's representation was also characterised by their plan to use forbidden weapons such as phosphorus and cluster munitions, and biological and chemical weapons.

- *The revealed information indicates that the government in Kyiv had seriously thought about using bioweapons against the people in Donbas and the Russian Federation.*[55]
- *We have reliable information that the Security Service of Ukraine plans to stage provocations using chemical weapons in Vysokopillia, Ivanivka, Kniazivka, Topolyne settlements of Kherson Oblast and Apostolove settlement in Dnipropetrovsk Oblast before 6 May 2022.*[56]

[53] Telegram post (*@VGA_Kherson*, forwarded from *@Stremousov_Kirill*, 7 October 2022), <https://t.me/VGA_Kherson/3900>.

[54] 'Counteroffensive: a meat grinder to account to the West', *Naddnepryanskaya Pravda*, no. 11, p. 4.

[55] Telegram post (*@VGA_Kherson*, forwarded from *@ukraina_ru*, 30 March 2022), <https://t.me/VGA_Kherson/651>.

[56] Telegram post (*@VGA_Kherson*, forwarded from *@mod_russia*, 6 May 2022), <https://t.me/VGA_Kherson/1719>.

- *Ukrainian nationalists and Banderites are already crossing all acceptable lines. According to experts, the AFU forces used white phosphorus ammunition near a Ukrainian settlement. Due to the highly toxic and lethal nature of the substance, its use is banned worldwide.*[57]

The word 'Banderites' appears in the last quote. Those known as 'Banderites' used to support one of the more radical factions of the Organization of Ukrainian Nationalists, formed in 1940 after a split within the group. Stepan Bandera became the leader of the faction. In 1941, at the start of the German invasion, the Banderites opted for an alliance with Germany. However, in 1943, they declared war on both the Third Reich and the USSR.[58] From the end of World War II until the 1950s, the Banderites actively participated in anti-Soviet resistance. Therefore, in Soviet propaganda, the term 'Banderite' was used with an extremely negative connotation. For a long time, many Ukrainians held negative associations with the term. But since 2014, Stepan Bandera has become the primary symbol of the fight against contemporary Russian aggressors. The image of Stepan Bandera has bifurcated—a historical character with a little-known biography to the wider public, on the other—a hero of mass culture, portrayed in songs, video clips, Internet memes, cartoons, patches, stickers, and more, who is often depicted as selflessly fighting against Russia. In 2022, a survey conducted by the Rating Sociological Group showed that Stepan Bandera is listed as one of the top five most influential Ukrainians ever. Amid the Russian-Ukrainian conflict, Bandera's popularity has increased from 4% in 2012 to 13% in 2022, which is more than

[57] Telegram post (*@VGA_Kherson*, forwarded from *@RKadyrov_95*, 25 March 2022), <https://t.me/VGA_Kherson/297>.

[58] P. P. Hai-Nyzhnyk, 'Orhanizatsiia ukrainskykh natsionalistiv (banderivtsiv)' [The Organisation of Ukrainian Nationalists (Banderites)], in I. M. Dziuba, A. I. Zhukovskyi, M. H. Zheleznìak et al. (eds), *Encyclopedia of Modern Ukraine [online resource]* (National Academy of Sciences of Ukraine, Shevchenko Scientific Society; Kyiv: Institute of Encyclopaedic Research of the National Academy of Sciences of Ukraine, 2022) <https://esu.com.ua/article-75640>.

three times the earlier figure.[59] In other words, the term 'Banderite' is no longer considered insulting to a large number of Ukrainians.

Other stories frequently seen in the media include reports of Ukrainian military torturing captured Russian soldiers, which can involve castration and is said to violate the Geneva Convention on the Treatment of Prisoners of War.[60] Additionally, there are reports of Ukrainian soldiers looting and extorting money from civilians.[61] There are even allegations of trading weapons and military equipment. Reports suggest that 'Ukrainian armed forces have been selling armoured vehicles and weapons to militias in the DPR. The exchange took place in the vicinity of the village of Snihurivka, Mykolaiv Oblast. It is telling that the Ukrainian military is willing to accept Russian rubles as payment'.[62]

The image of the Ukrainian military also includes cowardice. In particular, among the members of the National Guard: 'Based on the information available to me, Commander Prokopenko, who was leading the Azov regiment, abandoned his unit to their fate. He successfully fled the encirclement on one of the remaining helicopters and is currently in Kyiv'.[63] In fact, Denys Prokopenko, the commander of the Azov regiment, was captured and remained in captivity for four months before being exchanged.

Kherson occupation media has repeatedly released pieces depicting Ukrainian soldiers taking clothes from locals during the defence of Mariupol, supposedly in an attempt to conceal their presence.[64] According to those media sources they were even open to wearing women's clothes.[65]

[59] 'Popular Tops: The most prominent Ukrainians of all time (8–9 October 2022)'. Survey by the Rating Sociological Group (14 October 2022), <https://ratinggroup.ua/research/ukraine/narodnyy_top_vydayuschiesya_ukraincy_vseh_vremen_8-9_oktyabrya_2022.html>.

[60] Telegram post (*@VGA_Kherson*, forwarded from *@godeye_ru*, 5 April 2022), <https://t.me/VGA_Kherson/721>.

[61] Telegram post (*@VGA_Kherson*, 25 March 2022), <https://t.me/VGA_Kherson/306>.

[62] Telegram post (*@VGA_Kherson*, forwarded from *@ministryvs*, 15 April 2022), <https://t.me/VGA_Kherson/939>.

[63] Telegram post (*@VGA_Kherson*, forwarded from *@The_Kyva*, 12 April 2022), <https://t.me/VGA_Kherson/851>.

[64] Telegram post (*@VGA_Kherson*, 22 March 2022), <https://t.me/VGA_Kherson/143>.

[65] 'They killed his son and broke his grandson's legs in the driveway on March 14. On the following day, they got inside their flat via the balcony. Then, they asked his daughter-in-law for clothes they could wear to disguise themselves as civilians and avoid being spotted by Russian soldiers'. Telegram post (*@VGA_Kherson*, forwarded from *@donbassr*, 25 March 2022), <https://t.me/VGA_Kherson/305>.

A picture published in 2022 displays a scene of identifying an allegedly cross-dressing Ukrainian serviceman.[66]

Ukrainian Armed Forces are often compared to Nazis, the Wehrmacht and other military branches of Nazi Germany in the occupation media: 'People who carry out criminal orders to attack schools, kindergartens, churches, residential houses, or to harm children in Donbas are not legitimate military personnel. They should be seen as Nazis, regardless of their formal participation in any of the Ukrainian army's formations—Ukrainian Armed Forces, territorial defence units, or nationalist battalions'.[67]

One of the key elements of the enemy's discourse is the comparison of the current situation with World War II, specifically the Great Patriotic War. The 'special military operation' in this discourse serves as its continuation. The occupation authorities and the Russian Armed Forces frequently use the emblems of the Russian Empire and the Soviet Union to highlight this fact.[68] The media in annexed territories also feature articles where soldiers of the Ukrainian Armed Forces are portrayed as betrayers of their ancestors: 'Nationalists who had been fighting in Donbas since 2014 were invited to restore the monument commemorating the Soviet soldiers who liberated Beryslav, a city located in Kherson Oblast, from Nazi invaders. You have defamed the memory of your ancestors who fought in the Great Patriotic War for our Motherland. You now have a chance to make up for your mistakes by repairing this monument through your own hard work'.[69]

Reports often refer to the AFU as the 'Ukrowehrmacht' or 'Ukronazi Ze-Sonderkommandos' (relating to an election slogan of the Servant of the People Party 'Ze! team' [Zelenskyy supporters—*trans. note*]), 'Volkssturm' (referring to territorial defence units accused of conscripting children for service), and Ukraine as a whole as 'the Fourth Reich'. In the media of the occupying forces, having tattoos with Nazi symbols was a

[66] Mefody Gorbin, 'Ukrainian servicemen disguised as women try to escape from Mariupol', *Narodnye Novosti* (23 March 2022), <https://nation-news.ru/22080967-pereodetie_v_zhenschin_ukrainskie_voennie_pitayutsya_sbezhat_iz_mariupolya>.

[67] 'Insidious Strike', *Naddnepryanskaya Pravda*, no. 3, p. 4.

[68] Mykola Homanyuk, 'Reich, Union, Rossija. Die Symbolpolitik der Besatzer in der Ukraine', *Osteurope 11*, trans. Volker Weichsel (2022), <https://zeitschrift-osteuropa.de/blog/reich-union-rossija/>.

[69] Telegram post (*@VGA_Kherson*, forwarded from *@kgbchanel*, 5 May 2022), <https://t.me/VGA_Kherson/1640>.

common feature in the visual appearance of Ukrainian soldiers or supporters of Ukraine. Therefore, according to the occupying media, wartime laws do not apply to them, they are war criminals a priori. Most reports in the media controlled by the enemy accuse the AFU of committing crimes, which then lead to the conclusion that they should be completely destroyed: 'As you were informed a long time ago, we are at war with the Nazis. The AFU is heavily infused with these ideas. And the Nazi scum will have to be weeded out completely'.[70]

Enemy No. 4. 'Awaiters', Partisans, Provocateurs

During the first days of the occupation, civil protests began in Kherson Oblast. Besides the initial call-to-action by the director of Kherson Drama Theatre, Serhii Pavliuk, who chose the gathering time, no one was in charge of the subsequent daily rallies. Pro-Ukrainian rallies were held in all cities and urban-type settlements of Kherson Oblast on the same dates. It was truly a grassroots initiative. From March to April 2022, hundreds of people protested in Kherson on weekdays and thousands on weekends. During this time, protesters were frequently arrested, abducted after rallies, and subjected to beatings and torture.

On 21 March, the protesters were attacked by Russian Rosgvardiya officers.[71] The police began firing into the air, throwing light and stun, and tear gas grenades, hitting with batons and detaining the rally attendees. Several people were wounded. The media of the occupying force blamed troublemakers for this. The description beneath the photo of a man lying in a pool of blood reads: 'This is the outcome of troublemakers' and agitators' plotting. As a result, numerous residents of the city of Kherson have lost their lives. The time of Maidans and Nazism in Kherson Oblast is over. In the Maidans, they did not aim for the legs, but rather for the head, intending to kill'.[72]

The media occasionally covered the criminal acts committed by members of the resistance, who were referred to as terrorists, bandits, and mercenaries:

[70] Telegram post (*@VGA_Kherson*, forwarded from *@MedvedevVesti*, 27 March 2022), <https://t.me/VGA_Kherson/456>.

[71] Federal Service of National Guard Troops of the Russian Federation.

[72] Telegram post (*@VGA_Kherson*, 21 March 2022), <https://t.me/VGA_Kherson/123>.

- *The authorities arrested a Kherson resident for scattering 'petal mines'* [PFM-1 mines] *and location devices that the Armed Forces of Ukraine could use to destroy civilian objects. Remarkably, he was reported by his wife.*[73]
- *No 'partisan' was killed in Kherson. This is what the local terrorists call themselves while receiving allowances from their US handlers. Kherson scoundrels just kill people for money. I guess the CIA agents don't know the history of the partisan movement and the difference between a partisan and a mercenary. The Kherson 'partisans' are just ordinary bandits.*[74]

But it was the 'awaiters (or snorps)' who received the most attention. This is what the occupiers called the people living in Kherson Oblast who either refused to cooperate or waited for the AFU and, as per the invaders, looked forward to 'helping the AFU'. Kirill Stremousov even suggested a categorization of 'snorps': 'There are snorps who simply wait, being typical pacifists who don't care about any side winning. They constitute 90% of all snorps. Around 5–6% of snorps are willing to take action, ranging from passive to aggressive. And lastly, some snorps pose a serious threat because they are ready to harm civilians. Their future is not promising, because anyone who participates in terrorism will be punished according to the laws of the Russian Federation'.[75]

The occupying media also threatened the 'snorps' with the 'command given to the punitive units of the Ukro-Wehrmacht to destroy all civilians of the region regardless of whether they are "snorps" or sympathetic "pacifists"'.[76]

[73] 'Petal mines scattered in Donetsk. What should Khersonians expect', *Naddnepryanskaya Pravda*, no. 11, p. 8.

[74] 'Kherson residents respond to "partisans"', *Tochka Otscheta*, no.1 (16 August 2022), p. 4.

[75] Telegram post (*@Stremousov_Kirill*, 25 October 2022), <https://t.me/Stremousov_Kirill/617>.

[76] Telegram post (*@VGA_Kherson*, forwarded from *@Stremousov_Kirill*, 23 October 2022), <https://t.me/VGA_Kherson/4235>.

The Image of the Enemy in the Ukrainian Official Media of Kherson Oblast

To begin with, I will provide a short explanation and the overall attitude of the Telegram channel belonging to the Ukrainian Kherson Oblast Military Administration (OMA). The channel was created on 1 March 2022. By the beginning of 2023, the number of followers had reached 22,000. This channel offers a lot of shared content, but most of it comes from the official Telegram channels of the President of Ukraine, the AFU Commander-in-Chief, government ministers, and officials from the Presidential Administration. The channel uses a relatively dry language to provide information about the situation at the front and in occupied territories, as well as details on evacuation, delivery and distribution of humanitarian aid, and to debunk fakes. Reports like this often include the event's location and the age, sex, and social status of the people involved.

Russians are called invaders, occupiers, terrorists (terrorist state), and an aggressor country on the Kherson OMA's Telegram channel: 'The occupiers understand that we are getting stronger over time, and their goal is simply to exert pressure on our society, scare our people, and cause the most harm to Ukrainian cities as long as those terrorists have the capability to do so'.[77]

The term 'invaders' again brings to mind connections with World War II. For many years during Soviet times and currently in Ukraine, it has been commonly used in the term "German-fascist invaders". The posts on Victory Day and the Day of Ukraine's Liberation from the Nazis (28 October) show a clear parallel between World War II and the Russian attack on Ukraine. The Russian Federation is compared to Germany due to its failed 'blitzkrieg', and 'Ukraine and the democratic world' are compared to the anti-Hitler coalition:

- *Ukrainians who were part of the anti-Hitler alliance played a major role in defeating Nazism and its supporters. We are facing a cruel and cynical enemy once more. The Russian troops who have occupied Ukraine are committing true genocide against the Ukrainian people. They are intentionally destroying innocent civilians. Our native Ukrainian Kherson Oblast came under temporary occupation by the Ruscists.*

[77] Address of President Zelenskyy. Telegram post (*@OP_UA*, 15 July 2022), <https://t.me/OP_UA/7079>. Forwarded by *@khersonskaODA*.

However, we are protecting our own land and our homes. ...The entire democratic world is supporting us![78]

- *Just like 80 years ago, the people of Ukraine are standing up to defend their homeland. And the hostile blitzkrieg plan has become a failure. Evil always works the same way. It causes the death of innocent people while claiming to only attack military objectives. It sets up blockades and filtration camps. And Syrets*[79] *becomes Olenivka. It destroys towns and villages. And Koriukivka becomes Bucha.*[80]
- *Once again the enemy has set foot on our land. Today he is Russian. But we are the true descendants of the victors. We remember the feats of those who stopped the invasion at the cost of their own lives. Our warriors once again stand in defence of the entire civilized world. And the ruscists have no chance of becoming the masters on our land. To the eternal memory of the liberators of Ukraine who brought freedom from the Nazis in 1944!*[81]

Of the more expressive words, the most common is 'ruscists'—followers of the ruscist ideology. This term is made up of two words: 'Russia' and 'fascism':[82]

- *Ruscists have no respect for international prohibitions and human qualities. These pathetic occupiers are shooting civilians.*[83]

[78] Telegram post (*@khersonskaODA*, 8 May 2022), <https://t.me/khersonskaODA/363>.

[79] Syrets is the name of a concentration camp operated by the Nazis in Kyiv, and Olenivka is the name of a camp in the Donetsk region where Ukrainian prisoners of war were held during the Russian occupation. The Koriukivka tragedy is the name given to a tragic event that involved the killing of 6700 ordinary people in the village of Koriukivka during World War II. Bucha is a city in Ukraine where Russian soldiers committed war crimes in 2022.

[80] Telegram post (*@khersonskaODA*, forwarded from *@V_Zelenskiy_official*, 27 October 2022), <https://t.me/khersonskaODA/1433>.

[81] Telegram post (*@khersonskaODA*, 28 October 2022), <https://t.me/khersonskaODA/1435>.

[82] Timothy Snyder, 'The War in Ukraine Has Unleashed a New Word. In a creative play on three different languages, Ukrainians identify an enemy: "ruscism"', *The New York Times Magazine*, (22 April 2022), <https://www.nytimes.com/2022/04/22/magazine/ruscism-ukraine-russia-war.html>.

[83] Telegram post (*@khersonskaODA*, 1 March 2022), <https://t.me/khersonskaODA/14>.

- *The ruscists have once again stopped trucks carrying humanitarian aid into Kherson Oblast, thereby preventing residents from receiving much-needed supplies.*[84]

This term has been commonly used since the early days of the Russian Federation's full-scale attack on Ukraine, although it started being used in Ukraine as early as 2014, at the beginning of the Russian-Ukrainian conflict. The Verkhovna Rada of Ukraine approved a statement on 2 May 2023 defining Russia's current political regime as ruscism. The signs of ruscism, according to the statement, include militarism, the cult of the state's leader, and untouchable state institutions. Additionally, ruscism involves elevating the Russian Federation by violently oppressing or denying the existence of other peoples, propagating the ideas of the 'Russian world', continuously violating the norms and principles of international law, and infringing upon the sovereign rights, territorial integrity, and internationally recognised borders of other states.[85]

The term 'orcs' appears a few times. This word comes from the novel *The Lord of the Rings* by J. R. R. Tolkien: 'Irpin, Bucha, Hostomel and Borodianka, which are towns in the Kyiv Region, have been liberated. Hundreds of civilians were shot. Women were burned and raped. Children tied and shot dead. The scale of the atrocity is unmeasurable. Orcs, we will not forgive you for this'.[86]

This word is frequently used on Ukrainian social media, websites, and everyday conversations. It is frequently accompanied by other expressive terms like *rusnya*, *rusnyavye* (a garble of 'Russians'), and *chmobiki* (derived from '*chmo*'[87] and '*mobilized*').[88] While the novel and the movie trilogy based on it is more popular with younger people, even older people who

[84] Telegram post (*@khersonskaODA*, 19 March 2022), <https://t.me/khersonskaODA/195>.

[85] Bill record of 'Draft Resolution on the Statement of the Verkhovna Rada of Ukraine On Defining the Political Regime Existing in the Russian Federation as Ruscism and Condemning Its Ideological Foundations and Social Practices as Totalitarian and Hateful'. Web-portal of the Verkhovna Rada of Ukraine <https://itd.rada.gov.ua/billInfo/Bills/Card/41531>.

[86] Telegram post (*@khersonskaODA*, forwarded from *@tymoshenko_kyrylo*, 4 April 2022), <https://t.me/khersonskaODA/267>.

[87] Creep, pouge – *Dictionary of Ukrainian Slang*, <http://slovopedia.org.ua/57/53415/376410.html>.

[88] L. P. Kisliuk, 'The impact of the war on the activity of the Ukrainian community and the Ukrainian language system', online video recording, YouTube (Institute of Ukrainian Language of the National Academy of Sciences of Ukraine), <https://www.youtube.com/watch?v=ROF1V0plt08>.

do not know anything about Tolkien call Russians ‘orcs’.[89] However, the word was also used several times by Kherson’s official sources. One instance was a statement from the head of the Ukrainian National Security and Defense Council, Oleksii Danilov, who advised Ukrainians not to use new terms like *orc*, *rusnya*, or *svynosobaka* [pig-dog—*trans. note*] to avoid distorting the image of the enemy: ‘We should clearly write and document for history: russian people and russia[90] have killed, attacked, robbed, raped, destroyed, and lied’.[91]

More emotionally charged texts were posted on the Kherson Oblast Military Administration’s Telegram channel after another act of violence against civilians, including the destruction of Mariupol, cases of murder and rape in Bucha, the bombardment of Kramatorsk train station, and the attack on the Ukrainian prisoners of war camp in Olenivka.

- *The occupiers’ actions in Mariupol are extremely cruel. Europeans! Ukrainians! Mariupolians! We must unite today to denounce this Russian war crime, which reflects the evil the occupiers have brought to our land.*[92]
- *Bucharest, NATO summit declaration. April 3, 14 years ago. There was a chance to prevent it. To stop Russia from invading. Bucha, Kyiv Oblast. Now Russia is here. The mothers of Russian servicemen should pay close attention. Look at the bastards you’ve raised. Murderers,*

[89] Likewise, the term ‘Warriors of Light’, which originates from the fantasy genre, is frequently employed in Ukrainian social media and daily life to refer to Ukrainian soldiers. The song by Belarusian rock band Lyapis Trubetskoy with the same name made it popular in 2014, during the Euromaidan and the start of the Russian-Ukrainian war. A memorial plaque dedicated to everyone who died during the Anti-Terrorist Operation in Donbas was installed in Kherson Oblast. The plaque is titled *Warrior of Light*.

[90] Since the start of the full-scale invasion, Ukrainian media has mostly been referring to ‘Russia’, ‘Moscow’ and ‘Putin’ with a lowercase letter. In September 2023, the National Commission for State Language Standards in Ukraine stated that using lowercase letters for the names related to the aggressor country, specifically ‘russian federation’, ‘russia’, ‘moscow’, ‘moskovia’, ‘moscow tsardom’, ‘russian empire’, and ‘state duma of the russian federation’, does not violate the norms when in comes to non-formal texts, <https://mova.gov.ua/news/napysannia-nazv-rosiiska-federatsiia>.

[91] ‘Danilov explained why it is unreasonable to call Russians “orcs” and “svinosobaki”’, *IA Ukrinform* (25 December 2022), <https://www.ukrinform.ua/rubric-ato/3640814-danilov-poasniv-comu-nebazano-nazivati-rosian-orkami-ta-svinosobakami.html>.

[92] Address of President Zelenskyy. Telegram post (*@OP_UA*, 15 July 2022), <https://t.me/OP_UA/7079>. Forwarded by *@khersonska_ODA*.

looters, executioners [last sentence written in Russian instead of Ukrainian – *author's note*].[93]

- *The attackers targeted the Kramatorsk train station, where thousands of peaceful Ukrainians were waiting to be evacuated… Around 30 people died and approximately 100 others were injured, with varying degrees of severity. Russian villains don't give up their ways.*
- *Today, I received information about the invaders striking Olenivka in the Donetsk Oblast. This was a conscious Russian war crime, a deliberate mass murder of Ukrainian POWs.*[94]

The Kherson OMA Telegram channel describes the Russian army as both strong and weak. During the early months of the invasion, there was much talk about the power and superiority of the Russian army. This was mentioned in the context of how the Ukrainian Armed Forces managed to successfully confront them despite this technical and numerical advantage:

> We are aware that the occupiers prevail numerically. That they have more equipment. Russian leaders are trying to recruit more idiots among military cadets in all of Russia, those with combat experience and conscripts. They plan to send these soldiers to be killed in another offensive.[95]

The army that is occupying the region is shown to be weak, as soldiers of the Russian Armed Forces are facing difficulties with motivation. This is particularly prevalent among soldiers from national minorities and militia soldiers from DPR and LPR who were forcefully conscripted into the army: 'Marines captured 11 Russian soldiers on the Kherson front. The majority of them were mobilised in the occupied parts of Donetsk and Luhansk regions (ORDLO),[96] forcefully removed from their jobs, and threatened with criminal prosecution for any refusal'.[97]

[93] Address of President Zelenskyy. Telegram post (*@V_Zelenskiy_official*, 28 June 2022), <https://t.me/V_Zelenskiy_official/1074>. Forwarded by *@khersonska_ODA*.

[94] Telegram post (*@khersonskaODA*, forwarded from *@OP_UA*, 30 July 2022), <https://t.me/khersonskaODA/733>.

[95] Telegram post (*@khersonskaODA*, forwarded from *@OP_UA*, 6 April 2022), <https://t.me/khersonskaODA/275>.

[96] Acronym for the Isolated Areas of Donetsk and Luhansk Regions that were taken over in 2014. This abbreviation was used during the Minsk agreements.

[97] Telegram post (*@khersonskaODA*, 4 August 2022), <https://t.me/khersonskaODA/768>.

Some publications argue that Russian soldiers are not just being used as expendable manpower—Russia's attitude towards them is alleged to be a form of ethnic cleansing that aims to eliminate citizens of non-Russian origin: 'The repeated Russian attempts to advance in the Donetsk region are seen by some as one of the most cynical instances of the killing of their own in the history of warfare. Russia is using unlawful mobilisation tactics to not only prolong the suffering of people in Ukraine and worsen global instability, but also to destroy indigenous minority men who live in territories under Russian control'.[98]

The Telegram channel posts frequently called on the inhabitants of national republics to oppose mobilisation: 'I am urging the people of Caucasus, Siberia, and other native peoples of Russia to take action. Dagestanis should not die in Russia's despicable and shameful war. Nor should Chechens, Ingush, Ossetians, Circassians and any other people who found themselves under the Russian flag. …We acknowledge that you are standing up against illegal conscription, which is an attempt by the Russian authorities to mask their army's lack of success and their illegal orders. Fight to not die!' [the last sentence is written in Russian—*author's note*].[99]

Desertion was also mentioned as evidence of the Russian army's weakness: 'In Kherson Oblast, inspections are being intensified by the occupiers. They are not only searching for "Ukronazis", but also for their own soldiers who fled the battlefield and, disguised as civilians, are trying to find ways to escape'.[100]

This vulnerability affects not just the military, but the whole country, since 'asking for Iran's assistance implies the Kremlin acknowledges its military and political failure. For decades, they invested billions of money in their military-industrial complex, but in the end, they had to bow their heads to Tehran for relatively basic drones and missiles'.[101]

The Telegram channel focuses heavily on how the Russian military and special services treat the local population. The posts mention incidents

[98] Telegram post (*@khersonskaODA*, forwarded from *@V_Zelenskiy_official*, 25 September 2022), <https://t.me/khersonskaODA/1071>.

[99] Telegram post (*@khersonskaODA*, forwarded from *@V_Zelenskiy_official*, 29 September 2022), <https://t.me/khersonskaODA/1099>.

[100] Telegram post (*@khersonskaODA*, 18 September 2022), <https://t.me/khersonskaODA/1026>.

[101] Telegram post (*@khersonskaODA*, forwarded from *@V_Zelenskiy_official*, 18 October 2022), <https://t.me/khersonskaODA/1329>.

involving the deliberate targeting of passenger transport and housing resulting in casualties, abduction, and torture of people. It also refers to psychological pressure exerted on people, mining of civilian infrastructure, as well as planned provocations to falsely accuse the AFU:

- *Russian attackers in Kherson Oblast shot a minibus with ATGMs*[102] *near the village of Dovhove. The minibus was carrying evacuees from Starosillya. Two civilians were killed.*[103]
- *The occupiers fired cluster munitions at the village of Novopavlivka. No people were injured, but shells hit the cattle.*
- *In the de-occupied within Kherson Oblast, the invaders hid explosive devices in unexpected places like homes, parks, stores, and even dog crates.*[104]
- *A 66-year-old man was taken from his home in Biliaivka and held captive for over two months. He was subjected to torture during this period.*[105]
- *In Kherson Oblast, the invaders are offering to pay parents 10,000 rubles if they allow their children to attend the schools they have taken control of. Parents may be threatened with fines, confiscation of property, and loss of parental rights if they refuse to send their children to Russian educational programmes.*[106]
- *Ruscists are preparing bloody provocations in Kherson Oblast. Within the next few days, the Russian invaders will likely attack residential areas in the region and falsely accuse the Ukrainian Armed Forces.*[107]

Many posts on the Kherson OMA Telegram channel discussed the looting that the occupiers committed in the regions they controlled. There were many reports of looting, robbery, and theft from civilians. Large-scale theft from businesses was also commonly reported.

[102] Anti-tank guided missile.

[103] Telegram post (*@khersonskaODA*, 2 August 2022), <https://t.me/khersonskaODA/746>.

[104] Telegram post (*@khersonskaODA*, 7 August 2022), <https://t.me/khersonskaODA/777>.

[105] Telegram post (*@khersonskaODA*, 27 August 2022), <https://t.me/khersonskaODA/899>.

[106] Telegram post (*@khersonskaODA*, 28 August 2022), <https://t.me/khersonskaODA/904>.

[107] Telegram post (*@khersonskaODA*, 7 November 2022), <https://t.me/khersonskaODA/1546>.

- *As you all know, the city of Kherson is entirely surrounded by the enemy. During the night, there were many gunshots and explosions, and the SBU building was on fire.*[108] *The occupiers also looted many pharmacies and stores.*[109]
- *Russian invaders are stealing from garage cooperatives in Kherson with the help of local collaborators. Furthermore, incidents of Russians being relocated to captured flats are becoming commonplace.*[110]
- *In Kherson, occupants broke into homes on 1st Parkovy Lane and flats on Stroiteley Avenue to steal. Household and computer equipment, clothing and even carpets were taken. The items were taken away in APCs and KAMAZs.*[111]
- *Armed Russian military seized a cargo barge and a motor ship from the floodplain in the region's capital.*[112]
- *The occupants also 'evacuated' monuments to Ushakov and Suvorov from Kherson. All you need to know about the Russian military… They leave their soldiers behind on the battlefield, yet they remember to 'protect' the monuments.*[113]

Overall, the issue of looting by Russian soldiers was one of the most common topics discussed in Ukrainian social media. Lots of pictures, art, digital images, and tales are circulating online about this matter. An example that was discussed is the large-scale theft of household and computer equipment, clothes and toys by Russian soldiers who then send them to Russia. Numerous reported instances of theft and confiscation of utility companies' property, including cars, ambulances, special equipment, pieces of art and cultural artefacts can be found on the Telegram channel.

Besides the Russian army, Russian propagandists are also considered adversaries: 'I would like to talk about the duty of these Russian

[108] Security Service of Ukraine.

[109] Telegram post (*@khersonskaODA*, 2 March 2022), <https://t.me/khersonskaODA/18>.

[110] Telegram post (*@khersonskaODA*, 2 August 2022), <https://t.me/khersonskaODA/746>.

[111] Telegram post (*@khersonskaODA*, 13 September 2022), <https://t.me/khersonskaODA/986>.

[112] Telegram post (*@khersonskaODA*, 17 September 2022), <https://t.me/khersonskaODA/1014>.

[113] Telegram post (*@khersonskaODA*, 24 October 2022), <https://t.me/khersonskaODA/1379>.

propagandists who tell lies about these terrors and attempt to defend the actions of the Russian army. Each of you will also be punished'.[114]

When comparing the depiction of the enemy in Ukrainian and occupation sources, we notice that they tend to focus on the same subjects: these include World War II and its continuation in 2022, Nazis in Ukraine or Russia, ordinary soldiers being used as cannon fodder, portraying the enemy as terrorists who harm innocent civilians, concealing combat losses, using prohibited ammunition, looting, mentions of ethnicity (with the difference that the occupiers refer to the ethnic origin of leadership, while Ukraine refers to the nationality of the Russian Armed Forces soldiers), and propagandists being immoral. Both sides use lower-case letters for the names of adversarial states, government agencies and leaders. The only difference is that Ukrainian sources often report the news before Russian sources do. For instance, a Ukrainian source reported that Russian soldiers were involved in looting before it was mentioned in a Russian source. Later, a Russian source would allege that Ukrainian soldiers were also involved in looting.

Conclusions

Since its establishment, the media of the occupying forces have portrayed the Ukrainian political leadership and military as neo-Nazis and war criminals who act under the instructions of the 'collective West'. The "Special Military Operation" was frequently compared to World War II and presented as its continuation, with Russia—under the banners of both the USSR and the Russian Empire—fighting against Ukraine which was referred to as the 'Fourth Reich'. This discourse repeatedly highlighted the personal moral shortcomings of both the powerful leaders and the regular Ukrainian soldiers and activists, for instance, referring to drug addiction, homosexuality, brutality, and cowardice. The authors also tried to employ anti-Semitic themes. As an exception, captured Ukrainian soldiers were sometimes attributed human qualities, and this was done solely to display Russian mercy.

The media of the occupying force also attempted to alter the perception of who the aggressor is, to portray Ukraine as the perpetrator to the

[114] Address of President Zelenskyy. Telegram post (*@OP_UA*, 28 June 2022), <https://t.me/OP_UA/6951>. Forwarded by *@khersonska_ODA*.

people of Kherson Oblast. Starting in July 2022, they created a story that Ukraine is preventing the return of peacetime to the areas 'liberated' by Russia, spreading fear among the local population, and planning to completely destroy anyone who stayed in the occupied locations. The AFU counteroffensive was described as an attempt to occupy Kherson Oblast by the Nazis. The reversal of the aggressor was based on Ukrainian news coverage which included the bombardment of residential areas in Kharkiv and Mariupol by Russian troops, the killing of civilians in Bucha and Izyum, the use of cluster and phosphorus munitions, and the illegal relocation of children from the occupied territories. The occupation media authors used identical plots, with the only variation being the substitution Russia with Ukraine.

Spreading this narrative and portraying Ukraine as an adversary had multiple practical goals. The Russian leaders understood that they could not keep the territories on the right bank of the river Dnieper. So, they decided to take with them not only material resources and art treasures but also people. In Ukrainian and international news reports about the war, most of the focus was on territory—how much Russia has gained during the attack or how much Ukraine has liberated. Information about how many people were captured or how many were liberated is bypassed by the media. However, it is very important. Russia is experiencing a high level of population decline, and war is being used as one way to manage population levels. Only existential fear can force someone to abandon all they have and leave their homes forever. This was the goal of the occupation media from September to November 2022 when forming the image of the enemy.

The second task was to shape a negative attitude towards those residents of the occupied territories who took an active pro-Ukrainian stance, participated in protests and anti-Russian demonstrations, and passed on information about the deployment of Russian troops and military equipment. The goal was not just to create a negative attitude but also to encourage denunciations against those people.

Overall, it can be concluded that the information campaign aimed at portraying Ukraine as an enemy to the inhabitants of the occupied territories did not accomplish its objective. First, the occupation authorities failed to establish control over the information space. Second, the news items were mostly derivative in nature: a Russian version of the Ukrainian news agenda. Third, the occupiers failed to find credible speakers from the region. Fourth, all the people living in Kherson Oblast could inform themselves not only through phone and TV screens, but also through the windows of their houses and flats.

Acknowledgements The author is grateful for support from the Research Council of Norway, which provided an opportunity to work on this study in the frame of the project 'Contested Ukraine: Military Patriotism, Russian Influence and Implications for European Security', Project Number: 326509.

Afterword: Coming Home—Finding Form for Veterans' Life Stories of Enemy Encounter

Holly Furneaux and Matilda Greig

Words by Holly and Matilda. Incorporating thinking by Dave, Graham, Karin, Rach, Robbie

It's getting late on a Tuesday evening. On our Zoom screens, Robbie is describing a 'funny, bizarre' experience of 'sharing cups of coffee and a sticky bun, talking about the things we did during the war' with former enemies. He recalled how when he joined the Royal Marines in 1986, the training had been preparation against a Russian enemy, and how he had then come to Iraq towards the end of the Gulf War of 1990–1991. Working in 2003–2020 as a contractor in Iraq with a Russian oil company client, Robbie realised that the two men in his car were former Russian

H. Furneaux (✉)
Cardiff University, Cardiff, UK
e-mail: furneauxh@cardiff.ac.uk

M. Greig
National Army Museum, London, UK
e-mail: mgreig@nam.ac.uk

H. Furneaux, M. Greig (eds.), *Enemy Encounters in Modern Warfare*, https://doi.org/10.1007/978-3-031-56748-3_16

intelligence, while the driver was an Iraqi Gulf War veteran. An incident he'll always remember, Robbie described it as a 'lightbulb moment': 'I couldn't believe it. These are the people I was trained to kill'. Though having felt no human attachment when required to kill, years later Robbie felt this strange meeting gave a 'beating heart', 'brain', and 'humanness' to the former enemy.

We have been involved in these conversations through the partnership between the Enemy Encounters AHRC funded project and Re-Live, a Life Story Arts charity acclaimed for its work with veterans. We've been talking with a group of former British soldiers about our historical research on interactions across military sides, sharing first-hand and literary sources from the long nineteenth and twentieth centuries, and learning about how those relate to their myriad experiences of the enemy. The group comprises 3 men and 1 woman veteran each with different military experience from the 1970s onwards, as well as us and Karin Diamond, Re-Live Artistic Director. One thing that particularly stood out to the group was the widespread debate about and renegotiation of who the enemy was across the historical pieces. For some of the group, as for us, a favourite piece was from Leo Tolstoy's novel *War and Peace* (1869), where Nikolai Rostov, a formerly bellicose Russian officer, is discomforted by proximity to the French enemy he takes prisoner: 'This pale, mud-stained face of a fair-haired young man with a dimple on his chin and bright blue eyes had no business with battlefields; it was not the face of an enemy; it was a domestic, indoor face'.[1]

The 'enemy' means many things to different members of the Re-Live veterans group. While it can mean a military adversary or fought nation, the enemy might also be those who prosecute wars and feed the war machine. The group spoke of experiences of disconnection from civilian society on return from conflict and of feeling like an enemy within the context of legal prosecutions of soldiers and veterans and within the framework of widespread protest against unpopular wars, such as the Iraq War (2003–2011). They also reflected on moral injury, trauma, and the enemy within. Former soldiers' traumatic experiences can result in behaviours through which they become enemies within their families, and the inability or refusal of family members to listen and understand could also feel adversarial. As well as enemies within the self, and within the home, a hostile military environment, particularly to women personnel, meant that

[1] Leo Tolstoy, *War and Peace*, trans. Anthony Briggs (London: Penguin, 2007), p. 723.

members of the British forces and the patriarchal establishment became the enemy. The insufficiency of support within the military structure for the complex, long-silenced, and gendered challenges facing veterans is felt as a form of enmity. The group identified collective work, in community, that can communicate these experiences as essential to ensure that these enemies are not faced alone.

Personal insights from the veterans' group cover many of the central themes of this collection. We used our historical research material to anchor discussions of the group's experiences of the application of laws of war, with different rules of engagement regulating the use of force being applied in different conflicts as determined by the enemy and by the allies. In a conversation with particular resonance with Matilda's chapter on the long history of guerrilla warfare, the group reflected on the differences of contemporary war, less often between two clearly opposed, nationally identified uniformed groups, and the difficulties presented by less identifiable enemies including militias and insurgents. These distinct kinds of warfare present different challenges of processing for the soldiers involved. Discussions shared this book's concern with the contradictions within ideas of humanitarian war and the uneven application of laws designed to minimise war suffering. At times these stipulations might also be used for strategic advantage, such as a change to ammunition more likely to wound than kill, not to minimise war dead but to remove more people from the field in helping the casualties (taking 3 out rather than 1) and to commit the resources needed to evacuate and treat them.

Though technologies have made face-to-face combat rarer, the group rejected ideas that this depersonalises and desensitises war violence. They suggested, instead, that there is a very human reckoning to be gone through for those serving in the military across a range of roles. The introduction outlined the risk—through important work like Stahl's on the focus on the suffering of drone operatives in US and European popular culture—that attention to combatants' suffering might work to exonerate their nation from blame for its military policy. From our discussions of the personal cost of military service with these veterans, however, the opposite argument could also be made. A recognition of the widespread negative impact on military personnel and their families and communities might instead be used to hold countries to account for war violence to enemy nations and their own. As one group member put it, recognising the need for 'healing and mending is actually part of dealing with combat itself'.

As the story told by Robbie at the start of this piece suggests, there was a focus on the group's experience of shifting enmity and allegiance. In a raw parallel with the chapters by Tekseng and Aguirre Quiroga, a member of the group had lost a colleague after one member of a newly allied group they had trained turned on them. Though Captain Parquin's Napoleonic account of drinking brandy with the enemy before battle (discussed in the introduction) felt rather too 'gentlemanly' and 'polite' to the group, there were experiences of hospitality, as in Robbie's coffee and sticky buns, with former enemies years later. The group found especially moving an account from the Crimean War of a British Staff Officer's treating the Russian wounded and discovering commonality with them 'when around the necks of every one of these soldiers I found the cross and the Virgin and Child'.[2] While the immediacy of this reassessment, shortly after the battle, was distinct from their experiences, members of the group commented on the kinds of transformation of view of opposing forces over a longer time. As Graham put it, 'that definition of enemy I think will change from when you serve, to when you reflect, if you have the good fortune to reflect, years later'. He had had a direct experience of re-encountering the enemy 40 years later, having made contact with some of the Republicans and their families he'd been in opposition to when serving in Northern Ireland: 'I wanted to apologise for what I believed was wrong, but I also wanted to hear from them, and I wanted to meet them on a one-to-one human basis. Almost universally I was really welcomed into their homes and people wanted to talk, because we're all connected'. Robbie described the opportunity—also taken by many veterans of the Malvinas (Falklands) war on the recent 30 year anniversary—to reencounter a former enemy in peacetime and realise their similarity to yourself as 'part of the healing process of dealing with those demons and inner enemies'.

Re-Live uses a range of life-story telling techniques to give narrative form to complex experiences. In this project, our conversations about enemy encounters and broader life-story work helped to shape *Coming Home* (2022), a life story comic made in partnership with editor Steve Sullivan and professional illustrators. *Coming Home* gives an extraordinary rendering of four veterans' stories. Dave's and Claire's reflect on different facets of the enemy within. Dave's 'Hard Shoulder' explores the struggle to adjust to civilian society—'once you're out, you're on your own to deal with everything'—a sense of being alien to everyday life. It touches on the

[2] 'The Russian Wounded', *The Times*, 25 October 1854.

challenges to veterans' mental health and suicidal thoughts also explored in a stunning single page piece on 'Moral Injury' in the comic, in which artist Casey Raymond presents the work of repairing the heart as a form of kintsugi, a process by which broken ceramics are repieced with gold. There is growing attention to moral injury as a distinct form of trauma which may create different mental health challenges and require particular therapeutic approaches. A recent study led by Victoria Williamson identified moral injury in veterans who had 'experienced an event(s) during military service that challenged their view of who they are, the world they live in, or their sense of right and wrong'.[3] This built on a 2019 study of potentially morally injurious events described by UK military veterans, which included being ordered to break rules of engagement and disrespecting dead bodies, and highlighted concerns around mistreating the other side—civilians or enemy combatants (Fig. 1).[4]

Being set apart is also a feature of Claire's story, 'A Healthy Fighting Force', which shines a light on the harassment of women within the British forces. The narrative focuses on the abuse directed at a woman corporal when delivering a Force Health Protection brief. Male colleagues' insults at her ability and sexual taunts, unchecked by the sergeant major who looks on, demean her and prevent her from doing a job she is good at (Fig. 2). Emma Vieceli's art shows the absolute isolation of the woman soldier within the midst of her hostile 'own side'.

Robbie's story, 'Safe Haven' set in the Gulf War in 1991, recalls a traumatic experience of attachment, shifted allegiance, and betrayal. Tasked with providing displaced Kurdish civilians at risk of genocide with 'safe passage back to their homes and ongoing security', the Marine wonders, 'So this is a humanitarian mission?' It details the hospitality of the Kurdish people and the Marine's growing intimacy with Ahmed who declares him 'my adopted son, you are part of my family' before the order comes for British forces to move out (Fig. 3).

Graham's 'Stretcher Bearer Stan' directly explores the process of returning to the site of a particularly traumatic incident of enemy encounter. The

[3] Victoria Williamson, Dominic Murphy, Sharon A. M. Stevelink, Shannon Allen, Edgar Jones & Neil Greenberg, 'The impact of trauma exposure and moral injury on UK military veterans: a qualitative study', *European Journal of Psychotraumatology*, 11 (2020), pp. 1–10, p. 3.

[4] Victoria Williamson, Neil Greenberg & Dominic Murphy, 'Moral injury in UK armed forces veterans: A qualitative study', *European Journal of Psychotraumatology*, 10 (2019), pp. 1–8.

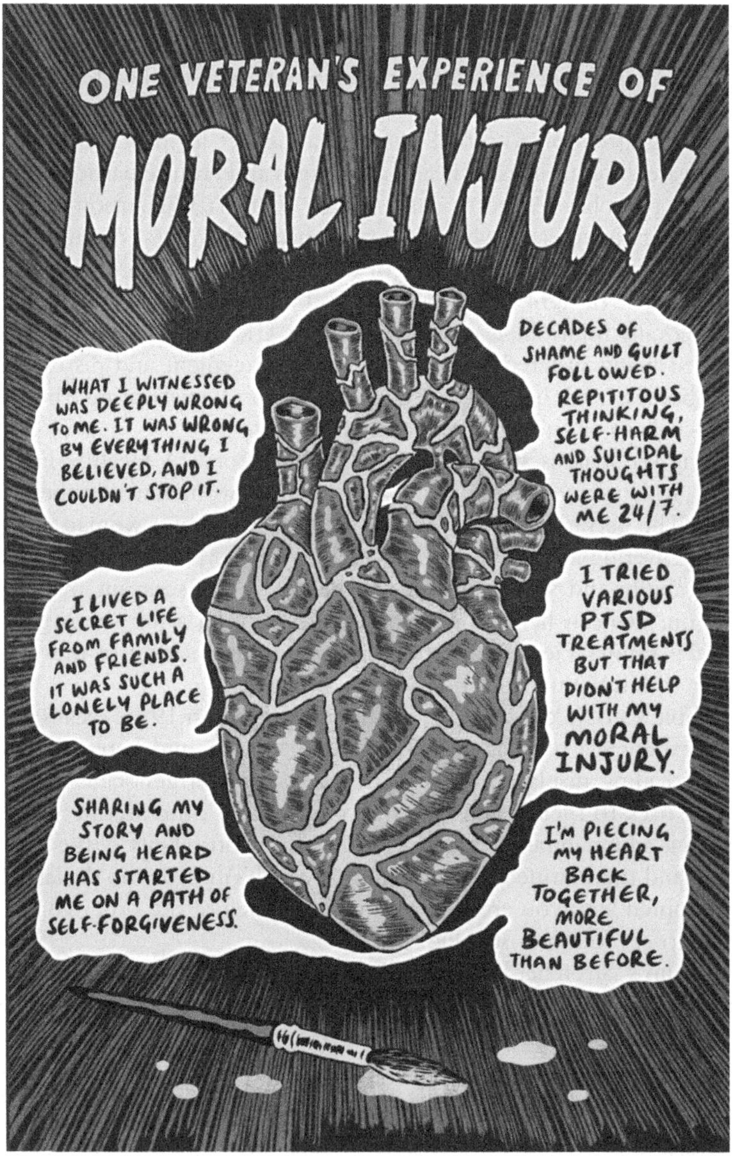

Fig. 1 'Moral Injury', words by Stan, art and lettering by Casey Raymond, *Coming Home*, issue 1 (2022). For more information on the *Coming Home* comic, please see www.re-live.org.uk

Fig. 2 'A Healthy Fighting Force', story by Claire, art and lettering by Emma Vieceli, *Coming Home*

narrative of a young bandsman working the gate at Long Kesh prison in Northern Ireland, which held Republican and Loyalist internees from 1971 to 1974, is intercut with the veteran's return decades later as he processes his experiences. The focus is on a prison riot in which Stan treats enemy wounded and is accused of lacking loyalty (Fig. 4).

The soldier's experience in 'Stretcher-bearer Stan' has resonance with Tolstoy's detailed exploration of Nikolai Rostov's disconcertment on coming face to face with his enemy, mentioned above. In both narratives cognitive and moral dissonance results from the gap between the enemy the soldier has been prepared to meet by his training and culture, and the actual individuals encountered. Our conversations with the Re-Live group suggest that these historical languages of strangeness and uneasiness in encountering an enemy very different to the one expected can be understood as earlier expressions of moral injury. In Tolstoy's novel, Rostov's difficult feeling about his behaviour to his enemy is described as 'a kind of

Fig. 3 'Safe Haven', story by Robbie, art by Keith Page, lettering by Josh Hicks, *Coming Home*

Fig. 4 'Stretcher-bearer Stan', story by Stan, art by Mike Donaldson, lettering by Josh Hicks, *Coming Home*

remorse'; he 'still felt embarrassed and somehow ashamed' (p. 724). In Williamson et al.'s study, a core identifier of moral injury was this kind of response, 'where the primary emotion expressed was of guilt/shame'.[5]

There is a repeated pattern in our historical materials, also visible in 'Stretcher-bearer Stan', where a bewildering change in status or power takes place in the movement of enemies from belligerents to suffering humans, as wounded and/or prisoners of war, in need of care. This shift

[5] Williamson et al., 'Moral injury', p. 3.

leads to a long and painful process of reassessment in which former certainties, identities, and allegiances are no longer reliable.

Together with members of the Re-Live group and the Museum of Military Medicine, we held an event on treating the enemy in 2022. Graham spoke to the profoundly life-changing experience of providing care to both sides after the riot at Long Kesh: 'in the course of that night I changed from a spotty teenager into someone who'd experienced some awful things in life'. His experience resonated with that described by volunteer nurse turned peace activist Asma Khalifa who had an experience of transformation in treating the enemy in Libya. Working as a field nurse for the opposition to Colonel Gaddafi in summer 2011, Asma put herself forward to care for Gaddafi's soldiers who had been neglected in military hospitals. Having lost her best friend in the conflict and with a brother and uncles fighting for the opposition, Asma maintained a strict demeanour on the wards, avoiding talking and eyecontact. On an early-morning antibiotics round, though, a soldier met her eyes and said 'you make me feel inhuman'. Asma describes this as 'one of the most impactful moments of my life, because when I looked at him, he seemed exactly the same as me ... the same pain, the same vulnerability, the same fear, the same understanding of this devastation we're going through'. She and the soldier prisoner cried together. Asma, who has gone on to become a peace activist and recipient of peace awards, described this moment as causing 'a fundamental shift': 'Ever since I don't think of anyone as an enemy'; 'my conditioning for accepting the enemy narrative has changed'. For Graham, in the midst of abuse and suspicion from both sides, treating the enemy felt instinctive: 'I looked on everyone as a human being.... They were suffering so I wanted to help'.[6]

The dialogue between Graham and Asma considered the dehumanisation and rehumanisation of the enemy and of the self. They strive to recount their stories in ways that might extend their own transformation. As Asma put it, 'I spoke when I felt it might do some good to speak'. For Graham, in working with Re-Live to tell his story and, crucially, to have it heard, the 'therapeutic process is enabling me to understand who I am – how this story affected me. Importantly, the change is that I can talk now without shaking or anything. I probably wouldn't have been able to do that this time last year'.

[6] 'Treating the Enemy', Museum of Military Medicine series, 12 April 2022.

Graham described how the process of making the comic had given a manageable, linear narrative shape to complex events which had previously been fragmented, circling, feeling almost unreal and impossible to order. Editing a long script into the compressed space of the short comic form 'helped to focus me on the parts of the story that truly mattered'. Thinking of the people who might one day read the comic and find a measure of healing in relating to it was an important motivation for him, as well as something that influenced how he told the story itself. 'Even though all the minutiae for me was so important... it wasn't actually that important for the potential reader'.

Graham's comments resonate with a long history of soldier narratives, with veterans of all ranks having for hundreds of years found it helpful and therapeutic to order their recollections into written or visual form, particularly from the early nineteenth century onwards. Even two hundred years ago, veterans also found that the collaborative editing and publishing process shaped the final format of their stories, gave them a sense of identity as an author apart from their military career, and connected them to other generations of soldiers.[7] In fact, although the group found the detached, old-fashioned tone of the Napoleonic-era writings we shared with them unfamiliar, the process of producing their own 'military memoir' of sorts via the comics brought them closer together with these historical counterparts, forming an unexpected but poignant link through time. For example, Graham's recollection of the intensive process of piecing back together his memories 'almost minute by minute', trying to concentrate on what emotions he might have felt at each moment aged eighteen, finds a powerful echo in the preface to a memoir written by a British rifleman from the Napoleonic Wars: 'I have... drawn on the reminiscences of my military life, and endeavoured faithfully to convey to the mind of the reader the impression which they made on my own at the time of their occurrence'.[8]

The re-opening and telling of stories has a consequence, even in a unified group. As Robbie put it, 'it is the regurgitation of events that has been laid dormant and tampered by years of emotion compression, and with it resurfaces feeling, emotion and trauma. Sometimes more than the actual

[7] Matilda Greig, *Dead Men Telling Tales: Napoleonic War Veterans and the Military Memoir Industry, 1808–1914* (Oxford: Oxford University Press, 2021).

[8] John Kincaid, *Adventures in the Rifle Brigade, in the Peninsula, France, and The Netherlands, from 1809 to 1815* (London: T. and W. Boone, 1830), p. vii.

event itself as there has been many years to evolve, question and ponder. The environment must be controlled at all times and that is where Karin comes in. Karin's steering and manoeuvring through such a brittle minefield took time, patience and a great deal of skill'. It takes courage to tell these stories, and many soldiers remain silenced by the possibilities of legal action, ostracisation, and reprisals.

From a practitioner's perspective, Karin noted that the narrative order brought by life story work has liberating effects for each individual in making space for engagement with new perspectives and allowing other stories to come to the fore. The mixed work of the project, with the dialogue on historical sources taking place alongside a releasing of contemporary life stories through the comic has had unique effects.

Reflecting on the reception of the *Coming Home* comic to date, Karin summed up the power of these life stories to release readers from their conditioning and preconceptions, by allowing 'people to reflect on an experience from someone else's point of view, which can be an incredibly informative place', providing new insights and education from which a change can begin.

Coming Home issue 1 was launched at The Lakes International Comic Art Festival, Windermere, UK, October 2022, where the veterans sold the comic to the comic reading public, and a panel discussion between veterans, their artists, and Re-Live about the process of creating the comic was held. The comic also featured at Thought Bubble Comic Art Festival in Harrogate, November 2022. It was released on sale internationally through comic shops in UK, Ireland, USA, and Australia. Additionally, it has been stocked by multiple libraries around Wales, UK. *Coming Home* has been shared with veterans' organisations in UK, including amongst veterans in the prison system. The comic has been shared and discussed with Help For Heroes' veterans' book club, January 2023. *Coming Home* has also been profiled and shared at academic conferences including Creative Brain Week organised by the Global Brain Health Institute at Trinity College, Dublin (February 2023) and Transatlantic Perspectives on the Present and Future of Moral Injury, International Centre For Moral Injury, Durham University (April 2023).

In sharing their stories, the group hopes to reach readers whose own experiences resonate, so that they no longer feel alone. This has been one effect of the connections our conversations have made between these historical materials and a history of reflection on enemy encounters, with this longer tradition of soldier experiences also helping to reduce the sense of alienation and isolation for veterans now. As Karin put it, to read that other people all throughout history have had similar experiences is immensely powerful: it creates the feeling that you have a 'community from the past'. Following this project, Re-Live intend to embed historical materials within their future life story work.

In opening up our interpretations of research material to those with direct military experience, we were continually required to revise and complicate our readings, recognising again that the order-imposing process of typical academic argument might not work for this archive and embracing more conflicting, oscillating forms that refuse to adjudicate between competing interpretations. The generosity of the group in engaging so openly and frankly with our research allowed us all to make new, unexpected connections between the materials. In a process similar, perhaps, to the comics, this Afterword has given cathartic narrative shape to a series of wide-ranging, emotionally testing discussions, and it has helped us to process our engagement with the group members' post-traumatic growth. We are deeply privileged to have worked with these remarkable individuals and to have their permission to share these conversations here. We hope that the shape we have given them does justice to the rich variety and individuality of each experience.

Index[1]

[1] Note: Page numbers followed by 'n' refer to notes.

H. Furneaux, M. Greig (eds.), *Enemy Encounters in Modern Warfare*, https://doi.org/10.1007/978-3-031-56748-3

B

SPRINGER NATURE

GPSR Compliance

The European Union's (EU) General Product Safety Regulation (GPSR) is a set of rules that requires consumer products to be safe and our obligations to ensure this.

If you have any concerns about our products, you can contact us on ProductSafety@springernature.com

In case Publisher is established outside the EU, the EU authorized representative is:

Springer Nature Customer Service Center GmbH
Europaplatz 3
69115 Heidelberg, Germany

The manufacturer's authorised representative in the EU is Springer Nature Customer Service Centre GmbH, Europaplatz 3, 69115 Heidelberg, Germany. If you have any concerns regarding our products, please contact ProductSafety@springernature.com

Printed and bound by CPI Group (UK) Ltd, Croydon, CR0 4YY

30/09/2026

02225583-0016